VOLUME ONE HUNDRED AND TWENTY SEVEN

ADVANCES IN
CANCER RESEARCH

VOLUME ONE HUNDRED AND TWENTY SEVEN

Advances in
CANCER RESEARCH

Edited by

PAUL B. FISHER

*Department of Human and Molecular Genetics,
VCU Institute of Molecular Medicine, and VCU Massey
Cancer Center, Virginia Commonwealth University,
School of Medicine, Richmond, Virginia, USA*

KENNETH D. TEW

*Department of Cell and Molecular Pharmacology,
Medical University of South Carolina, Charleston,
South Carolina, USA*

AMSTERDAM • BOSTON • HEIDELBERG • LONDON
NEW YORK • OXFORD • PARIS • SAN DIEGO
SAN FRANCISCO • SINGAPORE • SYDNEY • TOKYO

Academic Press is an imprint of Elsevier

Academic Press is an imprint of Elsevier
225 Wyman Street, Waltham, MA 02451, USA
525 B Street, Suite 1800, San Diego, CA 92101-4495, USA
125 London Wall, London, EC2Y 5AS, UK
The Boulevard, Langford Lane, Kidlington, Oxford OX5 1GB, UK

First edition 2015

© 2015 Elsevier Inc. All rights reserved.

No part of this publication may be reproduced or transmitted in any form or by any means, electronic or mechanical, including photocopying, recording, or any information storage and retrieval system, without permission in writing from the publisher. Details on how to seek permission, further information about the Publisher's permissions policies and our arrangements with organizations such as the Copyright Clearance Center and the Copyright Licensing Agency, can be found at our website: www.elsevier.com/permissions.

This book and the individual contributions contained in it are protected under copyright by the Publisher (other than as may be noted herein).

Notices
Knowledge and best practice in this field are constantly changing. As new research and experience broaden our understanding, changes in research methods, professional practices, or medical treatment may become necessary.

Practitioners and researchers must always rely on their own experience and knowledge in evaluating and using any information, methods, compounds, or experiments described herein. In using such information or methods they should be mindful of their own safety and the safety of others, including parties for whom they have a professional responsibility.

To the fullest extent of the law, neither the Publisher nor the authors, contributors, or editors, assume any liability for any injury and/or damage to persons or property as a matter of products liability, negligence or otherwise, or from any use or operation of any methods, products, instructions, or ideas contained in the material herein.

ISBN: 978-0-12-802920-6
ISSN: 0065-230X

For information on all Academic Press publications
visit our website at store.elsevier.com

CONTENTS

Contributors   ix

1. **Unravelling the Complexity and Functions of MTA Coregulators in Human Cancer**   1

   Da-Qiang Li and Rakesh Kumar

   1. Introduction   2
   2. Domain Architectures of MTA Proteins   4
   3. Regulation of Gene Expression by MTA Proteins   5
   4. Subcellular Localization of MTA Proteins   9
   5. Upstream Regulators of MTA Gene Products in Cancer   10
   6. Downstream Effectors of MTA Proteins in Cancer   21
   7. Expression and Clinical Significance of MTA Proteins in Cancer   24
   8. Therapeutic Targeting of MTA Family Members   31
   9. Conclusions and Prospective   33

   Acknowledgments   34
   References   35

2. **Examination of Epigenetic and other Molecular Factors Associated with *mda-9/Syntenin* Dysregulation in Cancer Through Integrated Analyses of Public Genomic Datasets**   49

   Manny D. Bacolod, Swadesh K. Das, Upneet K. Sokhi, Steven Bradley, David A. Fenstermacher, Maurizio Pellecchia, Luni Emdad, Devanand Sarkar, and Paul B. Fisher

   1. Background on *mda-9/Syntenin*   51
   2. Publicly Available Cancer Genomic Datasets   53
   3. Specific Datasets Examined for *mda-9* Analysis and the Analytical Approaches Employed   54
   4. The Patterns of *mda-9* Expression During Cancer Progression and Their Clinical Implications in Different Cancer Types   58
   5. *mda-9* Expression Correlates with Both Gene Copy Number and the Methylation Status of cg1719774   61
   6. *mda-9* is Highly Expressed in Melanoma   69
   7. Hypomethylation at cg1719774 is Associated with Histone Modifications Indicative of Higher Transcriptional Activity   71

		8.	Predicting Genes and Pathways Associated with *mda-9* Dysregulation in Glioma	74
		9.	Summary, Conclusion, and Future Directions	111
		Acknowledgments		115
		References		115

3. Exploitation of the Androgen Receptor to Overcome Taxane Resistance in Advanced Prostate Cancer — 123
Sarah K. Martin and Natasha Kyprianou

1. Introduction — 125
2. AR Signaling Finds Its Intracellular "Zip-Code" — 126
3. Can ADT Overcome AR Addiction in Prostate Tumors? — 128
4. Taxane Action in Prostate Cancer Cells: Up, Close, and "Personal" with Microtubules — 131
5. Taxane "Promiscuous" Actions Beyond Microtubule Stabilization — 133
6. Motor Proteins: An Intracellular Accomplice — 134
7. Mechanisms of Therapeutic Resistance — 139
8. Overcoming Taxane Resistance: The Translational Challenge — 143
9. Chemotherapy in Combination: "The Golden Fleece" of Cancer Treatment — 144
10. Conclusions — 147
References — 148

4. Stem Cell-Based Therapies for Cancer — 159
Deepak Bhere and Khalid Shah

1. Introduction — 159
2. Stem Cell Sources — 160
3. Stem Cell Homing and Migration — 161
4. Therapeutic Stem Cells for Cancer — 164
5. Encapsulated Stem Cells for Therapy — 176
6. Conclusions — 179
References — 179

5. Emerging Therapeutic Strategies for Overcoming Proteasome Inhibitor Resistance — 191
Nathan G. Dolloff

1. Introduction — 192
2. Mechanisms of Btz Resistance — 193
3. Approaches to Overcoming Btz Resistance — 200

4. Concluding Remarks	214
Acknowledgments	214
References	214

6. Influence of Bone Marrow Microenvironment on Leukemic Stem Cells: Breaking Up an Intimate Relationship   227
Puneet Agarwal and Ravi Bhatia

1. Introduction	228
2. Hematopoietic Stem Cells	229
3. Leukemia Stem Cells	230
4. Bone Marrow Microenvironment	231
5. Microenvironment Leukemia Crosstalk	236
6. In Vivo Models to Study BMM and Leukemia Interactions	241
7. Potential Therapeutic Avenue: Targeting the Seed and Soil?	242
8. Conclusions and Future Directions	243
References	245

7. Perspectives on Epidermal Growth Factor Receptor Regulation in Triple-Negative Breast Cancer: Ligand-Mediated Mechanisms of Receptor Regulation and Potential for Clinical Targeting   253
Carly Bess Williams, Adam C. Soloff, Stephen P. Ethier, and Elizabeth S. Yeh

1. Introduction	254
2. Regulation of EGFR Turnover and Signaling Outcomes	257
3. Common Molecular Characteristics of TNBC and Their Relationship to EGFR	264
4. Tumor Associated Macrophages in Breast Cancer Metastasis and Their Relationship with EGFR Signaling	267
5. Future Perspectives on Therapy	269
6. Conclusions	273
Acknowledgments	273
References	273

8. The Quest for an Effective Treatment for an Intractable Cancer: Established and Novel Therapies for Pancreatic Adenocarcinoma   283
Bridget A. Quinn, Nathaniel A. Lee, Timothy P. Kegelman, Praveen Bhoopathi, Luni Emdad, Swadesh K. Das, Maurizio Pellecchia, Devanand Sarkar, and Paul B. Fisher

1. Pancreatic Cancer	284
2. Current Pancreatic Cancer Therapies	285

3.	Novel Therapeutic Strategies	292
4.	Future Perspectives	301
	Acknowledgments	302
	References	302

Index *307*

CONTRIBUTORS

Puneet Agarwal
Division of Hematology-Oncology, Department of Medicine, University of Alabama Birmingham, Birmingham, Alabama, USA

Manny D. Bacolod
Department of Human and Molecular Genetics, and VCU Institute of Molecular Medicine, Virginia Commonwealth University, School of Medicine, Richmond, Virginia, USA

Ravi Bhatia
Division of Hematology-Oncology, Department of Medicine, University of Alabama Birmingham, Birmingham, Alabama, USA

Deepak Bhere
Massachusetts General Hospital, Harvard Medical School, Boston, Massachusetts, USA

Praveen Bhoopathi
Department of Human and Molecular Genetics, Virginia Commonwealth University, School of Medicine, Richmond, Virginia, USA

Steven Bradley
VCU Bioinformatics Program, School of Medicine, Virginia Commonwealth University, Richmond, Virginia, USA

Swadesh K. Das
Department of Human and Molecular Genetics; VCU Institute of Molecular Medicine, and VCU Massey Cancer Center, Virginia Commonwealth University, School of Medicine, Richmond, Virginia, USA

Nathan G. Dolloff
Department of Cellular and Molecular Pharmacology & Experimental Therapeutics, Medical University of South Carolina, Charleston, South Carolina, USA

Luni Emdad
Department of Human and Molecular Genetics; VCU Institute of Molecular Medicine, and VCU Massey Cancer Center, Virginia Commonwealth University, School of Medicine, Richmond, Virginia, USA

Stephen P. Ethier
Department of Pathology and Laboratory Medicine, Medical University of South Carolina, Charleston, South Carolina, USA

David A. Fenstermacher
VCU Massey Cancer Center, and Department of Biostatistics, Virginia Commonwealth University, School of Medicine, Richmond, Virginia, USA

Paul B. Fisher
Department of Human and Molecular Genetics; VCU Institute of Molecular Medicine, and VCU Massey Cancer Center, Virginia Commonwealth University, School of Medicine, Richmond, Virginia, USA

Timothy P. Kegelman
Department of Human and Molecular Genetics, Virginia Commonwealth University, School of Medicine, Richmond, Virginia, USA

Rakesh Kumar
Department of Biochemistry and Molecular Medicine, School of Medicine and Health Sciences, George Washington University, Washington, DC; Department of Molecular and Cellular Biology, Baylor College of Medicine, and Department of Molecular and Cellular Oncology, University of Texas M.D., Anderson Cancer Center, Houston, Texas, USA

Natasha Kyprianou
Department of Molecular and Cellular Biochemistry; Department of Urology; Department of Pathology and Toxicology, and Markey Cancer Center, University of Kentucky College of Medicine, Lexington, Kentucky, USA

Nathaniel A. Lee
Department of Human and Molecular Genetics, and Department of Surgery, Virginia Commonwealth University, School of Medicine, Richmond, Virginia, USA

Da-Qiang Li
Fudan University Shanghai Cancer Center and Institutes of Biomedical Sciences; Department of Oncology; Key Laboratory of Breast Cancer in Shanghai, and Key Laboratory of Epigenetics in Shanghai, Shanghai Medical College, Fudan University, Shanghai, China

Sarah K. Martin
Department of Molecular and Cellular Biochemistry, University of Kentucky College of Medicine, Lexington, Kentucky, USA

Maurizio Pellecchia
Sanford-Burnham Medical Research Institute, La Jolla, California, USA

Bridget A. Quinn
Department of Human and Molecular Genetics, Virginia Commonwealth University, School of Medicine, Richmond, Virginia, USA

Devanand Sarkar
Department of Human and Molecular Genetics; VCU Institute of Molecular Medicine, and VCU Massey Cancer Center, Virginia Commonwealth University, School of Medicine, Richmond, Virginia, USA

Khalid Shah
Massachusetts General Hospital, Harvard Medical School, Boston, Massachusetts, USA

Upneet K. Sokhi
Department of Human and Molecular Genetics, Virginia Commonwealth University, School of Medicine, Richmond, Virginia, USA

Adam C. Soloff
Department of Microbiology and Immunology, Medical University of South Carolina, Charleston, South Carolina, USA

Carly Bess Williams
Department of Cell and Molecular Pharmacology and Experimental Therapeutics, Medical University of South Carolina, Charleston, South Carolina, USA

Elizabeth S. Yeh
Department of Cell and Molecular Pharmacology and Experimental Therapeutics, Medical University of South Carolina, Charleston, South Carolina, USA

CHAPTER ONE

Unravelling the Complexity and Functions of MTA Coregulators in Human Cancer

Da-Qiang Li[*,†,‡,§,1], Rakesh Kumar[¶,||,#,1]

[*]Fudan University Shanghai Cancer Center and Institutes of Biomedical Sciences, Shanghai Medical College, Fudan University, Shanghai, China
[†]Department of Oncology, Shanghai Medical College, Fudan University, Shanghai, China
[‡]Key Laboratory of Breast Cancer in Shanghai, Shanghai Medical College, Fudan University, Shanghai, China
[§]Key Laboratory of Epigenetics in Shanghai, Shanghai Medical College, Fudan University, Shanghai, China
[¶]Department of Biochemistry and Molecular Medicine, School of Medicine and Health Sciences, George Washington University, Washington, DC, USA
[||]Department of Molecular and Cellular Biology, Baylor College of Medicine, Houston, Texas, USA
[#]Department of Molecular and Cellular Oncology, University of Texas M.D., Anderson Cancer Center, Houston, Texas, USA
[1]Corresponding authors: e-mail address: daqiangli1974@fudan.edu.cn; bcmrxk@gwu.edu

Contents

1. Introduction   2
2. Domain Architectures of MTA Proteins   4
3. Regulation of Gene Expression by MTA Proteins   5
      3.1 NuRD Corepressor Complexes   5
      3.2 Nucleosome Remodeling Factor Coactivator Complexes   6
4. Subcellular Localization of MTA Proteins   9
5. Upstream Regulators of MTA Gene Products in Cancer   10
      5.1 Transcription Factors and Coregulators   10
      5.2 MicroRNAs   16
      5.3 Growth Factors and Receptors   17
      5.4 Hormone and Hormone Receptors   18
      5.5 Enzymes and Protein Kinases   19
      5.6 Cell Stressors   20
6. Downstream Effectors of MTA Proteins in Cancer   21
      6.1 Transcriptional Targets of MTA Proteins   22
      6.2 Nontranscription Targets of MTA Proteins   23
7. Expression and Clinical Significance of MTA Proteins in Cancer   24
      7.1 Head and Neck Cancer   24
      7.2 Digestive System Cancer   29
      7.3 Respiratory System Cancer   30
      7.4 Hormone-Related and Reproductive System Cancer   30
8. Therapeutic Targeting of MTA Family Members   31

Advances in Cancer Research, Volume 127
ISSN 0065-230X
http://dx.doi.org/10.1016/bs.acr.2015.04.005

© 2015 Elsevier Inc.
All rights reserved.

9. Conclusions and Prospective	33
Acknowledgments	34
References	35

Abstract

Since the initial recognition of the metastasis-associated protein 1 (MTA1) as a metastasis-relevant gene approximately 20 years ago, our appreciation for the complex role of the MTA family of coregulatory proteins in human cancer has profoundly grown. MTA proteins consist of six family members with similar structural units and act as central signaling nodes for integrating upstream signals into regulatory chromatin-remodeling networks, leading to regulation of gene expression in cancer cells. Substantial experimental and clinical evidence demonstrates that MTA proteins, particularly MTA1, are frequently deregulated in a wide range of human cancers. The MTA family governs cell survival, the invasive and metastatic phenotypes of cancer cells, and the aggressiveness of cancer and the prognosis of patients with MTA1 overexpressing cancers. Our discussion here highlights our current understanding of the regulatory mechanisms and functional roles of MTA proteins in cancer progression and expands upon the potential implications of MTA proteins in cancer biology and cancer therapeutics.

1. INTRODUCTION

Distant metastasis represents a hallmark of human cancer and accounts for about 90% of cancer-related deaths (Hanahan & Weinberg, 2000). The metastatic cascade is a series of highly orchestrated biological processes that are driven by numerous gene products, including those important for cancer cell migration and invasion, angiogenesis, and cell survival and colonization of tumor cells at distant target organs (Hanahan & Weinberg, 2000; Nguyen, Bos, & Massague, 2009; Psaila & Lyden, 2009). One of such family of gene products with predominant roles in cancerous and metastatic process is the metastasis-associated proteins (MTA proteins). The MTA protein family consists of six members in vertebrates, including MTA1, MTA1s (MTA1 short form), ZG29p (zymogen granule 29 kDa protein), MTA2, MTA3, and MTA3L, which are coded by three distinct transcripts, namely, MTA1, MTA2, and MTA3 (Bowen, Fujita, Kajita, & Wade, 2004; Futamura et al., 1999; Kumar, Wang, & Bagheri-Yarmand, 2003; Manavathi & Kumar, 2007; Fig. 1).

MTA1, the founding member of the MTA1 gene family, was originally identified in 1994 by differential cDNA library screening prepared from the

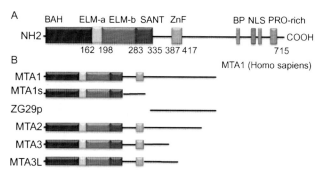

Figure 1 Structural domains of the MTA proteins. MTA proteins are highly conserved at the N-terminus but divergent at the C-terminus. NLS, nuclear localization signal; PRO-rich, proline-rich; ZnF, zinc finger. (See the color plate.)

13762NF rat mammary adenocarcinoma metastatic system (Toh, Pencil, & Nicolson, 1994). Subsequent studies mapped MTA1 to human chromosome 14q 31.2 (Martin et al., 2001). As MTA1's expression level closely correlates with the metastatic potential of breast cancer cells, the newly discovered gene was named as metastasis-associated protein 1 (Toh et al., 1994). Following this, two naturally occurring variants ZG29p and MTA1s were discovered (Kleene, Zdzieblo, Wege, & Kern, 1999; Kumar et al., 2002). The ZG29p encodes an N-terminally truncated form of MTA1 due to alternative transcription initiation and is exclusively expressed in pancreatic acinar cells (Kleene et al., 1999). In contrast, the MTA1s is a naturally occurring C-terminal truncated version of MTA1 generated by alternative splicing followed by addition of 33 novel animo acids due to a frame shift (Kumar et al., 2002).

In contrast to MTA1, MTA2 was identified in 1999 by sequencing of a 70-kDa polypeptide during analysis of the nucleosome remodeling and histone deacetylase (NuRD) complex (Zhang et al., 1999). Sequence alignment revealed that human MTA2 is 65% identical to human MTA1 (Yao & Yang, 2003; Zhang et al., 1999) and that both proteins are highly homologous in the N-terminal region but divergent in the C-terminal region (Yao & Yang, 2003). In general, MTA family members form distinct histone deacetylase (HDAC) containing complexes, and thus, potentially could target different genes (Yao & Yang, 2003).

Mouse MTA3 was initially cloned in 2001 (Simpson, Uitto, Rodeck, & Mahoney, 2001), followed by identification of human MTA3 as an estrogen-dependent component of the NuRD complex with a role in breast cancer invasion (Fujita et al., 2003). While MTA1 and MTA2 are primarily nuclear proteins, the expression of MTA3 is somewhat diffused. This suggests that the biologic functions of MTA3 are likely distinct as compared to MTA1 and MTA2 (Yao & Yang, 2003). In this context, the expression of MTA1 and MTA2, but not MTA3, increases during breast cancer progression in a MMTV-PyV-mT mouse model, wherein different stages of mammary tumorigenesis could be easily recognized (Zhang, Stephens, & Kumar, 2006).

During the past two decades, a large body of experimental and clinical studies demonstrates that MTA proteins are frequently deregulated in a wide range of human cancers and play a central role in tumor progression and metastasis (Li, Pakala, et al., 2012; Sen, Gui, & Kumar, 2014a). Moreover, MTA proteins localize at the nexus of multiple upstream oncogenic signaling pathways and regulate gene expression through a NuRD-dependent and -independent mechanism (Lai & Wade, 2011; Li, Pakala, et al., 2012).

2. DOMAIN ARCHITECTURES OF MTA PROTEINS

The fundamental unit of a protein is its structural domain, which is the building block of protein structure as well as a potential determinant for its putative function or functions. Accumulating structural and biochemical studies reveal that the MTA proteins, with the exception of two truncated version of MTA1, ZG29p, and MTA1s (Kleene et al., 1999; Kumar et al., 2002), contain three conserved structural domains, including the bromo-adjacent-homology (BAH) domain; the EGL-27 and MTA1 homology 2 (ELM2) domain; and the SWI3, ADA2, N-CoR, and TFIII-B (SANT) domain (Manavathi & Kumar, 2007; Fig. 1).

The BAH domain is commonly found in chromatin-associated proteins and plays critical roles in regulatory protein–protein interactions, nucleosome binding, and recognition of methylated histones (Kuo et al., 2012; Norris, Bianchet, & Boeke, 2008; Onishi, Liou, Buchberger, Walz, & Moazed, 2007; Yang & Xu, 2013). The ELM2 domain functions as a transcriptional repression domain via recruiting HDACs (Ding, Gillespie, & Paterno, 2003; Wang, Charroux, Kerridge, & Tsai, 2008). The ELM2 domain is also found transcription corepressors with proteins containing SANT domains (Ding et al., 2003). More recently, the ELM2-SANT combined domain has been shown to act as a binding scaffold for various transcriptional cofactors, and thus, allow proteins to modify chromatin structures leading to modulation of gene transcription (Wang et al., 2008). Consistent with this notion, recent structural studies of the HDAC1–MTA1 complex reveal that MTA1 makes extensive contacts with the HDAC through the ELM2 as well as SANT domains (Millard et al., 2013). The presence of multiple functional domains including the BAH domain, the ELM2, and the SANT domains in the MTA family members suggests a potential role of these coregulatory proteins in the regulation of target gene transcription.

3. REGULATION OF GENE EXPRESSION BY MTA PROTEINS

Cancer development and progression are tightly regulated by cell context-dependent transcriptional programs, which involves a dynamic fine balance between the packaging of regulatory sequences into chromatin and allowing transcriptional regulators to gain access to such sequences (Cairns, 2009; Voss & Hager, 2014). By its very nature, the highly condensed structure of chromatin generally limits the accessibility of transcription factors to the DNA, and thus, inhibits gene transcription. To overcome these barriers, cells use two major mechanisms to dynamically modulate chromatin structure for appropriate regulation of gene expression (Roberts & Orkin, 2004; Voss & Hager, 2014).

The first mechanism involves ATP-dependent chromatin-remodeling complexes, which hydrolyze ATP to alter histone–DNA contacts through nucleosome sliding, histone exchange, and/or nucleosome or histone eviction (Au, Rodriguez, Vincent, & Tsukiyama, 2011; Euskirchen, Auerbach, & Snyder, 2012; Lai & Wade, 2011; Wilson & Roberts, 2011). The second fundamental mechanism to remodel chromatin during gene transcription includes posttranslational modification of histones, which may directly influence the stability of nucleosomes through histone–DNA, histone–histone interactions (Tessarz & Kouzarides, 2014). For example, acetylation of histone tails by the histone acetyltransferases (HATs) is implicated to nucleosome remodeling and transcriptional activation, whereas HDAC-mediated histone deacetylation determines transcriptional repression through chromatin condensation (Johnstone, 2002). Hence, ATP-dependent chromatin-remodeling complexes and histone acetylation collectively play a central role in regulating gene transcription through modifying nucleosome structures in a dynamic manner.

3.1 NuRD Corepressor Complexes

To date, four classes of chromatin-remodeling complexes have been identified in eukaryotes based on its composition, enzymatic activities, and biological functions (Bao & Shen, 2007) One of such complexes is the NuRD (also known as Mi-2), which was independently identified in 1998 by several groups using different cell types and species (Tong, Hassig, Schnitzler, Kingston, & Schreiber, 1998; Wade, Jones, Vermaak, & Wolffe, 1998; Xue et al., 1998; Zhang, LeRoy, Seelig, Lane, & Reinberg, 1998). The NuRD complex is composed of multiple enzymatic subunits and possesses ATPase, histone deacetylase, and histone demethylase enzymatic activities (Wang et al., 2009; Xue et al., 1998; Zhang et al., 1999). In this regard, the chromodomain helicase DNA binding protein 3 (CHD3; also known as Mi2-α) and CHD4 (also known as Mi2-β) contain chromo (chromatin organization modifier) domains and SNF2-related helicase/ATPase domains, thus exerting ATP-dependent chromatin-remodeling activities (Hall & Georgel, 2007). In addition, HDAC1 and HDAC2 deacetylate histones and therefore facilitate the formation of a condensed and transcriptionally silenced chromatin (Marks et al., 2001). The lysine-specific demethylase 1 (LSD1) demethylates histone H3 to activate or repress gene expression in a context-dependent manner (Metzger et al., 2005; Wang et al., 2009). Other nonenzymatic subunits of the NuRD complex include retinoblastoma-binding protein P46 (RBAP46), RBAP48, methyl-CpG-binding domain protein 2 (MBD2), MBD3, MTA1, MTA2, and MTA3 (Fujita et al., 2003; Tong et al., 1998; Wade et al., 1998; Xue et al., 1998; Zhang et al., 1998, 1999). Being a core component of the NuRD complex (Tong et al., 1998; Wade et al., 1998; Xue et al., 1998; Zhang et al., 1998), the MTA family members have been implicated in the transcriptional repression of gene expression through recruitment of HDACs to target gene promoters via a NuRD-dependent mechanism (Lai & Wade, 2011; Li, Pakala, et al., 2012).

3.2 Nucleosome Remodeling Factor Coactivator Complexes

Although MTAs were originally discovered as a part of the NuRD complex, the roles of MTA proteins in gene regulation are not restricted to the ability of MTAs to form NuRD complexes. A case in point is the MTA1 which associates with the nucleosome remodeling factor (NURF) coactivator complex in a signaling-dependent manner (Nair, Li, & Kumar, 2013). The NURF complex was originally purified from Drosophila embryo extracts (Tsukiyama & Wu, 1995), which is composed of four subunits including the 140-kDa ATPase ISWI (NURF-140) (Tsukiyama, Daniel,

Tamkun, & Wu, 1995), the 301-kDa largest subunit (NURF-301/BPTF) (Xiao et al., 2001), the 55-kDa subunit (NURF-55 or p55 in fly and RbAp48/46 in human) (Martinez-Balbas, Tsukiyama, Gdula, & Wu, 1998; Nowak et al., 2011), and an uncharacterized 38-kDa subunit (Tsukiyama & Wu, 1995). Accumulating evidence suggests that the NURF complex enhances chromatin accessibility by perturbing nucleosome structure in an ATP-dependent manner (Tsukiyama & Wu, 1995), thus facilitating the binding of transcription factors to chromatin to activate gene transcription (Badenhorst, Voas, Rebay, & Wu, 2002; Mizuguchi, Tsukiyama, Wisniewski, & Wu, 1997; Tsukiyama & Wu, 1995). For example, the NURF complex promotes efficient expression of the wingless target genes (Song, Spichiger-Haeusermann, & Basler, 2009).

Previous studies have shown that MTA1 acts not only as a transcriptional repressor through a NuRD-dependent mechanism but also as a transcriptional activator for numerous target genes (Balasenthil et al., 2007; Gururaj, Singh, et al., 2006; Kumar, Balasenthil, Manavathi, Rayala, & Pakala, 2010; Li, Pakala et al., 2013; Reddy et al., 2011). However, the underlying mechanism for MTA1-mediated switch between gene activation and repression events remained mysterious, until recently when it was discovered that dynamic change from methylation to demethylation acts as a molecular switch between corepressor and coactivator function of MTA1 (Nair et al., 2013). In this context, MTA1's methylation at lysine 532 by the histone methyltransferase G9a directs orderly formation of the NuRD complex and this modification is essential for the assembly of a functional NuRD corepressor complex at the target gene chromatin such as $p21^{WAF1}$ (Li et al., 2010; Nair et al., 2013). In contrast, demethylation of MTA1 by the LSD1 demethylase is required for its coactivator function through an association with the NURF complex in cyclical- and -signaling-dependent manner (Nair et al., 2013).

It is being suggested that signaling-dependent modulation of combinatorial posttranslational modifications may initiate the dynamic process which ultimately drives the formation of distinct MTA1-containing chromatin-remodeling complexes in response to upstream signaling and/or intracellular cellular adaptations. In this context, MTA1 has been shown to undergo acetylation (Ohshiro et al., 2010), ubiquitination (Li, Ohshiro, et al., 2009), sumoylation (Cong, Pakala, Ohshiro, Li, & Kumar, 2011), and methylation (Nair et al., 2013; Fig. 2). The ability of these dynamic subcomplexes or complexes to regulate target gene chromatin is also influenced by the status and collaborative interactions with chromatin structures. Clearly, additional work is needed to fully understand the regulation of gene expression by

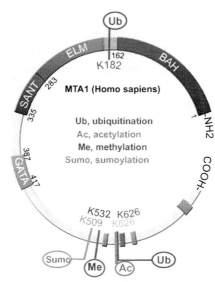

Figure 2 Posttranslational modification code of MTA1. MTA1 is posttranslationally modified by acetylation on lysine 626, ubiquitination on lysine 182 and 626, sumoylation on lysine 509, and methylation on lysine 532. Ac, acetylation; Me, methylation; Sumo, sumoylation; Ub, ubiquitination. Illustration is not on the scale of amino acid numbers. (See the color plate.)

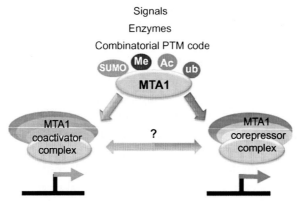

Figure 3 Combinatorial posttranslational modifications of MTA1 in response to upstream signaling mediate the formation of distinct complexes and, consequently, its coactivator versus corepressor function in gene transcription. (See the color plate.)

MTA1-containing coregulatory complexes and the underlying determinants of the formation, stability, and transition of such coactivator and corepressor complexes which ultimately regulate the expression of MTA1 target genes with roles in cancer (Fig. 3, Supplementary Fig. 1 on-line).

4. SUBCELLULAR LOCALIZATION OF MTA PROTEINS

The subcellular location of proteins provides clues about the predicted and putative protein's function. One of the best examples is the tumor suppressor phosphatase and tensin homolog (PTEN), which suppresses phosphatidylinositol 3-kinase (PI3K) signaling in the cytoplasm, whereas controls DNA repair and chromosomal stability in the nucleus (Bassi et al., 2013; Shen et al., 2007). Initial studies involving GFP-tagged MTA fusion proteins revealed that MTA1 and MTA2 are primarily localized in the nucleus (Simpson et al., 2001; Yao & Yang, 2003). Moreover, MTA1 nuclear localization depends on the presence of, at least, one nuclear localization signal (NLS) and one Src homology 3 (SH3) binding site (Simpson et al., 2001). With the development of new molecular technologies, recent evidence shows that MTA1 localizes to the nucleus, cytoplasm, and nuclear envelope (Liu et al., 2014). Furthermore, MTA1 localizes on the nuclear envelope in a translocated promoter region-dependent manner, while MTA1 exhibits an obvious localization on microtubules in the cytoplasm (Liu et al., 2014). In contrast, due to a lack of nuclear localization sequences, MTA1s predominately localizes in the cytoplasm, and thus, sequesters estrogen receptor α (ERα) in the cytoplasm to promote nongenomic responses of ERα (Kumar et al., 2002). In addition, consistent evidence shows that MTA3 has a diffused distribution and is present in both the nucleus and cytoplasm (Simpson et al., 2001; Yao & Yang, 2003).

In addition, subcellular localization of MTA1 may be closely associated with its tumor-promoting activities. For example, a higher MTA1 expression in colon cancer associates with MTA1 localization in the nucleus and cytoplasm (Liu et al., 2014), while only nuclear MTA1 localization correlates with suppression of cancer differentiation (Liu et al., 2014). In line with these observations, the nuclear-to-cytoplasmic ratio of MTA1 expression increases from prostatic intraepithelial neoplasia to metastatic prostate cancer, and nuclear MTA1 overexpression associates with aggressive prostate cancer, recurrence, and metastasis in African Americans (Dias et al., 2013). Similarly, nuclear overexpression of MTA1 correlates with poor survival in nasopharyngeal carcinoma (NPC) (Li, Liu, et al., 2012) and with the risk of high recurrence in breast cancer (Martin et al., 2006). Immunohistochemical analysis of MTA family proteins in a mouse model for human breast cancer also demonstrates an elevated nuclear MTA1 expression in ERα-negative tumors (Zhang et al., 2006). Furthermore, nuclear MTA1 staining associates with reduced E-cadherin as well as cytoplasmic β-catenin—both indicators of tumor invasion (Zhang et al., 2006). In endometrial adenocarcinomas, MTA1 is present in both the nucleus and cytoplasm compartments in grade 1 and grade 2 tumors, whereas MTA1 localizes in the cytoplasm only in grade 3 tumors, suggesting that different subcellular localization of MTA1 may play distinct roles during endometrial cancer progression (Balasenthil, Broaddus, & Kumar, 2006). In brief, subcellular localization of MTAs provides clues about the putative functions of MTA1 protein in cancer progression. However, the underlying mechanisms for regulating the distribution of MTA proteins in different subcellular compartments remain unknown at this time.

5. UPSTREAM REGULATORS OF MTA GENE PRODUCTS IN CANCER

MTA family of proteins acts as converging nodes for multiple upstream signaling pathways as well as master modifiers for various downstream effector factors. A comprehensive understanding of the regulatory mechanisms underlying the functions of MTA proteins will be important for developing effectively preventive and therapeutic strategies to block pathways contributing to MTA-mediated tumor progression and metastasis. To date, numerous upstream modifiers of MTA proteins (Table 1) and its downstream effectors and targets (Table 2) have been identified as discussed below.

5.1 Transcription Factors and Coregulators

Transcription factors constitute a large family of proteins, which bind to specific DNA sequences, and thereby, participate in the regulation of gene expression. The c-Myc oncoprotein is an enhancer box sequence-specific transcription factor that controls various biological processes related to cancer, such as cell-cycle progression and cellular transformation, through transcriptional regulation of its target genes (Adhikary & Eilers, 2005; van Riggelen, Yetil, & Felsher, 2010). Interestingly, c-Myc is an upstream transcriptional activator of MTA1 and that MTA1 is also required for c-Myc-mediated transformation (Zhang et al., 2005). In addition, increased expression of MTA1 by the eukaryotic initiation factor 5A2 (EIF5A2) in colorectal carcinoma cells partially depends on c-Myc (Zhu et al., 2012).

Table 1 Upstream Regulators of MTAs in Human Cancer

Genes	Regulators	Functions	References
MTA1	Akt	Invasion of prostate cancer	Wang, Fan, et al. (2012)
	ARF	MTA1-mediated oncogenesis	Li, Pakala, et al. (2011)
	HBx	HBX-associated hepatocellular carcinomas	Bui-Nguyen, Pakala, Sirigiri, Xia, et al. (2010); Lee, Na, Na, et al. (2012), Lee et al. (2014), and Yoo et al. (2008)
	Cigarette smoke	The smoke-related progress of NSCLC	Xu, Mao, Fan, and Zheng (2011)
	c-Myc	Cellular transformation	Zhang et al. (2005)
	COP1	MTA1 stability and DNA repair function	Li, Ohshiro, et al. (2009)
	eIF5A	Colorectal carcinoma aggressiveness	Zhu et al. (2012)
	G9a	MTA1 methylation at lysine 532	Nair et al. (2013)
	Heregulin	Transcriptional repression of ERα	Mazumdar et al. (2001)
	Hypoxia	Tumor progression	Yoo, Kong, and Lee (2006)
	IR	DNA repair and cell survival	Li, Ohshiro, et al. (2009)
	Leptin	Leptin-induced EMT of breast cancer cells	Yan, Avtanski, Saxena, and Sharma (2012)
	15-LOX-1	Invasive properties of colorectal carcinoma cells	Cimen, Tuncay, and Banerjee (2009)
	LPS	Inflammatory response	Pakala, Bui-Nguyen, et al. (2010)
	LSD1	MTA1 demethylation	Nair et al. (2013)
	miR-30c	Proliferation, migration/invasion of endometrial cancer	Kong et al. (2014) and Zhou et al. (2012)
		Lung cancer invasion	Xia et al. (2013)
	miR-661	Breast cancer growth and invasion	Reddy, Pakala, Ohshiro, Rayala, and Kumar (2009)

Continued

Table 1 Upstream Regulators of MTAs in Human Cancer—cont'd

Genes	Regulators	Functions	References
	NF-κB	Inflammatory response	Pakala, Bui-Nguyen, et al. (2010)
	P300	Oncogenic activity of MTA1	Ohshiro et al. (2010)
	P53	Angiogenesis and invasion	Lee, Na, Kim, et al. (2012)
	PIAS	MTA1 sumoylation	Cong et al. (2011)
	SENP	MTA1 desumoylation	Cong et al. (2011)
	TGF-β1	TGF-β1-induced EMT phenotypes in breast cancer	Pakala et al. (2011)
	VEGF	Breast cancer angiogenesis and metastasis	Nagaraj, Shilpa, Rachaiah, and Salimath (2013)
	Ultraviolet	DNA repair and cell survival	Li, Ohshiro, Khan, and Kumar (2010)
MTA1s	CKI γ2	Nongenomic action of estrogen in breast cancer cells	Mishra, Yang, et al. (2004)
MTA2	hBD-3	Colon cancer progression	Uraki et al. (2014)
	miR-146a	Pancreatic cancer cell invasion	Li, Vandenboom, et al. (2010)
	P300	Colorectal cancer cells growth and migration	Zhou et al. (2014)
	Rho GDIα	Metastasis and tamoxifen resistance	Barone et al. (2011)
	SP1	Gastric cancer cells invasion	Zhou et al. (2013)
MTA3	Estrogen and ERα	Breast cancer growth	Fujita, Kajita, Taysavang, and Wade (2004) and Mishra, Talukder, et al. (2004)
	miR-495	Lung cancer growth and migration	Chu et al. (2014)
	SP1	Breast cancer growth	Fujita et al. (2004)

Table 2 Downstream Effectors of MTA Proteins in Human Cancer

Genes	Effectors	Functions	References
MTA1	ARF	MTA1-mediated oncogenesis	Li, Pakala, et al. (2011)
	BCAS3	Breast cancer progression and resistance to tamoxifen	Gururaj, Holm, et al. (2006) and Gururaj, Singh, et al. (2006)
	BRCA1	Abnormal centrosome number and chromosomal instability	Molli, Singh, Lee, and Kumar (2008)
	CXCR2	Ovarian cancer progression	Dannenmann et al. (2008)
	E-cadherin	Esophageal cancer invasion	Weng, Yin, Zhang, Qiu, and Wang (2014)
		Epithelial-to-mesenchymal transition of breast cancer	Pakala et al. (2011)
		Metastasis and epithelial-to-mesenchymal transition of colon cancer	Tuncay Cagatay, Cimen, Savas, and Banerjee (2013)
		Invasiveness of prostate cancer cells	Wang, Fan, et al. (2012)
		Prostate cancer metastasis	Fan et al. (2012)
	ERα	Endocrinal resistance to tamoxifen of breast cancer	Kang et al. (2014)
		Breast cancer growth and progression	Mazumdar et al. (2001)
	ERβ	Ovarian cancer progression	Dannenmann et al. (2008)
	FosB	EMT phenotype of breast cancer	Pakala et al. (2011)
	Gαi2	Cell transformation	Ohshiro et al. (2010)
	Hedgehog	Aggressive phenotypes of nasopharyngeal carcinoma	Song et al. (2013)
	HIF1α	Angiogenesis	Moon et al., 2006; Yoo et al., 2006

Continued

Table 2 Downstream Effectors of MTA Proteins in Human Cancer—cont'd

Genes	Effectors	Functions	References
	iNOS	HBx stimulation of iNOS expression and activity	Bui-Nguyen, Pakala, Sirigiri, Martin, et al. (2010)
	miR-125b	Promoting invasion and migration of NSCLC cells	Li, Chao, et al. (2013)
	MMP9	Unknown	Yan, Wang, Toh, and Boyd (2003)
	MyD88	Inflammatory response	Pakala, Reddy, et al. (2010)
	NR4A1	Docetaxel resistance in prostate cancer cells	Yu, Su, Zhao, Wang, and Li (2013)
	P21	P53-independent DNA repair	Li, Pakala, et al. (2010)
	P53	DNA damage response	Li, Divijendra Natha Reddy, et al. (2009)
		Inhibition of p53-induced apoptosis	Moon, Cheon, and Lee (2007)
	Pax5	Lymphomagenesis	Balasenthil et al. (2007)
	PTEN	Proliferation in prostate cancer	Dhar, Kumar, Li, Tzivion, and Levenson (2014)
		Stimulation of the PI3K/AKT signaling in breast cancer	Reddy et al. (2012)
	Rho GTPases	Aggressive phenotypes of nasopharyngeal carcinoma	Song, Li, et al. (2013)
	SMAD7	Tumourigenesis and metastasis of breast caner	Salot and Gude (2013)
	Snail and Slug	Metastasis and epithelial-to-mesenchymal transition of colon cancer	Tuncay Cagatay et al. (2013)
	STAT3	Breast cancer metastasis	Pakala et al. (2013)
	TGM2	Inflammatory response	Ghanta, Pakala (2011)

Table 2 Downstream Effectors of MTA Proteins in Human Cancer—cont'd

Genes	Effectors	Functions	References
	VEGF and Flt-1	Tumor angiogenesis and metastasis	Nagaraj et al. (2013)
	Vimentin	Metastasis and epithelial-to-mesenchymal transition of colon cancer	Tuncay Cagatay et al. (2013)
	Wnt1	Mammary gland development and tumorigenesis	Kumar, Balasenthil, Manavathi, et al. (2010)
	Wnt/β-catenin	NSCLC cell growth and invasion	Lu, Wei, and Xi (2014)
MTA1s	ERα	Nongenomic responses of ERα	Kumar et al. (2002)
	Wnt1	Breast cancer progression	Kumar, Balasenthil, Manavathi, et al. (2010) and Kumar, Balasenthil, Pakala, et al. (2010)
MTA2	ERα	Breast cancer growth	Cui et al. (2006)
	P53	Cell growth and apoptosis	Luo, Su, Chen, Shiloh, and Gu (2000)

Consequently, MTA1 may be responsible for EIF5A2-induced epithelial-to-mesenchymal transition (EMT) and invasiveness of colorectal carcinoma cells (Zhu et al., 2012). There is another regulatory connectivity between the c-Myc and MTA1; c-Myc is regulated by the histone acetyltransferase p300 coactivator via acetylation-dependent control of its protein turnover and coactivation of Myc-induced transcription (Faiola et al., 2005); similarly, p300-dependent acetylation of MTA1 at lysine 626 activates the Ras–Raf pathway and determines oncogenic activity of MTA1 in cellular transformation (Ohshiro et al., 2010). Interestingly, p300 also binds to and acetylates MTA2 at lysine 152 to promote the growth of colorectal cancer cells (Zhou et al., 2014).

The specificity protein 1 (SP1) protein is a zinc finger transcription factor that directly binds to GC-rich motifs in gene promoters and transactivates its transcription (Safe & Abdelrahim, 2005). Growing evidence shows that SP1 protein plays a critical role in the growth and metastasis of cancer cells by regulating expression of its downstream target genes such as MTA proteins (Fujita et al., 2004; Li, Pakala, et al., 2011; Safe & Abdelrahim, 2005; Xia & Zhang, 2001; Zhou et al., 2013). For example, SP1 is recruited to the MTA1 promoter and activates its transcription (Li, Pakala, et al., 2011). Furthermore, the tumor suppressor alternative reading frame (ARF) promotes the proteasomal degradation of SP1 by enhancing its interaction with proteasome subunit Rpt6, and thus, abrogating SP1-mediated transactivation of MTA1 (Li, Pakala, et al., 2011). In addition to MTA1, SP1 also regulates the expression of MTA2 and MTA3 (Fujita et al., 2004; Zhou et al., 2013). In this context, SP1 binds to the *MTA2* gene promoter and transactivates its expression in gastric cancer cells (Xia & Zhang, 2001; Zhou et al., 2013). Consistently, MTA2 expression in gastric cancer tissues is related with the level of SP1 expression (Zhou et al., 2013). Similarly, SP1 also regulates MTA3 expression in breast cancer cells (Fujita et al., 2004).

In addition to transactivation of MTA genes by transcription factors, the expression of MTA gene products is also negatively regulated by specific transcription factors. For example, the p53 transcription factor, a profound modulator of transcription under a variety of stress signals (Riley, Sontag, Chen, & Levine, 2008), represses the expression of the *MTA1* gene via p53-binding elements located in its upstream promoter and this regulatory process is tightly regulated by poly(ADP-ribose)ylation of p53 by poly(ADP-ribose) polymerase 1 (PARP-1) (Lee, Na, Kim, Lee, & Lee, 2012). Thus, p53 and PARP-1 repress the MTA1-mediated expression of hypoxia inducible factor alpha (HIF1α) and vascular endothelial growth factor (VEGF), and thereby, regulate angiogenesis (Lee, Na, Kim, et al., 2012). Given the fact that more than 50% of human tumors contain mutations or deletions of the p53 gene (Hollstein, Sidransky, Vogelstein, & Harris, 1991), the loss of p53 in cancer cells could potentially contribute to generally noticed MTA1 overexpression and MTA-induced oncogenesis and tumor progression in human cancers. However, this hypothesis has yet to be experimentally tested.

5.2 MicroRNAs

MicroRNAs (miRNAs) are evolutionarily conserved small noncoding RNAs that play crucial roles in regulating gene expression at the posttranscriptional level (Li & Rana, 2014). Not surprisingly, miRNAs have also emerged as key upstream regulators for MTA gene expression in human cancer. The first identified miRNA for regulating MTA family is the miR-661, which inhibits MTA1 expression and thereby MTA1-mediated tumor-promoting functions in breast cancer cells (Reddy et al., 2009). More recently, several studies demonstrated that miR-30c negatively regulate MTA1 expression, thus acting as a tumor suppressor in endometrial cancer and non-small-cell lung cancer cells (Kong et al., 2014; Xia et al., 2013; Zhou et al., 2012). Similarly, miR-146a also inhibits, directly or indirectly, the expression of MTA2 in pancreatic cancer cells, leading to the suppression of cancer cell invasion and metastasis (Li, Vandenboom, et al., 2010). Genome-wide characterization of miR-34a targets revealed MTA2 is a potential target of miR-34a (Kaller et al., 2011). In addition, miR-495 regulates the proliferation and migration of non-small-cell lung cancer cells by targeting MTA3 (Chu et al., 2014). These observations are in line with the growing notion of using miRNA-based anticancer therapies to improve disease response and increase cure rates (Garzon, Marcucci, & Croce, 2010).

5.3 Growth Factors and Receptors

Growth factors and its receptors have been widely recognized as regulators of abnormal cancer growth and progression through activating downstream signaling cascades, and thus, several agents targeting growth factor receptors and signaling are being developed as targeted cancer therapeutics. For example, transforming growth factor β (TGFβ) plays a mechanistic role in the regulation of tumor development and progression, and therefore, many drugs are being developed to target TGFβ signaling in cancer (Akhurst & Hata, 2012). In this context, TGFβ1 regulates EMT via stimulating the expression of MTA1 in experimental model systems (Pakala et al., 2011). In turn, MTA1 status determines TGFβ1-induced EMT phenotypes through transcriptional repression of E-cadherin expression (Pakala et al., 2011).

The VEGF-mediated angiogenesis is another turning point in cancer development (Hanahan & Weinberg, 2011). Consequently, antiangiogenic drugs targeting the VEGF pathway have shown signs of clinical benefits in patients with advanced-stage cancer (Ellis & Hicklin, 2008). Interestingly, MTA1 is a potent pro-angiogenic molecule that regulates tumor angiogenesis and metastasis through a crosstalk between the VEGF and MTA1 signaling pathways (Nagaraj et al., 2013). To this point, VEGF induces the expression of MTA1 and also promotes its phosphorylation on tyrosine residues through VEGF receptor 2 and p38 mitogen-activated protein kinase (Nagaraj et al., 2013). In turn, MTA1 upregulates the expression of VEGF and its receptor Flt-1 genes (Du et al., 2011; Nagaraj et al., 2013). Thus, MTA1 and VEGF may form a double-positive feedback to promote tumor angiogenesis and progression. In line with these observations, targeting MTA1 in prostate cancer cells effectively inhibits angiogenesis and cancer aggressiveness (Kai et al., 2011).

The ErbB family of the cell surface receptors consists of four receptor tyrosine kinases, including EGFR, ErbB2, ErbB3, and ErbB4. Deregulation of ErbB signaling is associated with the development of a wide variety of types of human cancer. Mahoney et al. reported that treatment of human keratinocytes with EGFR selective inhibitor or EGFR-specific antibody results in a significant downregulation of MTA1 expression, indicating an EGFR-dependent MTA1 expression in keratinocytes (Mahoney et al., 2002). Previous studies have shown that heregulin-β1 (HRG) is a combinatorial ligand for ErbB3 and ErbB4 and transactivates ErbB2, whose activation controls the development of aggressive phenotypes of human cancer (Lewis et al., 1996; Mazumdar et al., 2001). Interestingly, MTA1 is a target of HRG and represses ERα-mediated transcription in breast cancer cells by recruiting HDACs (Mazumdar et al., 2001). Subsequent studies further demonstrate that HRG stimulates MTA1 expression in a heat shock factor 1-dependent manner (Khaleque et al., 2008).

5.4 Hormone and Hormone Receptors

As a class of signaling molecules, a hormone regulates gene expression through binding to and activating its receptor in the target cells, thus exerting its molecular functions in various biological processes including cancer. For instance, the ERα is an estrogen-dependent transcription factor that plays a critical role in the development and progression of hormone-related cancers through regulating the expression of genes involved in cell proliferation, invasion, and metastasis (Klinge, 2001; Thomas & Gustafsson, 2011). Emerging evidence suggests that estrogen upregulates the expression of MTA1 in endometrial cancer cells through an ER-dependent and ER-independent mechanism (Kong et al., 2014). The Rho GDP dissociation inhibitor α (Rho GDIα) directly binds to ERα and alters the interaction between the ERα and ERE elements in the target genes, and thus, influences the ability of ERα to regulate transcription of estrogen-responsive genes (El Marzouk et al., 2007). It was recently shown that Rho GDIα negatively regulates the expression of MTA2 in ER-positive breast cancer cells and the Rho GDIα-MTA2 is implicated in breast cancer metastasis and resistance to tamoxifen (Barone et al., 2011). In addition, *MTA3* is an ER-regulated gene in mammary epithelial cells as estrogen positively regulates the expression of *MTA3* through functional ERE half-sites located in the vicinity of AP1 or SP1 binding sites in the *MTA3* promoter (Fujita et al., 2003, 2004; Mishra, Talukder, et al., 2004). Consistently, SP1 and ERα synergistically drive high-level transcription of the MTA3 gene in breast cancer cells (Fujita et al., 2004). Further, ER-mediated transcriptional regulation of MTA3 is tightly regulated by the dynamic interplay of ER coregulators, such as MTA1 (Mishra, Talukder, et al., 2004).

In addition to the estrogen signaling, the expression of MTA1 is also under control of androgen pathway, and accordingly, MTA1 is also implicated in the androgen-dependent prostate cancer growth and progression (Ma et al., 2010). Interestingly, a recent study suggests that leptin, a hormone synthesized by fat cells to regulate the amount of fat stored in the body, enhances the expression of MTA1, and in turn, MTA1 mechanistically participates in leptin-induced Wnt1 upregulation and EMT in breast cancer cells (Yan et al., 2012).

5.5 Enzymes and Protein Kinases

An abundance of evidence has firmly established that MTA1 is widely upregulated in human cancer, but the underlying mechanisms governing the stability of MTA1 proteins remains elusive, until Li et al. found that a RING-finger ubiquitin-protein ligase constitutive photomorphogenic 1 (COP1) regulates MTA1 stability via the ubiquitin-proteasome pathway (Li, Ohshiro, et al., 2009). In addition, the tumor suppressor protein ARF, encoded by a gene located in the Ink4a/ARF gene that is frequently mutated in human cancer (Sherr, 2001), also physically associates with MTA1 and affects its protein stability (Li, Pakala, et al., 2011). Moreover, ARF suppresses MTA1 transcription by abrogating the transcription factor SP1-mediated transactivation of MTA1 (Li, Pakala, et al., 2011).

In addition to ubiquitination, MTA proteins are also posttranslationally modified by acetylation, methylation, and sumoylation (Fig. 2). In this context, MTA1 undergoes acetylation at lysine 626 by the p300 HAT, and acetylated MTA1 activates the Ras–Raf pathway and thereby stimulates cell transformation (Ohshiro et al., 2010). As discussed above, methylation of MTA1 at lysine 532 by the G9a methyltransferase and its demethylation by the LSD1 demethylase governs its coactivator versus corepressor activity through multivalent reading of nucleosome codes in a signaling-dependent manner (Nair et al., 2013). MTA1 also undergoes sumoylation, while sentrin/SUMO-specific protease 1 (SENP1) and 2 deSUMOylate MTA1 (Cong et al., 2011). Furthermore, MTA1 sumoylation at lysine 509 along with SUMO-interacting motif (SIM) at its C-termini participates in the regulation of MTA1's corepressor activity (Cong et al., 2011). In addition, casein kinase I-gamma 2 (CKI-γ2), a ubiquitously expressed cytoplasmic kinase, phosphorylates MTA1s and thereby potentiates the ability of MTA1s to sequester ERα in the cytosol, thus resulting in suppression of transactivation functions of ERα in the nucleus (Mishra, Yang, et al., 2004).

5.6 Cell Stressors

Cells are constantly exposed to various exogenous and endogenous stressors, such as infections, hypoxia, nutrition deprivation, heat shock, and DNA damage. The inability of cells to timely and effectively respond to these stressors can cause apoptosis, cell death, and even neoplasm. Accumulating evidence suggests that MTA1 is a stress response protein, which is upregulated and facilitates cell survival in response to various stresses (Wang, 2014).

Hepatitis B virus (HBV) infection is a leading cause for the pathogenesis of hepatocellular carcinoma (HCC), in which HBV-encoded X protein (HBx) plays a central role (Trepo, Chan, & Lok, 2014). HBx has been shown to induce the expression of MTA1, but not MTA2 and MTA3 (Bui-Nguyen, Pakala, Sirigiri, Xia, et al., 2010; Lee, Na, Na, et al., 2012; Yoo et al., 2008). One of the mechanisms involves the nuclear factor kappa-B (NF-κB) signaling; HBx recruits the HBx/NF-κB p65 complex to the NF-κB consensus motif at the *MTA1* promoter, and thereby, stimulates MTA1 transcription (Bui-Nguyen, Pakala, Sirigiri, Xia, et al., 2010). Another proposed mechanism includes HBx induction of CpG island methylation in the *MTA1* promoter via recruiting DNA methyltransferases DNMT3a and DNMT3b, and in turn, abrogating p53-mediated transcriptional repression of MTA1 (Lee, Na, Na, et al., 2012). In turn, MTA1 drives development and progression of HBx-related hepatocarcinoma through multiple effector pathways. For instance, MTA1 contributes to HBx regulation of NF-κB signaling and hence, regulation of NF-κB target gene products with roles in inflammation (Bui-Nguyen, Pakala, Sirigiri, Xia, et al., 2010). In addition, MTA1 mediates HBx-induced stabilization of HIF1α, and thus, contributes to the promotion of angiogenesis and metastasis of HBV-associated liver cancer (Yoo et al., 2008). HBx–MTA1 complex is also recruited onto the inducible nitric oxide synthase (iNOS) promoter in a NF-κB-dependent manner and positively regulates iNOS transcription (Bui-Nguyen, Pakala, Sirigiri, Martin, et al., 2010). Accordingly, depletion of MTA1 in HBx-expressing cells severely impairs the ability of HBx to modulate the endogenous levels of iNOS and nitrite production

(Bui-Nguyen, Pakala, Sirigiri, Martin, et al., 2010), which play a key role in HBV-related hepatocarcinoma development and progression through crosstalk with other signaling cascades (Park et al., 2013). Further MTA1 level is a good indicator of microvascular invasion and poor survival of HCC patients infected with HBV (Ryu et al., 2008).

Hypoxia is a shared phenotypic feature of most solid tumors and promotes angiogenesis, metastasis, and resistance to therapy (Wouters & Koritzinsky, 2008). HIF1α is a stress-responsive transcription factor, which regulates the expression of genes important for adaption to hypoxia and angiogenesis (Ayrapetov et al., 2011). Emerging evidence shows that hypoxia is a strong inducer of MTA1 expression in breast cancer cell lines (Yoo et al., 2006). MTA1, in turn, induces the deacetylation of HIF1α by increasing the expression of HDAC1, resulting in enhanced the transcriptional activity and stability of HIF1α and thereby HIF1α-induced tumor angiogenesis (Moon et al., 2006; Yoo et al., 2006). In addition, MTA1 and HDAC1 are also involved in the stabilization of HIF1α induced by HBx (Yoo et al., 2008).

Cancer cells appear to utilize another mechanism to stabilize MTA1 following ionizing radiation (IR) involving disruption of the COP1-mediated proteolysis of MTA1 (Li, Ohshiro, et al., 2009). Interestingly, MTA1 also is required for optimal repair of the double-strand DNA break and cellular survival following IR treatment (Li, Ohshiro, et al., 2009). MTA1 is also stabilized by ultraviolet (UV) and such posttranslational stability of MTA1 contributes to UV-induced DNA damage pathway and facilitates cell survival (Li, Ohshiro, Khan, & Kumar, 2010). In addition, MTA1 protects cells from heat-stress-induced apoptosis (Li et al., 2008; Li, Wu, et al., 2011).

6. DOWNSTREAM EFFECTORS OF MTA PROTEINS IN CANCER

As a multifaceted transcriptional coregulator (Manavathi & Kumar, 2007), roles of MTA proteins in cancer progression largely involve its ability to regulate the transcription of downstream target genes that encode effector proteins involved in tumor growth and progression (Table 2). As key components of MTA-containing coregulatory complexes (Fujita et al., 2003; Xue et al., 1998; Zhang et al., 1998, 1999), MTA proteins are generally associated with transcriptional modulation by recruiting HDACs or HATs onto their target genes (Manavathi & Kumar, 2007).

7. EXPRESSION AND CLINICAL SIGNIFICANCE OF MTA PROTEINS IN CANCER

A large body of evidence over the last two decades has established that MTA1 and MTA2 are upregulated in a wide range of human cancer, and their expression levels are closely associated with disease progression and poor prognosis (Kaur, Gupta, & Dutt, 2014; Li, Pakala, et al., 2012; Luo, Li, Yao, Hu, & He, 2014; Sen et al., 2014a; Toh & Nicolson, 2009, 2014). Here, we briefly summarize the expression patterns and its clinical significance of MTA proteins in human cancer (Table 3).

7.1 Head and Neck Cancer

In general, head and neck cancer includes carcinomas arising from the mucosal epithelia of the head and neck region and various cell types of salivary glands and the thyroid (Kang, Kiess, & Chung, 2015; Marzook, Deivendran, Kumar, & Pillai, 2014). Among them, nasopharyngeal cancer (NPC) represents one of the common malignancies (Zhang, Chen, Liu, Tang, & Mai, 2013). Immunohistochemistry staining of paraffin-embedded NPC tissues reveals that nuclear overexpression of MTA1 is present in about one-half of NPC tissues, and nuclear overexpression of MTA1 positively correlates with clinical stage, distant metastasis, and poor survival of patients (Li, Liu, et al., 2012; Yuan et al., 2014). Mechanistically, MTA1 participates in the reorganization of actin cytoskeleton as well as NPC metastasis via Rho GTPases and Hedgehog signaling (Song, Li, et al., 2013). Consistently, MTA1-induced invasion and migration of cancer cells could be inhibited by specific inhibitors targeting Hedgehog signaling and Rho GTPases (Song, Li, et al., 2013). In addition, MTA1-associated growth of NPC tumor cells may be accompanied by accelerated progress of the cell cycle through the G1-S phase (Song, Zhang, et al., 2013).

Similarly, high levels of MTA1 are also present oral squamous cell carcinomas and closely associates with tumor progression, angiogenesis and invasion, and lymph node metastasis (Kawasaki et al., 2008). Immunohistochemical analysis of tonsillar cancer tissues for MTA1 demonstrates that MTA1 overexpression also correlates with lymph node metastasis (Park et al., 2011). In head and neck squamous cell carcinoma, however, MTA1 has been reported to be reduced expression in the lymph node metastases as compared with the matched primary tumors (Roepman et al., 2006).

Table 3 Expression and Significance of MTA Proteins in Human Cancer

Genes	Cancer Types	Detection Methods	Expression Pattern	Clinico-Pathologic Correlations	References
MTA1	Breast	qPCR	Upregulation	Tumor stage and lymph node metastasis	Cheng et al. (2012)
		IHC	Upregulation	Tumor grade and angiogenesis	Jang, Paik, Chung, Oh, and Kong (2006)
	Cervical	IHC	55.3%	Tumor progression and metastasis	Liu et al. (2013)
	Colon	IHC	51.4%	Tumor invasion and poorer prognosis	Higashijima et al. (2011)
	Colorectal	RT-PCR	38.9%	Invasion and lymphatic metastasis	Toh et al. (1997)
		IHC	Upregulation	Metastasis and EMT	Tuncay Cagatay et al. (2013)
	Endometrial	IHC	75.7%	Unknown	Balasenthil et al. (2006)
	Esophageal	IHC	44.4%	Shorter disease-free interval	Li, Wang, et al. (2009)
		qPCR	34%	Invasion and metastasis	Toh, Kuwano, Mori, Nicolson, and Sugimachi (1999)
	Esophageal squamous carcinoma	IHC	Upregulation	Tumor stage and survival	Song, Wang, and Liu (2012)
		IHC	43.5%	Tumor progression, angiogenesis, and poor survival	Li, Tian, et al. (2012)
		IHC	42.9%	Invasion and metastasis	Toh et al. (2004)
		RT-PCR	34.0%	Invasion and metastasis	Toh et al. (1999)
	Gastric	RT-PCR	38.2%	Invasion and lymphatic metastasis	Toh et al. (1997)
		IHC	36.04%	Angiogenesis and poor prognosis	Deng et al. (2013)

Continued

Table 3 Expression and Significance of MTA Proteins in Human Cancer—cont'd

Genes	Cancer Types	Detection Methods	Expression Pattern	Clinico-Pathologic Correlations	References
	Head and neck	DNA microarray	Reduced in lymph node tissue	Lymph node metastasis	Roepman et al. (2006)
	Hepatocellular	IHC	69%	Tumor growth and vascular invasion	Moon, Chang, and Tarnawski (2004)
		IHC	17%	Microvascular invasion, recurrence, and poor survival	Ryu et al. (2008)
		RT-PCR	42%	Poor survival	Hamatsu et al. (2003)
	Nasopharyngeal	IHC	48.6%	Poor survival	Li, Liu, et al. (2012)
		IHC	Upregulation	Clinical stage, metastasis, and poor survival	Yuan et al. (2014)
		ISH	Upregulation	Metastasis, recurrence, and poor survival	Deng, Zhou, Ye, Zeng, and Yin (2012)
	Non-small-cell lung	IHC	40.2%	Tumor angiogenesis and poor survival	Li, Tian, et al. (2011)
		IHC	61.0%	Tumor progression and clinical outcome	Zhu, Guo, Li, Ding, and Chen (2010)
		IHC	63.5%	Lymph node metastasis and TNM stage	Xu et al. (2011)
		IHC	36.7%	Tumor differentiation and size and poor survival	Yu, Wang, Zhang, Liu, and Zhang (2011)
		RT-PCR	87.8%	Invasiveness and metastasis	Sasaki et al. (2002)

Oral squamous	IHC	97.7%	Metastasis and angiogenesis	Andishehtadbir et al. (2015)
	IHC	86.8%	Invasion and metastasis	Kawasaki, Yanamoto, Yoshitomi, Yamada, and Mizuno (2008)
Ovarian	IHC	Upregulation	Progression, poor response to treatment, and survival	Prisco et al. (2012)
Osteosarcoma	IHC	81.25%	Hematogenous metastasis	Park et al. (2005)
Osteosarcomas, jaw	IHC	77.4%	Tumor grade	Park et al. (2009)
Pancreatic	IHC	31%	Poor prognosis	Miyake et al. (2008)
	IHC	Upregulation	Recurrent disease and metastasis	Dias et al. (2013)
Prostate	IHC	Upregulation	Invasiveness	Wang, Fan, et al. (2012)
	IHC	Upregulation	Prostate cancer progression	Hofer et al. (2004)
Pancreatic endocrine	IHC	Upregulation	Malignant behavior	Hofer, Chang, Hirko, Rubin, and Nose (2009)
Tonsil cancer	IHC	41.9%	Lymph node metastasis	Park, Jung, Sun, Joo, and Kim (2011)
Thymoma	RT-PCR	Upregulation	Invasion	Sasaki et al. (2001)

Continued

Table 3 Expression and Significance of MTA Proteins in Human Cancer—cont'd

Genes	Cancer Types	Detection Methods	Expression Pattern	Clinico-Pathologic Correlations	References
MTA2	ESCC	IHC	40.1%	Tumor progression and poor prognosis	Liu, Shan, Wang, and Ma (2012)
	Gastric cancer	IHC	55.9%	Tumor invasion, lymph nodes metastasis, and TNM staging	Zhou et al. (2013)
	Hepatocellular	IHC	96.2%	Tumor size and differentiation	Lee et al. (2009)
	Non-small-cell lung	IHC	66.4%	TNM stages, tumor size, and lymph node metastasis	Liu, Han, et al. (2012)
	Pancreatic ductal adenocarcinoma	qPCR and IHC	Upregulation	Poor tumor differentiation, TNM stage, lymph node metastasis, and poor survival	Chen, Fan, Li, and Jiang (2013)
	Thymomas	IHC	70.8% (nuclear)	Histological type and tumor stage	Wang, Li, Li, Xie, and Wang (2012)
MTA3	Endometrial	IHC	Downregulation	Endometrial carcinogenesis	Bruning et al. (2010)
	Gastroesophageal junction	IHC	Downregulation	EMT and disease progression	Dong et al. (2013)
	Non-small-cell lung	qPCR and IHC	59.32%	Lymph node metastasis and survival	Zheng et al. (2013)
		IHC	57.4%	TNM stage, nodal metastasis, and poor prognosis	Li, Sun, et al. (2013)
	Osteosarcoma	IHC	81.25%	Progression and metastasis	Park et al. (2005)

Notes: EMT, epithelial-mesenchymal transition; IHC, immunohistochemistry; ISH, *in situ* hybridization; qPCR, quantitative PCR; RT-PCR, reverse transcription PCR; and TNM, tumor, lymph node, metastasis.

7.2 Digestive System Cancer

Esophageal cancer is the eighth most common cancer globally and has two main subtypes, namely esophageal squamous cell carcinoma and esophageal adenocarcinoma (Siegel, Ma, Zou, & Jemal, 2014). Interestingly, overexpression of MTA1 mRNA was found to be overexpressed in about one-third of esophageal carcinomas, and such tumors show a significantly higher frequency of adventitial invasion and lymph node metastasis (Toh et al., 1999). Likewise, MTA1 is overexpressed in over one-third of esophageal cancer tissues and correlates well with a shorter disease-free interval of patients (Li, Wang, & Liu, 2009). In addition to a poor prognosis, MTA1 overexpression in esophageal squamous cell carcinoma also associates with the degree of invasion and lymph node metastasis (Song et al., 2012; Toh et al., 2004). Consistently, RNA interference-mediated MTA1 downregulation inhibits invasiveness of esophageal squamous cell lines (Qian et al., 2005). In addition, MTA2 overexpression is also detected in esophageal squamous cell carcinoma and significantly correlates with the tumor stage, lymphatic and blood-vessel invasion, distant metastasis, and poor prognosis (Liu, Shan, Wang, & Ma, 2012).

Globally, gastric cancer is the fifth leading cause of cancer and the third leading cause of death from cancer (Siegel et al., 2014). In this context, MTA1 mRNA is overexpressed in about one-third of gastric cancer and correlates with lymph node metastasis (Toh et al., 1997). MTA2 is also upregulated in more than one-half of gastric cancer tissues analyzed in a recent study, and its status associates with tumor invasion and lymph nodes metastasis (Zhou et al., 2013). Accordingly, MTA2 knockdown inhibits the growth of xenografts and lung metastasis in animal models (Zhou et al., 2013). In contrast to MTA1 and MTA2, the expression of MTA3 is reduced in gastroesophageal junction adenocarcinoma at both mRNA and protein levels, and patients with lower expression of MTA3 had poorer outcomes (Dong et al., 2013).

HCC is one of the most common tumors worldwide and is characterized by MTA1 overexpression and increased vascularity and frequent metastasis (Moon et al., 2004). For example, MTA1 is overexpressed in over two-third of HCC and correlates well with the growth and vascular invasion (Moon et al., 2004). In line with these observations, high expression of the *MTA1* gene in over one-third of HCC samples in a study involving mRNA expression in HCC and paired nontumor liver tissues (Hamatsu et al., 2003). In addition, immunohistochemical staining of MTA2 in human HCC tissues suggest that MTA2 is easily detectable expression in most of HCC samples, and MTA2 overexpression correlates with tumor size and differentiation; however, surprisingly, MTA2 expression does not associate with the survival rate of patients and cancer recurrence (Lee et al., 2009).

In colon cancer, MTA1 is overexpressed in over one-half of cancer tissues and MTA1 expression tends to correlate with tumor status, vascular invasion, and decreased patient survival (Higashijima et al., 2011). In addition, several lines of evidence show that MTA1 is upregulated in colorectal cancer and associates with invasion and metastasis as well as EMT (Toh et al., 1997; Tuncay Cagatay et al., 2013; Zhao et al., 2007).

7.3 Respiratory System Cancer

The most common types of cancer in males are lung cancer (Siegel et al., 2014), which is histologically classified into two main primary types, small-cell lung carcinoma and non-small-cell lung carcinoma (NSCLC). The expression of MTA1 mRNA is easily detectable in over 80% of NSCLC tumors and its expression levels closely relate with tumor progression and invasive and metastatic potential of tumors (Sasaki et al., 2002). Consistently, MTA1 expression is also upregulated in NSCLC and associates with smoking history, tumor angiogenesis, lymph node metastasis, and poor survival (Li, Tian, et al., 2011; Xu et al., 2011). In addition, MTA1 expression correlates well with cigarette smoke in NSCLC, raising the possibility of MTA1 involvement in smoked-linked NSCLC (Xu et al., 2011). MTA1 upregulation also correlates with the invasion of NSCLC cells via negatively impacting the level of miR-125b (Li, Chao, et al., 2013). Similar to MTA1, nuclear MTA2 expression in about two-third of NSCLC and correlates with advanced tumor stages, tumor size, lymph node metastasis, higher Ki-67 proliferation index, and poor overall survival (Liu, Han, et al., 2012). Although MTA3 expression inversely correlates with MTA1 in breast cancer, MTA3 mRNA is also overexpressed and believed to be a risk factor of lymph node metastasis in NSCLC patients (Zheng et al., 2013).

7.4 Hormone-Related and Reproductive System Cancer

In the United States, breast cancer accounts for about 29% of all new cancers and 15% cancer-related death among women (Siegel et al., 2014). Molecular and genetic mouse model studies have demonstrated that MTA1 represents a strong candidate for breast cancer metastasis-promoting gene (Martin et al., 2001; Pakala et al., 2013). Clinically, MTA1 expression in breast cancer strongly correlates with tumor grade and increased tumor angiogenesis, but not with tumor stage, status of hormone receptors, and axillary lymph node metastasis (Jang et al., 2006). More interestingly, MTA1 overexpression in the nucleus may predict an early disease relapse, and such breast tumors may be sensitive to systemic therapies (Martin et al., 2006). However, a more recent study suggests that a higher expression of MTA1 mRNA also associates with poorer clinical outcomes (Cheng et al., 2012). In addition, MTA2 expression also promotes metastatic behavior and associates with a poor outcome in ER-negative breast cancer (Covington et al., 2013).

In prostate cancer, multiple lines of evidence suggest that MTA1 expression also associates with cancer progression (Dias et al., 2013; Hofer et al., 2004). Specifically, elevated MTA1 expression in hormone-refractory metastatic prostate cancer inversely correlates with cancer recurrence after radical prostatectomy (Hofer et al., 2004). In general, MTA1 overexpression correlates well with the aggressive and recurrence of prostate cancer in African American patients (Dias et al., 2013).

In cervical cancer, MTA1 overexpression could be detected in over one-half of patients, and a higher MTA1 status correlates with histologic grade, lymph node metastasis, recurrence, and poor survival (Liu et al., 2013). MTA1 expression is also upregulated in about three-fourth of endometrial adenocarcinomas in a small study involving 70 endometrial adenocarcinomas (Balasenthil et al., 2006). In ovarian cancer, high levels of MTA1 expression also associates with advanced stage, worse response to cancer treatment, and poor disease-free survival (Prisco et al., 2012). In support of these findings, MTA1 expression regulates the expression of several tumor promoting and metastasis-facilitating factors, such as of E-cadherin and its upstream regulators SNAIL and SLUG and growth-regulated oncogene during ovarian cancer progression (Dannenmann et al., 2008). In line with emerging role for MTA1 in DNA damage repair (Li & Kumar, 2010; Li, Pakala, et al., 2012; Li, Yang, & Kumar, 2014), MTA1 has been also proposed to promotes the proliferation of epithelial ovarian cancer cells by enhancing DNA repair (Yang et al., 2014).

8. THERAPEUTIC TARGETING OF MTA FAMILY MEMBERS

Given the functional importance of MTA1 in tumor growth and metastasis, this group of coregulatory proteins have emerged a potential targets for developing anticancer strategies. In this context, natural products have emerged as potent anticancer agents over the past 30 years (Mann, 2002).

Interestingly, natural compound such as curcumin—a naturally occurring phenolic compound—suppresses proliferation and invasion in non-small-cell lung cancer by modulation of MTA1-mediated Wnt/β-catenin pathway (Lu et al., 2014). Another naturally occurring phytochemical resveratrol found in grapes, peanuts and berries, also downregulates the expression of MTA1, and thus, enhances acetylation and activation of p53 by destabilizing the MTA1/NuRD complex (Kai, Samuel, & Levenson, 2010). Consistently, experimental depletion of MTA1 in prostate cancer cells sensitizes cancer cells to resveratrol-induced apoptosis (Kai et al., 2010). In addition, resveratrol promotes the acetylation status and reactivation of PTEN via inhibiting the MTA1/HDAC complex, and in turn, leads to inhibition of the Akt pathway as well as prostate cancer progression (Dhar et al., 2014). In support of these observations, pterostilbene (PTER)—a dimethylether analogue of resveratrol that is naturally present in blueberries, are a potent inhibitor of MTA1 expression in prostate cancer cells and also inhibits the growth, progression, and metastasis of cancer cells in experimental models (Li, Dias, et al., 2013). More interestingly, PTER is more potent than resveratrol in its ability to promote p53 acetylation through inhibition of MTA1 (Li, Dias, et al., 2013). In addition, β-elemene, a noncellular antineoplastic agent, downregulates the expression of MTA1 in bladder cancer cells, and leads to tumor inhibitory effects in model systems (Chen et al., 2012). Interestingly, β-elemene also inhibits cell invasion by upregulating E-cadherin expression through controlling the ERα/MTA3/Snail signaling pathway in human MCF-7 breast cancer cells (Zhang, Zhang, & Li, 2013). Likewise, phytoestrogen genistein, found in soybeans, inhibits choriocarcinoma cell invasion by triggering the MTA3/Snail/E-cadherin regulatory pathway by binding with ERβ (Liu et al., 2011). Taken together, these exciting findings suggest that these natural compounds inhibit tumor growth and progression by targeting MTA signaling (Table 4).

As MTA1 overexpression in numerous cancer model systems has been shown to confer resistance to growth inhibitory and anticancer activities of IR and chemotherapeutic agents, MTA1 could also be targeted by agents other than natural compounds. For example, MTA1 status is a good determinant of radiosensitivity of cancer cells to radiation induced growth inhibition (Chou et al., 2010; Li, Ohshiro, et al., 2009). In another study, MTA1 status has been shown to be a good indicator of chemosensitivity of prostate cancer to Docetaxel (Yu et al., 2013). Interestingly, MTA1 overexpression in nasopharyngeal cancer cells confers resistance to cisplatin as experimental downregulation of MTA1 leads to growth inhibition and increased susceptibility of tumor xenografts to cisplatin's anticancer activity (Feng et al., 2014).

Table 4 Natural Compounds Targeting MTA Family to Inhibit Tumor Growth and Progression

Compounds	Targets	Cancer Type	Effects on Cancer	References
β-elemene	MTA1	Bladder cancer	Induction of apoptosis	Chen et al. (2012)
	MTA3	Breast cancer	Inhibition of cell migration and invasion	Zhang, Chen, Liu, Tang, and Mai (2013) and Zhang, Zhang, and Li (2013)
Curcumin	MTA1	Lung cancer	Inhibition of cell growth and invasion	Lu et al. (2014)
Genistein	MTA3	Choriocarcinoma	Inhibition of cell invasion	Liu et al. (2011)
Pterostilbene	MTA1	Prostate cancer	Inhibition of tumor growth, progression and metastasis	Li, Dias, et al. (2013)
Portulacerebroside A	MTA1	Liver cancer	Inhibition of invasion and metastasis	Ji et al. (2014)
Resveratrol	MTA1	Prostate cancer	Inhibition of cell proliferation and survival	Dhar et al. (2014)
			Induction of apoptosis	Kai et al. (2010)

9. CONCLUSIONS AND PROSPECTIVE

Since the initial discovery of the founding member of the MTA family of genes in 1994, the scientific community has witnessed an incredible progress in our understanding of the regulatory mechanisms and molecular functions of MTA proteins in cancer cells. A timeline of major advances in MTA family of coregulators during the last two decades is illustrated in Fig. 4. As a result, the MTA1 is now recognized as one of the most

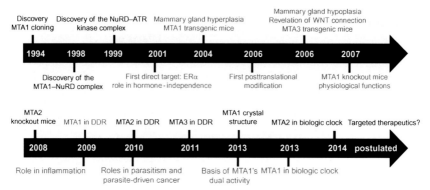

Figure 4 Timeline of major advances in the field of MTA family members in cancer research. Contributions from authors' laboratories are in dark blue. (See the color plate.)

upregulated gene products in human cancer and that this family of proteins represents a major modifiers of gene expression due to its role in chromatin remodeling. In addition, we have begun to appreciate cytoplasmic functions of previously thought to be nuclear proteins. Lessons from the MTA family will also allow us to bring out the concept of dynamic proteome of the same molecule as a function of time in response to a given signal. Based on the molecular insights of MTA1 action and preliminary clues in the literature, the authors hope to further unearth the role of MTA protein in immunologic responses, a converging integrator of upstream signaling, and epigenetic biology. Excitingly, recent studies discovered that MTA signaling pathways could be effectively targeted by natural compounds to suppress the growth and tumor progression in multiple model systems (Chen et al., 2012; Dhar et al., 2014; Kai et al., 2010; Li, Dias, et al., 2013; Liu et al., 2011; Lu et al., 2014; Zhang, Zhang, & Li, 2013). However, given the fact that MTA1 gene is widely expressed in normal tissues and participates in essential physiologic functions such as circadian rhythm, morphogenesis, and development (Chen & Han, 2001; Li, Pakala, et al., 2013; Manavathi et al., 2007; Reddy et al., 2011; Sen et al., 2014b), identifying selective inhibitors that specifically kill cancer cells without significant toxicity against normal cells is likely to be the next major thrust in the field. Evidently, deeper insights into the regulatory networks of MTA proteins in human cancer will provide an opportunity to develop MTA-directed anticancer therapies in the near future.

ACKNOWLEDGMENTS

We are indebted to many of our colleagues in this field whose original work could not be cited here due to space limitations. The authors thank the previous and current lab members of Dr. Kumar's Laboratory for their contributions to the MTA field, and specially, Dr. Bin Gui for his exceptional efforts in preparing the figures and Dr. Li's Laboratory for assisting in the preparation of this manuscript. We gratefully appreciate the support by the National Institutes of Health grants CA098823 (to R.K.), the National Natural Science Foundation of China (No. 81372847), the Program for Professor of Special Appointment (Eastern Scholar) at Shanghai Institutions of Higher Learning (No. 2013–06), and the Innovation Program of Shanghai Municipal Education Commission (No. 2015ZZ007) (to D.Q.L.).

Conflict of Interest Statement: The authors declare no any potential conflict of interest.

REFERENCES

Adhikary, S., & Eilers, M. (2005). Transcriptional regulation and transformation by Myc proteins. *Nature Reviews. Molecular Cell Biology, 6*, 635–645.

Akhurst, R. J., & Hata, A. (2012). Targeting the TGFbeta signalling pathway in disease. *Nature Reviews Drug Discovery, 11*, 790–811.

Andishehtadbir, A., Najvani, A. D., Pardis, S., Ashkavandi, Z. J., Ashraf, M. J., Khademi, B., et al. (2015). Metastasis-associated protein 1 expression in oral squamous cell carcinomas: correlation with metastasis and angiogenesis. *Turk Patoloji Derg, 31*, 9–15.

Au, T. J., Rodriguez, J., Vincent, J. A., & Tsukiyama, T. (2011). ATP-dependent chromatin remodeling factors tune S phase checkpoint activity. *Molecular and Cellular Biology, 31*, 4454–4463.

Ayrapetov, M. K., Xu, C., Sun, Y., Zhu, K., Parmar, K., D'Andrea, A. D., et al. (2011). Activation of Hif1alpha by the prolylhydroxylase inhibitor dimethyoxalyglycine decreases radiosensitivity. *PloS One, 6*, e26064.

Badenhorst, P., Voas, M., Rebay, I., & Wu, C. (2002). Biological functions of the ISWI chromatin remodeling complex NURF. *Genes & Development, 16*, 3186–3198.

Balasenthil, S., Broaddus, R. R., & Kumar, R. (2006). Expression of metastasis-associated protein 1 (MTA1) in benign endometrium and endometrial adenocarcinomas. *Human Pathology, 37*, 656–661.

Balasenthil, S., Gururaj, A. E., Talukder, A. H., Bagheri-Yarmand, R., Arrington, T., Haas, B. J., et al. (2007). Identification of Pax5 as a target of MTA1 in B-cell lymphomas. *Cancer Research, 67*, 7132–7138.

Bao, Y., & Shen, X. (2007). SnapShot: Chromatin remodeling complexes. *Cell, 129*, 632.

Barone, I., Brusco, L., Gu, G., Selever, J., Beyer, A., Covington, K. R., et al. (2011). Loss of Rho GDIalpha and resistance to tamoxifen via effects on estrogen receptor alpha. *Journal of the National Cancer Institute, 103*, 538–552.

Bassi, C., Ho, J., Srikumar, T., Dowling, R. J., Gorrini, C., Miller, S. J., et al. (2013). Nuclear PTEN controls DNA repair and sensitivity to genotoxic stress. *Science, 341*, 395–399.

Bieging, K. T., Mello, S. S., & Attardi, L. D. (2014). Unravelling mechanisms of p53-mediated tumour suppression. *Nature Reviews Cancer, 14*, 359–370.

Bowen, N. J., Fujita, N., Kajita, M., & Wade, P. A. (2004). Mi-2/NuRD: Multiple complexes for many purposes. *Biochimica et Biophysica Acta, 1677*, 52–57.

Bruning, A., Juckstock, J., Blankenstein, T., Makovitzky, J., Kunze, S., & Mylonas, I. (2010). The metastasis-associated gene MTA3 is downregulated in advanced endometrioid adenocarcinomas. *Histology and Histopathology, 25*, 1447–1456.

Bui-Nguyen, T. M., Pakala, S. B., Sirigiri, D. R., Martin, E., Murad, F., & Kumar, R. (2010). Stimulation of inducible nitric oxide by hepatitis B virus transactivator protein HBx requires MTA1 coregulator. *The Journal of Biological Chemistry, 285*, 6980–6986.

Bui-Nguyen, T. M., Pakala, S. B., Sirigiri, R. D., Xia, W., Hung, M. C., Sarin, S. K., et al. (2010). NF-kappaB signaling mediates the induction of MTA1 by hepatitis B virus transactivator protein HBx. *Oncogene, 29*, 1179–1189.

Cairns, B. R. (2009). The logic of chromatin architecture and remodelling at promoters. *Nature, 461*, 193–198.

Chen, D. W., Fan, Y. F., Li, J., & Jiang, X. X. (2013). MTA2 expression is a novel prognostic marker for pancreatic ductal adenocarcinoma. *Tumour Biology: The journal of the International Society for Oncodevelopmental Biology and Medicine, 34*, 1553–1557.

Chen, Z., & Han, M. (2001). Role of C. elegans lin-40 MTA in vulval fate specification and morphogenesis. *Development, 128*, 4911–4921.

Chen, X., Wang, Y., Luo, H., Luo, Z., Zhang, T., Yang, N., et al. (2012). beta-elemene acts as an antitumor factor and downregulates the expression of survivin, Bcl-xL and Mta-1. *Molecular Medicine Reports, 6*, 989–995.

Cheng, C. W., Liu, Y. F., Yu, J. C., Wang, H. W., Ding, S. L., Hsiung, C. N., et al. (2012). Prognostic significance of cyclin D1, beta-catenin, and MTA1 in patients with invasive ductal carcinoma of the breast. *Annals of Surgical Oncology, 19*, 4129–4139.

Chou, D. M., Adamson, B., Dephoure, N. E., Tan, X., Nottke, A. C., Hurov, K. E., et al. (2010). A chromatin localization screen reveals poly (ADP ribose)-regulated recruitment of the repressive polycomb and NuRD complexes to sites of DNA damage. *Proceedings of the National Academy of Sciences of the United States of America, 107*, 18475–18480.

Chu, H., Chen, X., Wang, H., Du, Y., Wang, Y., Zang, W., et al. (2014). MiR-495 regulates proliferation and migration in NSCLC by targeting MTA3. *Tumour Biology: The journal of the International Society for Oncodevelopmental Biology and Medicine, 35*, 3487–3494.

Cimen, I., Tuncay, S., & Banerjee, S. (2009). 15-Lipoxygenase-1 expression suppresses the invasive properties of colorectal carcinoma cell lines HCT-116 and HT-29. *Cancer Science, 100*, 2283–2291.

Cong, L., Pakala, S. B., Ohshiro, K., Li, D. Q., & Kumar, R. (2011). SUMOylation and SUMO-interacting motif (SIM) of metastasis tumor antigen 1 (MTA1) synergistically regulate its transcriptional repressor function. *The Journal of Biological Chemistry, 286*, 43793–43808.

Covington, K. R., Brusco, L., Barone, I., Tsimelzon, A., Selever, J., Corona-Rodriguez, A., et al. (2013). Metastasis tumor-associated protein 2 enhances metastatic behavior and is associated with poor outcomes in estrogen receptor-negative breast cancer. *Breast Cancer Research and Treatment, 141*, 375–384.

Cui, Y., Niu, A., Pestell, R., Kumar, R., Curran, E. M., Liu, Y., et al. (2006). Metastasis-associated protein 2 is a repressor of estrogen receptor alpha whose overexpression leads to estrogen-independent growth of human breast cancer cells. *Molecular Endocrinology, 20*, 2020–2035.

Dannenmann, C., Shabani, N., Friese, K., Jeschke, U., Mylonas, I., & Bruning, A. (2008). The metastasis-associated gene MTA1 is upregulated in advanced ovarian cancer, represses ERbeta, and enhances expression of oncogenic cytokine GRO. *Cancer Biology & Therapy, 7*, 1460–1467.

Deng, X., Du, L., Wang, C., Yang, Y., Li, J., Liu, H., et al. (2013). Close association of metastasis-associated protein 1 overexpression with increased angiogenesis and poor survival in patients with histologically node-negative gastric cancer. *World Journal of Surgery, 37*, 792–798.

Deng, Y. F., Zhou, D. N., Ye, C. S., Zeng, L., & Yin, P. (2012). Aberrant expression levels of MTA1 and RECK in nasopharyngeal carcinoma: Association with metastasis, recurrence, and prognosis. *The Annals of Otology, Rhinology, and Laryngology, 121*, 457–465.

Dhar, S., Kumar, A., Li, K., Tzivion, G., & Levenson, A. S. (2014). Resveratrol regulates PTEN/Akt pathway through inhibition of MTA1/HDAC unit of the NuRD complex in prostate cancer. *Biochimica et Biophysica Acta, 1853*, 265–275.

Dias, S. J., Zhou, X., Ivanovic, M., Gailey, M. P., Dhar, S., Zhang, L., et al. (2013). Nuclear MTA1 overexpression is associated with aggressive prostate cancer, recurrence and metastasis in African Americans. *Scientific Reports, 3*, 2331.

Ding, Z., Gillespie, L. L., & Paterno, G. D. (2003). Human MI-ER1 alpha and beta function as transcriptional repressors by recruitment of histone deacetylase 1 to their conserved ELM2 domain. *Molecular and Cellular Biology, 23*, 250–258.

Dong, H., Guo, H., Xie, L., Wang, G., Zhong, X., Khoury, T., et al. (2013). The metastasis-associated gene MTA3, a component of the Mi-2/NuRD transcriptional repression complex, predicts prognosis of gastroesophageal junction adenocarcinoma. *PloS One, 8*, e62986.

Du, B., Yang, Z. Y., Zhong, X. Y., Fang, M., Yan, Y. R., Qi, G. L., et al. (2011). Metastasis-associated protein 1 induces VEGF-C and facilitates lymphangiogenesis in colorectal cancer. *World Journal of Gastroenterology: WJG, 17*, 1219–1226.

El Marzouk, S., Schultz-Norton, J. R., Likhite, V. S., McLeod, I. X., Yates, J. R., & Nardulli, A. M. (2007). Rho GDP dissociation inhibitor alpha interacts with estrogen receptor alpha and influences estrogen responsiveness. *Journal of Molecular Endocrinology, 39*, 249–259.

Ellis, L. M., & Hicklin, D. J. (2008). VEGF-targeted therapy: Mechanisms of anti-tumour activity. *Nature Reviews Cancer, 8*, 579–591.

Euskirchen, G., Auerbach, R. K., & Snyder, M. (2012). SWI/SNF chromatin-remodeling factors: Multiscale analyses and diverse functions. *Journal of Biological Chemistry, 287*, 30897–30905.

Faiola, F., Liu, X., Lo, S., Pan, S., Zhang, K., Lymar, E., et al. (2005). Dual regulation of c-Myc by p300 via acetylation-dependent control of Myc protein turnover and coactivation of Myc-induced transcription. *Molecular and Cellular Biology, 25*, 10220–10234.

Fan, L., Wang, H., Xia, X., Rao, Y., Ma, X., Ma, D., et al. (2012). Loss of E-cadherin promotes prostate cancer metastasis via upregulation of metastasis-associated gene 1 expression. *Oncology Letters, 4*, 1225–1233.

Feng, X., Zhang, Q., Xia, S., Xia, B., Zhang, Y., Deng, X., et al. (2014). MTA1 overexpression induces cisplatin resistance in nasopharyngeal carcinoma by promoting cancer stem cells properties. *Molecules and Cells, 37*, 699–704.

Fujita, N., Jaye, D. L., Kajita, M., Geigerman, C., Moreno, C. S., & Wade, P. A. (2003). MTA3, a Mi-2/NuRD complex subunit, regulates an invasive growth pathway in breast cancer. *Cell, 113*, 207–219.

Fujita, N., Kajita, M., Taysavang, P., & Wade, P. A. (2004). Hormonal regulation of metastasis-associated protein 3 transcription in breast cancer cells. *Molecular Endocrinology, 18*, 2937–2949.

Futamura, M., Nishimori, H., Shiratsuchi, T., Saji, S., Nakamura, Y., & Tokino, T. (1999). Molecular cloning, mapping, and characterization of a novel human gene, MTA1-L1, showing homology to a metastasis-associated gene, MTA1. *Journal of Human Genetics, 44*, 52–56.

Garzon, R., Marcucci, G., & Croce, C. M. (2010). Targeting microRNAs in cancer: Rationale, strategies and challenges. *Nature Reviews Drug Discovery, 9*, 775–789.

Ghanta, K. S., Pakala, S. B., Reddy, S. D., Li, D. Q., Nair, S. S., & Kumar, R. (2011). MTA1 coregulation of transglutaminase 2 expression and function during inflammatory response. *The Journal of Biological Chemistry, 286*, 7132–7138.

Gururaj, A. E., Holm, C., Landberg, G., & Kumar, R. (2006). Breast cancer-amplified sequence 3, a target of metastasis-associated protein 1, contributes to tamoxifen resistance in premenopausal patients with breast cancer. *Cell Cycle, 5*, 1407–1410.

Gururaj, A. E., Singh, R. R., Rayala, S. K., Holm, C., den Hollander, P., Zhang, H., et al. (2006). MTA1, a transcriptional activator of breast cancer amplified sequence 3. *Proceedings of the National Academy of Sciences of the United States of America, 103*, 6670–6675.

Hall, J. A., & Georgel, P. T. (2007). CHD proteins: A diverse family with strong ties. *Biochemistry and Cell Biology (Biochimie et Biologie Cellulaire), 85*, 463–476.

Hamatsu, T., Rikimaru, T., Yamashita, Y., Aishima, S., Tanaka, S., Shirabe, K., et al. (2003). The role of MTA1 gene expression in human hepatocellular carcinoma. *Oncology Reports, 10*, 599–604.

Hanahan, D., & Weinberg, R. A. (2000). The hallmarks of cancer. *Cell, 100*, 57–70.

Hanahan, D., & Weinberg, R. A. (2011). Hallmarks of cancer: The next generation. *Cell, 144*, 646–674.

Higashijima, J., Kurita, N., Miyatani, T., Yoshikawa, K., Morimoto, S., Nishioka, M., et al. (2011). Expression of histone deacetylase 1 and metastasis-associated protein 1 as prognostic factors in colon cancer. *Oncology Reports, 26*, 343–348.

Hofer, M. D., Chang, M. C., Hirko, K. A., Rubin, M. A., & Nose, V. (2009). Immunohistochemical and clinicopathological correlation of the metastasis-associated gene 1 (MTA1) expression in benign and malignant pancreatic endocrine tumors. *Modern Pathology: An Official Journal of the United States and Canadian Academy of Pathology, Inc. 22*, 933–939.

Hofer, M. D., Kuefer, R., Varambally, S., Li, H., Ma, J., Shapiro, G. I., et al. (2004). The role of metastasis-associated protein 1 in prostate cancer progression. *Cancer Research, 64*, 825–829.

Hollstein, M., Sidransky, D., Vogelstein, B., & Harris, C. C. (1991). p53 mutations in human cancers. *Science, 253*, 49–53.

Jang, K. S., Paik, S. S., Chung, H., Oh, Y. H., & Kong, G. (2006). MTA1 overexpression correlates significantly with tumor grade and angiogenesis in human breast cancers. *Cancer Science, 97*, 374–379.

Ji, Q., Zheng, G. Y., Xia, W., Chen, J. Y., Meng, X. Y., Zhang, H., et al. (2014). Inhibition of invasion and metastasis of human liver cancer HCCLM3 cells by portulacerebroside A. *Pharmaceutical Biology*, 1–8.

Johnstone, R. W. (2002). Histone-deacetylase inhibitors: Novel drugs for the treatment of cancer. *Nature Reviews Drug Discovery, 1*, 287–299.

Kai, L., Samuel, S. K., & Levenson, A. S. (2010). Resveratrol enhances p53 acetylation and apoptosis in prostate cancer by inhibiting MTA1/NuRD complex. *International Journal of Cancer. Journal International du Cancer, 126*, 1538–1548.

Kai, L., Wang, J., Ivanovic, M., Chung, Y. T., Laskin, W. B., Schulze-Hoepfner, F., et al. (2011). Targeting prostate cancer angiogenesis through metastasis-associated protein 1 (MTA1). *The Prostate, 71*, 268–280.

Kaller, M., Liffers, S. T., Oeljeklaus, S., Kuhlmann, K., Roh, S., Hoffmann, R., et al. (2011). Genome-wide characterization of miR-34a induced changes in protein and mRNA expression by a combined pulsed SILAC and microarray analysis. *Molecular & Cellular Proteomics: MCP, 10*(M111), 010462.

Kang, H., Kiess, A., & Chung, C. H. (2015). Emerging biomarkers in head and neck cancer in the era of genomics. *Nature Reviews. Clinical Oncology, 12*, 11–26.

Kang, H. J., Lee, M. H., Kang, H. L., Kim, S. H., Ahn, J. R., Na, H., et al. (2014). Differential regulation of estrogen receptor alpha expression in breast cancer cells by metastasis-associated protein 1. *Cancer Research, 74*, 1484–1494.

Kaur, E., Gupta, S., & Dutt, S. (2014). Clinical implications of MTA proteins in human cancer. *Cancer Metastasis Reviews, 33*, 1017–1024.

Kawasaki, G., Yanamoto, S., Yoshitomi, I., Yamada, S., & Mizuno, A. (2008). Overexpression of metastasis-associated MTA1 in oral squamous cell carcinomas: Correlation with metastasis and invasion. *International Journal of Oral and Maxillofacial Surgery, 37*, 1039–1046.

Khaleque, M. A., Bharti, A., Gong, J., Gray, P. J., Sachdev, V., Ciocca, D. R., et al. (2008). Heat shock factor 1 represses estrogen-dependent transcription through association with MTA1. *Oncogene, 27*, 1886–1893.

Kleene, R., Zdzieblo, J., Wege, K., & Kern, H. F. (1999). A novel zymogen granule protein (ZG29p) and the nuclear protein MTA1p are differentially expressed by alternative transcription initiation in pancreatic acinar cells of the rat. *Journal of Cell Science, 112*(Pt 15), 2539–2548.

Klinge, C. M. (2001). Estrogen receptor interaction with estrogen response elements. *Nucleic Acids Research, 29*, 2905–2919.

Kong, X., Xu, X., Yan, Y., Guo, F., Li, J., Hu, Y., et al. (2014). Estrogen regulates the tumour suppressor MiRNA-30c and its target gene, MTA-1, in endometrial cancer. *PloS One, 9*, e90810.

Kumar, R., Balasenthil, S., Manavathi, B., Rayala, S. K., & Pakala, S. B. (2010). Metastasis-associated protein 1 and its short form variant stimulates Wnt1 transcription through promoting its derepression from Six3 corepressor. *Cancer Research, 70*, 6649–6658.

Kumar, R., Balasenthil, S., Pakala, S. B., Rayala, S. K., Sahin, A. A., & Ohshiro, K. (2010). Metastasis-associated protein 1 short form stimulates Wnt1 pathway in mammary epithelial and cancer cells. *Cancer Research, 70*, 6598–6608.

Kumar, R., Wang, R. A., & Bagheri-Yarmand, R. (2003). Emerging roles of MTA family members in human cancers. *Seminars in Oncology, 30*, 30–37.

Kumar, R., Wang, R. A., Mazumdar, A., Talukder, A. H., Mandal, M., Yang, Z., et al. (2002). A naturally occurring MTA1 variant sequesters oestrogen receptor-alpha in the cytoplasm. *Nature, 418*, 654–657.

Kuo, A. J., Song, J., Cheung, P., Ishibe-Murakami, S., Yamazoe, S., Chen, J. K., et al. (2012). The BAH domain of ORC1 links H4K20me2 to DNA replication licensing and Meier-Gorlin syndrome. *Nature, 484*, 115–119.

Lai, A. Y., & Wade, P. A. (2011). Cancer biology and NuRD: A multifaceted chromatin remodelling complex. *Nature Reviews Cancer, 11*, 588–596.

Lee, M. H., Na, H., Kim, E. J., Lee, H. W., & Lee, M. O. (2012). Poly(ADP-ribosyl)ation of p53 induces gene-specific transcriptional repression of MTA1. *Oncogene, 31*, 5099–5107.

Lee, M. H., Na, H., Na, T. Y., Shin, Y. K., Seong, J. K., & Lee, M. O. (2012). Epigenetic control of metastasis-associated protein 1 gene expression by hepatitis B virus X protein during hepatocarcinogenesis. *Oncogenesis, 1*, e25.

Lee, M. H., Na, H., Na, T. Y., Shin, Y. K., Seong, J. K., & Lee, M. O. (2014). Epigenetic control of metastasis-associated protein 1 gene expression by hepatitis B virus X protein during hepatocarcinogenesis. *Oncogenesis, 3*, e88.

Lee, H., Ryu, S. H., Hong, S. S., Seo, D. D., Min, H. J., Jang, M. K., et al. (2009). Overexpression of metastasis-associated protein 2 is associated with hepatocellular carcinoma size and differentiation. *Journal of Gastroenterology and Hepatology, 24*, 1445–1450.

Lewis, G. D., Lofgren, J. A., McMurtrey, A. E., Nuijens, A., Fendly, B. M., Bauer, K. D., et al. (1996). Growth regulation of human breast and ovarian tumor cells by heregulin: Evidence for the requirement of ErbB2 as a critical component in mediating heregulin responsiveness. *Cancer Research, 56*, 1457–1465.

Li, W., Bao, W., Ma, J., Liu, X., Xu, R., Wang, R. A., et al. (2008). Metastasis tumor antigen 1 is involved in the resistance to heat stress-induced testicular apoptosis. *FEBS Letters, 582*, 869–873.

Li, Y., Chao, Y., Fang, Y., Wang, J., Wang, M., Zhang, H., et al. (2013). MTA1 promotes the invasion and migration of non-small cell lung cancer cells by downregulating miR-125b. *Journal of Experimental & Clinical Cancer Research: CR, 32*, 33.

Li, K., Dias, S. J., Rimando, A. M., Dhar, S., Mizuno, C. S., Penman, A. D., et al. (2013). Pterostilbene acts through metastasis-associated protein 1 to inhibit tumor growth, progression and metastasis in prostate cancer. *PloS One, 8*, e57542.

Li, D. Q., Divijendra Natha Reddy, S., Pakala, S. B., Wu, X., Zhang, Y., Rayala, S. K., et al. (2009). MTA1 coregulator regulates p53 stability and function. *The Journal of Biological Chemistry, 284*, 34545–34552.

Li, D. Q., & Kumar, R. (2010). Mi-2/NuRD complex making inroads into DNA-damage response pathway. *Cell Cycle, 9*, 2071–2079.

Li, W. F., Liu, N., Cui, R. X., He, Q. M., Chen, M., Jiang, N., et al. (2012). Nuclear overexpression of metastasis-associated protein 1 correlates significantly with poor survival in nasopharyngeal carcinoma. *Journal of Translational Medicine, 10*, 78.

Li, D. Q., Ohshiro, K., Khan, M. N., & Kumar, R. (2010). Requirement of MTA1 in ATR-mediated DNA damage checkpoint function. *The Journal of Biological Chemistry, 285*, 19802–19812.

Li, D. Q., Ohshiro, K., Reddy, S. D., Pakala, S. B., Lee, M. H., Zhang, Y., et al. (2009). E3 ubiquitin ligase COP1 regulates the stability and functions of MTA1. *Proceedings of the National Academy of Sciences of the United States of America, 106*, 17493–17498.

Li, D. Q., Pakala, S. B., Nair, S. S., Eswaran, J., & Kumar, R. (2012). Metastasis-associated protein 1/nucleosome remodeling and histone deacetylase complex in cancer. *Cancer Research, 72*, 387–394.

Li, D. Q., Pakala, S. B., Reddy, S. D., Ohshiro, K., Peng, S. H., Lian, Y., et al. (2010). Revelation of p53-independent function of MTA1 in DNA damage response via modulation of the p21 WAF1-proliferating cell nuclear antigen pathway. *The Journal of Biological Chemistry, 285*, 10044–10052.

Li, D. Q., Pakala, S. B., Reddy, S. D., Ohshiro, K., Zhang, J. X., Wang, L., et al. (2011). Bidirectional autoregulatory mechanism of metastasis-associated protein 1-alternative reading frame pathway in oncogenesis. *Proceedings of the National Academy of Sciences of the United States of America, 108*, 8791–8796.

Li, D. Q., Pakala, S. B., Reddy, S. D., Peng, S., Balasenthil, S., Deng, C. X., et al. (2013). Metastasis-associated protein 1 is an integral component of the circadian molecular machinery. *Nature Communications, 4*, 2545.

Li, Z., & Rana, T. M. (2014). Therapeutic targeting of microRNAs: Current status and future challenges. *Nature Reviews Drug Discovery, 13*, 622–638.

Li, H., Sun, L., Xu, Y., Li, Z., Luo, W., Tang, Z., et al. (2013). Overexpression of MTA3 correlates with tumor progression in non-small cell lung cancer. *PloS One, 8*, e66679.

Li, S. H., Tian, H., Yue, W. M., Li, L., Gao, C., Li, W. J., et al. (2012). Metastasis-associated protein 1 nuclear expression is closely associated with tumor progression and angiogenesis in patients with esophageal squamous cell cancer. *World Journal of Surgery, 36*, 623–631.

Li, S. H., Tian, H., Yue, W. M., Li, L., Li, W. J., Chen, Z. T., et al. (2011). Overexpression of metastasis-associated protein 1 is significantly correlated with tumor angiogenesis and poor survival in patients with early-stage non-small cell lung cancer. *Annals of Surgical Oncology, 18*, 2048–2056.

Li, Y., Vandenboom, T. G., 2nd, Wang, Z., Kong, D., Ali, S., Philip, P. A., et al. (2010). miR-146a suppresses invasion of pancreatic cancer cells. *Cancer Research, 70*, 1486–1495.

Li, S. H., Wang, Z., & Liu, X. Y. (2009). Metastasis-associated protein 1 (MTA1) overexpression is closely associated with shorter disease-free interval after complete resection of histologically node-negative esophageal cancer. *World Journal of Surgery, 33*, 1876–1881.

Li, W., Wu, Z. Q., Zhao, J., Guo, S. J., Li, Z., Feng, X., et al. (2011). Transient protection from heat-stress induced apoptotic stimulation by metastasis-associated protein 1 in pachytene spermatocytes. *PloS One, 6*, e26013.

Li, D. Q., Yang, Y., & Kumar, R. (2014). MTA family of proteins in DNA damage response: Mechanistic insights and potential applications. *Cancer Metastasis Reviews, 33*, 993–1000.

Liu, S. L., Han, Y., Zhang, Y., Xie, C. Y., Wang, E. H., Miao, Y., et al. (2012). Expression of metastasis-associated protein 2 (MTA2) might predict proliferation in non-small cell lung cancer. *Targeted Oncology, 7,* 135–143.

Liu, X., Li, X., Yin, L., Ding, J., Jin, H., & Feng, Y. (2011). Genistein inhibits placental choriocarcinoma cell line JAR invasion through ERbeta/MTA3/Snail/E-cadherin pathway. *Oncology Letters, 2,* 891–897.

Liu, Y. P., Shan, B. E., Wang, X. L., & Ma, L. (2012). Correlation between MTA2 overexpression and tumour progression in esophageal squamous cell carcinoma. *Experimental and Therapeutic Medicine, 3,* 745–749.

Liu, J., Xu, D., Wang, H., Zhang, Y., Chang, Y., Zhang, J., et al. (2014). The subcellular distribution and function of MTA1 in cancer differentiation. *Oncotarget, 5,* 5153–5164.

Liu, T., Yang, M., Yang, S., Ge, T., Gu, L., & Lou, G. (2013). Metastasis-associated protein 1 is a novel marker predicting survival and lymph nodes metastasis in cervical cancer. *Human Pathology, 44,* 2275–2281.

Lu, Y., Wei, C., & Xi, Z. (2014). Curcumin suppresses proliferation and invasion in non-small cell lung cancer by modulation of MTA1-mediated Wnt/beta-catenin pathway. *In Vitro Cellular & Developmental Biology: Animal, 50,* 840–850.

Luo, H., Li, H., Yao, N., Hu, L., & He, T. (2014). Metastasis-associated protein 1 as a new prognostic marker for solid tumors: A meta-analysis of cohort studies. *Tumour Biology: The Journal of the International Society for Oncodevelopmental Biology and Medicine, 35,* 5823–5832.

Luo, J., Su, F., Chen, D., Shiloh, A., & Gu, W. (2000). Deacetylation of p53 modulates its effect on cell growth and apoptosis. *Nature, 408,* 377–381.

Ma, L., Li, W., Zhu, H. P., Li, Z., Sun, Z. J., Liu, X. P., et al. (2010). Localization and androgen regulation of metastasis-associated protein 1 in mouse epididymis. *PloS One, 5,* e15439.

Mahoney, M. G., Simpson, A., Jost, M., Noe, M., Kari, C., Pepe, D., et al. (2002). Metastasis-associated protein (MTA)1 enhances migration, invasion, and anchorage-independent survival of immortalized human keratinocytes. *Oncogene, 21,* 2161–2170.

Manavathi, B., & Kumar, R. (2007). Metastasis tumor antigens, an emerging family of multifaceted master coregulators. *The Journal of Biological Chemistry, 282,* 1529–1533.

Manavathi, B., Peng, S., Rayala, S. K., Talukder, A. H., Wang, M. H., Wang, R. A., et al. (2007). Repression of Six3 by a corepressor regulates rhodopsin expression. *Proceedings of the National Academy of Sciences of the United States of America, 104,* 13128–13133.

Mann, J. (2002). Natural products in cancer chemotherapy: Past, present and future. *Nature Reviews Cancer, 2,* 143–148.

Mao, X. Y., Chen, H., Wang, H., Wei, J., Liu, C., Zheng, H. C., et al. (2012). MTA1 expression correlates significantly with ER-alpha methylation in breast cancer. *Tumour Biology: The Journal of the International Society for Oncodevelopmental Biology and Medicine, 33,* 1565–1572.

Marks, P., Rifkind, R. A., Richon, V. M., Breslow, R., Miller, T., & Kelly, W. K. (2001). Histone deacetylases and cancer: Causes and therapies. *Nature Reviews Cancer, 1,* 194–202.

Martin, M. D., Fischbach, K., Osborne, C. K., Mohsin, S. K., Allred, D. C., & O'Connell, P. (2001). Loss of heterozygosity events impeding breast cancer metastasis contain the MTA1 gene. *Cancer Research, 61,* 3578–3580.

Martin, M. D., Hilsenbeck, S. G., Mohsin, S. K., Hopp, T. A., Clark, G. M., Osborne, C. K., et al. (2006). Breast tumors that overexpress nuclear metastasis-associated 1 (MTA1) protein have high recurrence risks but enhanced responses to systemic therapies. *Breast Cancer Research and Treatment, 95,* 7–12.

Martinez-Balbas, M. A., Tsukiyama, T., Gdula, D., & Wu, C. (1998). Drosophila NURF-55, a WD repeat protein involved in histone metabolism. *Proceedings of the National Academy of Sciences of the United States of America, 95,* 132–137.

Marzook, H., Deivendran, S., Kumar, R., & Pillai, M. R. (2014). Role of MTA1 in head and neck cancers. *Cancer and Metastasis Reviews, 33*(4), 953–964.

Mazumdar, A., Wang, R. A., Mishra, S. K., Adam, L., Bagheri-Yarmand, R., Mandal, M., et al. (2001). Transcriptional repression of oestrogen receptor by metastasis-associated protein 1 corepressor. *Nature Cell Biology, 3*, 30–37.

Metzger, E., Wissmann, M., Yin, N., Muller, J. M., Schneider, R., Peters, A. H., et al. (2005). LSD1 demethylates repressive histone marks to promote androgen-receptor-dependent transcription. *Nature, 437*, 436–439.

Millard, C. J., Watson, P. J., Celardo, I., Gordiyenko, Y., Cowley, S. M., Robinson, C. V., et al. (2013). Class I HDACs share a common mechanism of regulation by inositol phosphates. *Molecular Cell, 51*, 57–67.

Mishra, S. K., Mazumdar, A., Vadlamudi, R. K., Li, F., Wang, R. A., Yu, W., et al. (2003). MICoA, a novel metastasis-associated protein 1 (MTA1) interacting protein coactivator, regulates estrogen receptor-alpha transactivation functions. *The Journal of Biological Chemistry, 278*, 19209–19219.

Mishra, S. K., Talukder, A. H., Gururaj, A. E., Yang, Z., Singh, R. R., Mahoney, M. G., et al. (2004). Upstream determinants of estrogen receptor-alpha regulation of metastatic tumor antigen 3 pathway. *The Journal of Biological Chemistry, 279*, 32709–32715.

Mishra, S. K., Yang, Z., Mazumdar, A., Talukder, A. H., Larose, L., & Kumar, R. (2004). Metastatic tumor antigen 1 short form (MTA1s) associates with casein kinase I-gamma2, an estrogen-responsive kinase. *Oncogene, 23*, 4422–4429.

Miyake, K., Yoshizumi, T., Imura, S., Sugimoto, K., Batmunkh, E., Kanemura, H., et al. (2008). Expression of hypoxia-inducible factor-1alpha, histone deacetylase 1, and metastasis-associated protein 1 in pancreatic carcinoma: Correlation with poor prognosis with possible regulation. *Pancreas, 36*, e1–e9.

Mizuguchi, G., Tsukiyama, T., Wisniewski, J., & Wu, C. (1997). Role of nucleosome remodeling factor NURF in transcriptional activation of chromatin. *Molecular Cell, 1*, 141–150.

Molli, P. R., Singh, R. R., Lee, S. W., & Kumar, R. (2008). MTA1-mediated transcriptional repression of BRCA1 tumor suppressor gene. *Oncogene, 27*, 1971–1980.

Moon, W. S., Chang, K., & Tarnawski, A. S. (2004). Overexpression of metastatic tumor antigen 1 in hepatocellular carcinoma: Relationship to vascular invasion and estrogen receptor-alpha. *Human Pathology, 35*, 424–429.

Moon, H. E., Cheon, H., Chun, K. H., Lee, S. K., Kim, Y. S., Jung, B. K., et al. (2006). Metastasis-associated protein 1 enhances angiogenesis by stabilization of HIF-1alpha. *Oncology Reports, 16*, 929–935.

Moon, H. E., Cheon, H., & Lee, M. S. (2007). Metastasis-associated protein 1 inhibits p53-induced apoptosis. *Oncology Reports, 18*, 1311–1314.

Nagaraj, S. R., Shilpa, P., Rachaiah, K., & Salimath, B. P. (2013). Crosstalk between VEGF and MTA1 signaling pathways contribute to aggressiveness of breast carcinoma. *Molecular Carcinogenesis, 54*, 333–350. http://dx.doi.org/10.1002/mc.22104.

Nair, S. S., Li, D. Q., & Kumar, R. (2013). A core chromatin remodeling factor instructs global chromatin signaling through multivalent reading of nucleosome codes. *Molecular Cell, 49*, 704–718.

Nguyen, D. X., Bos, P. D., & Massague, J. (2009). Metastasis: From dissemination to organ-specific colonization. *Nature Reviews Cancer, 9*, 274–284.

Norris, A., Bianchet, M. A., & Boeke, J. D. (2008). Compensatory interactions between Sir3p and the nucleosomal LRS surface imply their direct interaction. *PLoS Genetics, 4*, e1000301.

Nowak, A. J., Alfieri, C., Stirnimann, C. U., Rybin, V., Baudin, F., Ly-Hartig, N., et al. (2011). Chromatin-modifying complex component Nurf55/p55 associates with histones H3 and H4 and polycomb repressive complex 2 subunit Su(z)12 through partially overlapping binding sites. *The Journal of Biological Chemistry, 286*, 23388–23396.

Ohshiro, K., Rayala, S. K., Wigerup, C., Pakala, S. B., Natha, R. S., Gururaj, A. E., et al. (2010). Acetylation-dependent oncogenic activity of metastasis-associated protein 1 co-regulator. *EMBO Reports, 11*, 691–697.

Onishi, M., Liou, G. G., Buchberger, J. R., Walz, T., & Moazed, D. (2007). Role of the conserved Sir3-BAH domain in nucleosome binding and silent chromatin assembly. *Molecular Cell, 28*, 1015–1028.

Pakala, S. B., Bui-Nguyen, T. M., Reddy, S. D., Li, D. Q., Peng, S., Rayala, S. K., et al. (2010). Regulation of NF-kappaB circuitry by a component of the nucleosome remodeling and deacetylase complex controls inflammatory response homeostasis. *The Journal of Biological Chemistry, 285*, 23590–23597.

Pakala, S. B., Rayala, S. K., Wang, R. A., Ohshiro, K., Mudvari, P., Reddy, S. D., et al. (2013). MTA1 promotes STAT3 transcription and pulmonary metastasis in breast cancer. *Cancer Research, 73*, 3761–3770.

Pakala, S. B., Reddy, S. D., Bui-Nguyen, T. M., Rangparia, S. S., Bommana, A., & Kumar, R. (2010). MTA1 coregulator regulates LPS response via MyD88-dependent signaling. *The Journal of Biological Chemistry, 285*, 32787–32792.

Pakala, S. B., Singh, K., Reddy, S. D., Ohshiro, K., Li, D. Q., Mishra, L., et al. (2011). TGF-beta1 signaling targets metastasis-associated protein 1, a new effector in epithelial cells. *Oncogene, 30*, 2230–2241.

Park, H. R., Cabrini, R. L., Araujo, E. S., Paparella, M. L., Brandizzi, D., & Park, Y. K. (2009). Expression of ezrin and metastatic tumor antigen in osteosarcomas of the jaw. *Tumori, 95*, 81–86.

Park, H. R., Jung, W. W., Kim, H. S., Bacchini, P., Bertoni, F., & Park, Y. K. (2005). Overexpression of metastatic tumor antigen in osteosarcoma: Comparison between conventional high-grade and central low-grade osteosarcoma. *Cancer Research and Treatment: Official Journal of Korean Cancer Association, 37*, 360–364.

Park, J. O., Jung, C. K., Sun, D. I., Joo, Y. H., & Kim, M. S. (2011). Relationships between metastasis-associated protein (MTA) 1 and lymphatic metastasis in tonsil cancer. *European Archives of Oto-Rhino-Laryngology: Official Journal of the European Federation of Oto-Rhino-Laryngological Societies, 268*, 1329–1334.

Park, Y. H., Shin, H. J., Kim, S. U., Kim, J. M., Kim, J. H., Bang, D. H., et al. (2013). iNOS promotes HBx-induced hepatocellular carcinoma via upregulation of JNK activation. *Biochemical and Biophysical Research Communications, 435*, 244–249.

Prisco, M. G., Zannoni, G. F., De Stefano, I., Vellone, V. G., Tortorella, L., Fagotti, A., et al. (2012). Prognostic role of metastasis tumor antigen 1 in patients with ovarian cancer: A clinical study. *Human Pathology, 43*, 282–288.

Psaila, B., & Lyden, D. (2009). The metastatic niche: Adapting the foreign soil. *Nature Reviews Cancer, 9*, 285–293.

Qian, H., Lu, N., Xue, L., Liang, X., Zhang, X., Fu, M., et al. (2005). Reduced MTA1 expression by RNAi inhibits in vitro invasion and migration of esophageal squamous cell carcinoma cell line. *Clinical & Experimental Metastasis, 22*, 653–662.

Reddy, S. D., Pakala, S. B., Molli, P. R., Sahni, N., Karanam, N. K., Mudvari, P., et al. (2012). Metastasis-associated protein 1/histone deacetylase 4-nucleosome remodeling and deacetylase complex regulates phosphatase and tensin homolog gene expression and function. *The Journal of Biological Chemistry, 287*, 27843–27850.

Reddy, S. D., Pakala, S. B., Ohshiro, K., Rayala, S. K., & Kumar, R. (2009). MicroRNA-661, a c/EBPalpha target, inhibits metastatic tumor antigen 1 and regulates its functions. *Cancer Research, 69*, 5639–5642.

Reddy, S. D., Rayala, S. K., Ohshiro, K., Pakala, S. B., Kobori, N., Dash, P., et al. (2011). Multiple coregulatory control of tyrosine hydroxylase gene transcription. *Proceedings of the National Academy of Sciences of the United States of America, 108*, 4200–4205.

Riley, T., Sontag, E., Chen, P., & Levine, A. (2008). Transcriptional control of human p53-regulated genes. *Nature Reviews. Molecular Cell Biology, 9*, 402–412.

Roberts, C. W., & Orkin, S. H. (2004). The SWI/SNF complex—Chromatin and cancer. *Nature Reviews Cancer, 4*, 133–142.

Roepman, P., de Jager, A., Groot Koerkamp, M. J., Kummer, J. A., Slootweg, P. J., & Holstege, F. C. (2006). Maintenance of head and neck tumor gene expression profiles upon lymph node metastasis. *Cancer Research, 66*, 11110–11114.

Ryu, S. H., Chung, Y. H., Lee, H., Kim, J. A., Shin, H. D., Min, H. J., et al. (2008). Metastatic tumor antigen 1 is closely associated with frequent postoperative recurrence and poor survival in patients with hepatocellular carcinoma. *Hepatology, 47*, 929–936.

Safe, S., & Abdelrahim, M. (2005). Sp transcription factor family and its role in cancer. *European Journal of Cancer, 41*, 2438–2448.

Salot, S., & Gude, R. (2013). MTA1-mediated transcriptional repression of SMAD7 in breast cancer cell lines. *European Journal of Cancer, 49*, 492–499.

Sasaki, H., Moriyama, S., Nakashima, Y., Kobayashi, Y., Yukiue, H., Kaji, M., et al. (2002). Expression of the MTA1 mRNA in advanced lung cancer. *Lung Cancer, 35*, 149–154.

Sasaki, H., Yukiue, H., Kobayashi, Y., Nakashima, Y., Kaji, M., Fukai, I., et al. (2001). Expression of the MTA1 mRNA in thymoma patients. *Cancer Letters, 174*, 159–163.

Semenza, G. L. (2003). Targeting HIF-1 for cancer therapy. *Nature Reviews Cancer, 3*, 721–732.

Sen, N., Gui, B., & Kumar, R. (2014a). Role of MTA1 in cancer progression and metastasis. *Cancer Metastasis Reviews, 33*, 879–889.

Sen, N., Gui, B., & Kumar, R. (2014b). Physiological functions of MTA family of proteins. *Cancer and Metastasis Reviews, 33*, 869–877.

Shen, W. H., Balajee, A. S., Wang, J., Wu, H., Eng, C., Pandolfi, P. P., et al. (2007). Essential role for nuclear PTEN in maintaining chromosomal integrity. *Cell, 128*, 157–170.

Sherr, C. J. (2001). The INK4a/ARF network in tumour suppression. *Nature Reviews. Molecular Cell Biology, 2*, 731–737.

Siegel, R., Ma, J., Zou, Z., & Jemal, A. (2014). Cancer statistics, 2014. *CA: A Cancer Journal for Clinicians, 64*, 9–29.

Simpson, A., Uitto, J., Rodeck, U., & Mahoney, M. G. (2001). Differential expression and subcellular distribution of the mouse metastasis-associated proteins Mta1 and Mta3. *Gene, 273*, 29–39.

Singh, R. R., Barnes, C. J., Talukder, A. H., Fuqua, S. A., & Kumar, R. (2005). Negative regulation of estrogen receptor alpha transactivation functions by LIM domain only 4 protein. *Cancer Research, 65*, 10594–10601.

Song, Q., Li, Y., Zheng, X., Fang, Y., Chao, Y., Yao, K., et al. (2013). MTA1 contributes to actin cytoskeleton reorganization and metastasis of nasopharyngeal carcinoma by modulating Rho GTPases and Hedgehog signaling. *The International Journal of Biochemistry & Cell Biology, 45*, 1439–1446.

Song, H., Spichiger-Haeusermann, C., & Basler, K. (2009). The ISW1-containing NURF complex regulates the output of the canonical Wingless pathway. *EMBO Reports, 10*, 1140–1146.

Song, L., Wang, Z., & Liu, X. (2012). MTA1: A prognosis indicator of postoperative patients with esophageal carcinoma. *The Thoracic and Cardiovascular Surgeon, 61*, 479–485.

Song, Q., Zhang, H., Wang, M., Song, W., Ying, M., Fang, Y., et al. (2013). MTA1 promotes nasopharyngeal carcinoma growth in vitro and in vivo. *Journal of Experimental & Clinical Cancer Research: CR, 32*, 54.

Talukder, A. H., Gururaj, A., Mishra, S. K., Vadlamudi, R. K., & Kumar, R. (2004). Metastasis-associated protein 1 interacts with NRIF3, an estrogen-inducible nuclear receptor coregulator. *Molecular and Cellular Biology, 24*, 6581–6591.

Talukder, A. H., Mishra, S. K., Mandal, M., Balasenthil, S., Mehta, S., Sahin, A. A., et al. (2003). MTA1 interacts with MAT1, a cyclin-dependent kinase-activating kinase complex ring finger factor, and regulates estrogen receptor transactivation functions. *The Journal of Biological Chemistry, 278*, 11676–11685.

Tessarz, P., & Kouzarides, T. (2014). Histone core modifications regulating nucleosome structure and dynamics. *Nature Reviews. Molecular Cell Biology, 15*, 703–708.

Thomas, C., & Gustafsson, J. A. (2011). The different roles of ER subtypes in cancer biology and therapy. *Nature Reviews Cancer, 11*, 597–608.

Toh, Y., Kuwano, H., Mori, M., Nicolson, G. L., & Sugimachi, K. (1999). Overexpression of metastasis-associated MTA1 mRNA in invasive oesophageal carcinomas. *British Journal of Cancer, 79*, 1723–1726.

Toh, Y., & Nicolson, G. L. (2009). The role of the MTA family and their encoded proteins in human cancers: Molecular functions and clinical implications. *Clinical & Experimental Metastasis, 26*, 215–227.

Toh, Y., & Nicolson, G. L. (2014). Properties and clinical relevance of MTA1 protein in human cancer. *Cancer Metastasis Reviews, 33*, 891–900.

Toh, Y., Ohga, T., Endo, K., Adachi, E., Kusumoto, H., Haraguchi, M., et al. (2004). Expression of the metastasis-associated MTA1 protein and its relationship to deacetylation of the histone H4 in esophageal squamous cell carcinomas. *International Journal of Cancer. Journal International du Cancer, 110*, 362–367.

Toh, Y., Oki, E., Oda, S., Tokunaga, E., Ohno, S., Maehara, Y., et al. (1997). Overexpression of the MTA1 gene in gastrointestinal carcinomas: Correlation with invasion and metastasis. *International Journal of Cancer. Journal International du Cancer, 74*, 459–463.

Toh, Y., Pencil, S. D., & Nicolson, G. L. (1994). A novel candidate metastasis-associated gene, mta1, differentially expressed in highly metastatic mammary adenocarcinoma cell lines. cDNA cloning, expression, and protein analyses. *The Journal of Biological Chemistry, 269*, 22958–22963.

Tong, J. K., Hassig, C. A., Schnitzler, G. R., Kingston, R. E., & Schreiber, S. L. (1998). Chromatin deacetylation by an ATP-dependent nucleosome remodelling complex. *Nature, 395*, 917–921.

Trepo, C., Chan, H. L., & Lok, A. (2014). Hepatitis B virus infection. *Lancet, 384*, 2053–2063.

Tsukiyama, T., Daniel, C., Tamkun, J., & Wu, C. (1995). ISWI, a member of the SWI2/SNF2 ATPase family, encodes the 140 kDa subunit of the nucleosome remodeling factor. *Cell, 83*, 1021–1026.

Tsukiyama, T., & Wu, C. (1995). Purification and properties of an ATP-dependent nucleosome remodeling factor. *Cell, 83*, 1011–1020.

Tuncay Cagatay, S., Cimen, I., Savas, B., & Banerjee, S. (2013). MTA-1 expression is associated with metastasis and epithelial to mesenchymal transition in colorectal cancer cells. *Tumour Biology, 34*(2), 1189–1204. http://dx.doi.org/10.1007/s13277-013-0662-x.

Uraki, S., Sugimoto, K., Shiraki, K., Tameda, M., Inagaki, Y., Ogura, S., et al. (2014). Human beta-defensin-3 inhibits migration of colon cancer cells via downregulation of metastasis-associated 1 family, member 2 expression. *International Journal of Oncology, 45*, 1059–1064.

van Riggelen, J., Yetil, A., & Felsher, D. W. (2010). MYC as a regulator of ribosome biogenesis and protein synthesis. *Nature Reviews Cancer, 10*, 301–309.

Voss, T. C., & Hager, G. L. (2014). Dynamic regulation of transcriptional states by chromatin and transcription factors. *Nature Reviews Genetics, 15*, 69–81.

Wade, P. A., Jones, P. L., Vermaak, D., & Wolffe, A. P. (1998). A multiple subunit Mi-2 histone deacetylase from Xenopus laevis cofractionates with an associated Snf2 superfamily ATPase. *Current Biology: CB, 8*, 843–846.

Wang, R. A. (2014). MTA1-a stress response protein: A master regulator of gene expression and cancer cell behavior. *Cancer Metastasis Reviews, 33*, 1001–1009.

Wang, L., Charroux, B., Kerridge, S., & Tsai, C. C. (2008). Atrophin recruits HDAC1/2 and G9a to modify histone H3K9 and to determine cell fates. *EMBO Reports, 9*, 555–562.

Wang, H., Fan, L., Wei, J., Weng, Y., Zhou, L., Shi, Y., et al. (2012). Akt mediates metastasis-associated gene 1 (MTA1) regulating the expression of E-cadherin and promoting the invasiveness of prostate cancer cells. *PloS One, 7*, e46888.

Wang, Y., Li, L., Li, Q., Xie, C., & Wang, E. (2012). Expression of P120 catenin, Kaiso, and metastasis tumor antigen-2 in thymomas. *Tumour Biology : The Journal of the International Society for Oncodevelopmental Biology and Medicine, 33*, 1871–1879.

Wang, Y., Zhang, H., Chen, Y., Sun, Y., Yang, F., Yu, W., et al. (2009). LSD1 is a subunit of the NuRD complex and targets the metastasis programs in breast cancer. *Cell, 138*, 660–672.

Weng, W., Yin, J., Zhang, Y., Qiu, J., & Wang, X. (2014). Metastasis-associated protein 1 promotes tumor invasion by downregulation of E-cadherin. *International Journal of Oncology, 44*, 812–818.

Wilson, B. G., & Roberts, C. W. (2011). SWI/SNF nucleosome remodellers and cancer. *Nature Reviews Cancer, 11*, 481–492.

Wouters, B. G., & Koritzinsky, M. (2008). Hypoxia signalling through mTOR and the unfolded protein response in cancer. *Nature Reviews Cancer, 8*, 851–864.

Xia, Y., Chen, Q., Zhong, Z., Xu, C., Wu, C., Liu, B., et al. (2013). Down-regulation of miR-30c promotes the invasion of non-small cell lung cancer by targeting MTA1. *Cellular Physiology and Biochemistry: International Journal of Experimental Cellular Physiology, Biochemistry, and Pharmacology, 32*, 476–485.

Xia, L., & Zhang, Y. (2001). Sp1 and ETS family transcription factors regulate the mouse Mta2 gene expression. *Gene, 268*, 77–85.

Xiao, H., Sandaltzopoulos, R., Wang, H. M., Hamiche, A., Ranallo, R., Lee, K. M., et al. (2001). Dual functions of largest NURF subunit NURF301 in nucleosome sliding and transcription factor interactions. *Molecular Cell, 8*, 531–543.

Xu, L., Mao, X. Y., Fan, C. F., & Zheng, H. C. (2011). MTA1 expression correlates significantly with cigarette smoke in non-small cell lung cancer. *Virchows Archiv: An International Journal of Pathology, 459*, 415–422.

Xue, Y., Wong, J., Moreno, G. T., Young, M. K., Cote, J., & Wang, W. (1998). NURD, a novel complex with both ATP-dependent chromatin-remodeling and histone deacetylase activities. *Molecular Cell, 2*, 851–861.

Yan, D., Avtanski, D., Saxena, N. K., & Sharma, D. (2012). Leptin-induced epithelial-mesenchymal transition in breast cancer cells requires beta-catenin activation via Akt/GSK3- and MTA1/Wnt1 protein-dependent pathways. *The Journal of Biological Chemistry, 287*, 8598–8612.

Yan, C., Wang, H., Toh, Y., & Boyd, D. D. (2003). Repression of 92-kDa type IV collagenase expression by MTA1 is mediated through direct interactions with the promoter via a mechanism, which is both dependent on and independent of histone deacetylation. *The Journal of Biological Chemistry, 278*, 2309–2316.

Yang, Q. Y., Li, J. H., Wang, Q. Y., Wu, Y., Qin, J. L., Cheng, J. J., et al. (2014). MTA1 promotes cell proliferation via DNA damage repair in epithelial ovarian cancer. *Genetics and Molecular Research: GMR, 13*, 10269–10278.

Yang, N., & Xu, R. M. (2013). Structure and function of the BAH domain in chromatin biology. *Critical Reviews in Biochemistry and Molecular Biology, 48*, 211–221.

Yao, Y. L., & Yang, W. M. (2003). The metastasis-associated proteins 1 and 2 form distinct protein complexes with histone deacetylase activity. *The Journal of Biological Chemistry, 278*, 42560–42568.

Yoo, Y. G., Kong, G., & Lee, M. O. (2006). Metastasis-associated protein 1 enhances stability of hypoxia-inducible factor-1alpha protein by recruiting histone deacetylase 1. *The EMBO Journal, 25*, 1231–1241.

Yoo, Y. G., Na, T. Y., Seo, H. W., Seong, J. K., Park, C. K., Shin, Y. K., et al. (2008). Hepatitis B virus X protein induces the expression of MTA1 and HDAC1, which enhances hypoxia signaling in hepatocellular carcinoma cells. *Oncogene*, *27*, 3405–3413.

Yu, L., Su, Y. S., Zhao, J., Wang, H., & Li, W. (2013). Repression of NR4A1 by a chromatin modifier promotes docetaxel resistance in PC-3 human prostate cancer cells. *FEBS Letters*, *587*, 2542–2551.

Yu, Y., Wang, Z., Zhang, M. Y., Liu, X. Y., & Zhang, H. (2011). Relation between prognosis and expression of metastasis-associated protein 1 in stage I non-small cell lung cancer. *Interactive Cardiovascular and Thoracic Surgery*, *12*, 166–169.

Yuan, T., Zhang, H., Liu, B., Zhang, Q., Liang, Y., Zheng, R., et al. (2014). Expression of MTA1 in nasopharyngeal carcinoma and its correlation with prognosis. *Medical Oncology*, *31*, 330.

Zhang, L., Chen, Q. Y., Liu, H., Tang, L. Q., & Mai, H. Q. (2013). Emerging treatment options for nasopharyngeal carcinoma. *Drug Design, Development and Therapy*, *7*, 37–52.

Zhang, X. Y., DeSalle, L. M., Patel, J. H., Capobianco, A. J., Yu, D., Thomas-Tikhonenko, A., et al. (2005). Metastasis-associated protein 1 (MTA1) is an essential downstream effector of the c-MYC oncoprotein. *Proceedings of the National Academy of Sciences of the United States of America*, *102*, 13968–13973.

Zhang, Y., LeRoy, G., Seelig, H. P., Lane, W. S., & Reinberg, D. (1998). The dermatomyositis-specific autoantigen Mi2 is a component of a complex containing histone deacetylase and nucleosome remodeling activities. *Cell*, *95*, 279–289.

Zhang, Y., Ng, H. H., Erdjument-Bromage, H., Tempst, P., Bird, A., & Reinberg, D. (1999). Analysis of the NuRD subunits reveals a histone deacetylase core complex and a connection with DNA methylation. *Genes & Development*, *13*, 1924–1935.

Zhang, H., Stephens, L. C., & Kumar, R. (2006). Metastasis tumor antigen family proteins during breast cancer progression and metastasis in a reliable mouse model for human breast cancer. *Clinical Cancer Research: An Official Journal of the American Association for Cancer Research*, *12*, 1479–1486.

Zhang, X., Zhang, Y., & Li, Y. (2013). beta-elemene decreases cell invasion by upregulating E-cadherin expression in MCF-7 human breast cancer cells. *Oncology Reports*, *30*, 745–750.

Zhao, L., Liu, L., Wang, S., Zhang, Y. F., Yu, L., & Ding, Y. Q. (2007). Differential proteomic analysis of human colorectal carcinoma cell lines metastasis-associated proteins. *Journal of Cancer Research and Clinical Oncology*, *133*, 771–782.

Zheng, S., Du, Y., Chu, H., Chen, X., Li, P., Wang, Y., et al. (2013). Analysis of MAT3 gene expression in NSCLC. *Diagnostic Pathology*, *8*, 166.

Zhou, C., Ji, J., Cai, Q., Shi, M., Chen, X., Yu, Y., et al. (2013). MTA2 promotes gastric cancer cells invasion and is transcriptionally regulated by Sp1. *Molecular Cancer*, *12*, 102.

Zhou, H., Xu, X., Xun, Q., Yu, D., Ling, J., Guo, F., et al. (2012). microRNA-30c negatively regulates endometrial cancer cells by targeting metastasis-associated gene-1. *Oncology Reports*, *27*, 807–812.

Zhou, J., Zhan, S., Tan, W., Cheng, R., Gong, H., & Zhu, Q. (2014). P300 binds to and acetylates MTA2 to promote colorectal cancer cells growth. *Biochemical and Biophysical Research Communications*, *444*, 387–390.

Zhu, W., Cai, M. Y., Tong, Z. T., Dong, S. S., Mai, S. J., Liao, Y. J., et al. (2012). Overexpression of EIF5A2 promotes colorectal carcinoma cell aggressiveness by upregulating MTA1 through C-myc to induce epithelial-mesenchymaltransition. *Gut*, *61*, 562–575.

Zhu, X., Guo, Y., Li, X., Ding, Y., & Chen, L. (2010). Metastasis-associated protein 1 nuclear expression is associated with tumor progression and clinical outcome in patients with non-small cell lung cancer. *Journal of Thoracic Oncology: Official Publication of the International Association for the Study of Lung Cancer*, *5*, 1159–1166.

CHAPTER TWO

Examination of Epigenetic and other Molecular Factors Associated with *mda-9/Syntenin* Dysregulation in Cancer Through Integrated Analyses of Public Genomic Datasets

Manny D. Bacolod*,[†], Swadesh K. Das*,[†,‡], Upneet K. Sokhi*, Steven Bradley[§], David A. Fenstermacher[‡,¶], Maurizio Pellecchia[∥], Luni Emdad*,[†,‡], Devanand Sarkar*,[†,‡], Paul B. Fisher*,[†,‡,1]

*Department of Human and Molecular Genetics, Virginia Commonwealth University, School of Medicine, Richmond, Virginia, USA
[†]VCU Institute of Molecular Medicine, Virginia Commonwealth University, School of Medicine, Richmond, Virginia, USA
[‡]VCU Massey Cancer Center, Virginia Commonwealth University, School of Medicine, Richmond, Virginia, USA
[§]VCU Bioinformatics Program, School of Medicine, Virginia Commonwealth University, Richmond, Virginia, USA
[¶]Department of Biostatistics, School of Medicine, Virginia Commonwealth University, Richmond, Virginia, USA
[∥]Sanford-Burnham Medical Research Institute, La Jolla, California, USA
[1]Corresponding author: e-mail address: pbfisher@vcu.edu

Contents

1. Background on *mda-9/Syntenin*   51
2. Publicly Available Cancer Genomic Datasets   53
3. Specific Datasets Examined for *mda-9* Analysis and the Analytical Approaches Employed   54
   3.1 Publicly Available Genomic Datasets   54
   3.2 Analytical Tools   57
4. The Patterns of *mda-9* Expression During Cancer Progression and Their Clinical Implications in Different Cancer Types   58
   4.1 Glioma   58
   4.2 Melanoma, Prostate Cancer, Liver Hepatocellular Carcinoma, and Kidney Renal Papillary Carcinoma   59
   4.3 *mda-9* Expression in Metastasis Samples   60
5. *mda-9* Expression Correlates with Both Gene Copy Number and the Methylation Status of cg1719774   61
   5.1 mda-9 RNA Expression Generated Illumina HiSeq Data   61

    5.2 mda-9 Copy Number as Determined by Affymetrix SNP Array   61
    5.3 *mda-9* CpG Sites Interrogated in Illumina 450K Array   61
    5.4 Which Among the Two CpG Sites Influence *mda-9* Expression?   62
    5.5 Glioma   65
    5.6 Melanoma, Prostate Cancer, Liver Hepatocellular Carcinoma, and Kidney Renal Papillary Carcinoma   67
    5.7 Combination of the Six Cancer Datasets   68
6. *mda-9* is Highly Expressed in Melanoma   69
    6.1 mda-9 RNA Expression Across all Cancer Types (PANCAN Dataset)   69
    6.2 MDA-9 Protein Expression Across all Cancer Types (The Human Protein Atlas)   69
7. Hypomethylation at cg1719774 is Associated with Histone Modifications Indicative of Higher Transcriptional Activity   71
    7.1 Information from the ENCODE Project   71
    7.2 How CpG Methylation is Associated with Differential Binding of Modified Histones   72
8. Predicting Genes and Pathways Associated with *mda-9* Dysregulation in Glioma   74
    8.1 Comparing *mda-9*-High and *mda-9*-Low Subtypes of Glioma   74
    8.2 Genes that are Most Highly Dysregulated in *mda-9*-High Glioma   76
    8.3 Pathways and Gene Groups Identified Through GSEA   99
    8.4 A Network of GSEA-Identified Gene Sets   109
9. Summary, Conclusion, and Future Directions   111
Acknowledgments   115
References   115

Abstract

mda-9/Syntenin (melanoma differentiation-associated gene 9) is a PDZ domain containing, cancer invasion-related protein. In this study, we employed multiple integrated bioinformatic approaches to identify the probable epigenetic factors, molecular pathways, and functionalities associated with *mda-9* dysregulation during cancer progression. Analyses of publicly available genomic data (e.g., expression, copy number, methylation) from TCGA, GEO, ENCODE, and Human Protein Atlas projects led to the following observations: (a) *mda-9* expression correlates with both copy number and methylation level of an intronic CpG site (cg1719774) located downstream of the CpG island, (b) cg1719774 methylation is a likely prognostic marker in glioma, (c) among 22 cancer types, melanoma exhibits the highest *mda-9* level, and lowest level of methylation at cg1719774, (d) cg1719774 hypomethylation is also associated with histone modifications (at the *mda-9* locus) indicative of more active transcription, (e) using Gene Set Enrichment Analysis (GSEA), and the *Virtual Gene Overexpression or Repression* (VIGOR) analytical scheme, we were able to predict *mda-9*'s association with extracellular matrix organization (e.g., MMPs, collagen, integrins), IGFBP2 and NF-κB signaling pathways, phospholipid metabolism, cytokines (e.g., interleukins), CTLA-4, and components of complement cascade pathways. Indeed, previous publications have shown that many of the aforementioned genes and pathways are associated with *mda-9*'s functionality.

1. BACKGROUND ON *MDA-9/SYNTENIN*

mda-9/Syntenin (melanoma differentiation-associated gene-9), also known as *SDCBP* (*Syndecan binding protein*), is a human gene transcribed in the plus strand of the 8q12.1 region. The gene consists of nine exons (including the UTR regions), and has five transcript variants coding for three protein isoforms. Isoforms 1, 2, and 3, code for proteins with 298, 292, and 297 amino acids, respectively. As indicated in the UCSC Genome Browser (genome.ucsc.edu), there is a possible 10th exon located 3′ of the 5′ UTR exon, which may result in an even longer isoform (318 aa). The prominent features of the protein (isoform 1 is 33 kDa) are its two PDZ domains (PDZ1 and PDZ2). PDZ domains, due to their repertoire of possible interactions (C-terminal peptide recognition, interactions with internal peptide ligands, PDZ–PDZ interactions, PDZ–phospholipid interactions) (Ivarsson, 2012) are known to mediate a wide array of signaling pathways and cellular functions. Among the molecules, the protein interacts with are phosphoinositides, IL-5 receptor α (IL5RA), proTGFα, syndecan, eIF5A, Schwannomin, and CD6 (see Beekman & Coffer, 2008; Sarkar, Boukerche, Su, & Fisher, 2008 for reviews). A short list of cellular processes involving MDA-9 are: axon outgrowth (Tomoda, Kim, Zhan, & Hatten, 2004), chemotaxis (Sala-Valdes et al., 2012), HIV-entry (Gordon-Alonso et al., 2012), development of neuronal membrane architecture (Hirbec, Martin, & Henley, 2005), protein-cell surface localization (Fernandez-Larrea, Merlos-Suarez, Urena, Baselga, & Arribas, 1999), cell adhesion (Zimmermann et al., 2001), and pro-metastatic and pro-angiogenic activities (Boukerche et al., 2005; Das et al., 2013).

mda-9, as numerous studies have already shown, also plays an important role in cancer progression, particularly during the invasion/metastasis stage (reviewed in Kegelman et al., 2015). The earliest report describing this gene in our laboratory dates back to 1996 (Lin, Jiang, & Fisher, 1998), when it was first cloned using a subtraction hybridization approach developed in our laboratory (Jiang, Lin, Su, Goldstein, & Fisher, 1995) using melanoma cells induced to terminally differentiate by treatment with interferon β (IFN-β) and the antileukemic compound mezerein (Fisher, Prignoli, Hermo, Weinstein, & Pestka, 1985). Subsequent studies have demonstrated that MDA-9 is a positive regulator of metastasis in melanoma partially attributed to its interaction with c-Src, which eventually leads to the activation of the transcription factor NF-κB (Boukerche et al., 2010; Boukerche, Su, Prevot,

Sarkar, & Fisher, 2008; Das et al., 2012). These changes induce an increase in the transcription of matrix metalloproteinases (MMPs), necessary for the degradation of extracellular matrix (ECM) during invasion (Das et al., 2012). MDA-9's interaction with c-Src can also lead to transcriptional activation of insulin growth factor binding protein 2 (IGFBP2), which can promote angiogenesis in melanoma (Das et al., 2013). As our group recently reported, *mda-9*'s invasion-promoting property is also evident in glioma, in which its overexpression led to activation of c-Src, p38 MAPK, and NF-κB, and eventually the elevated expression of MMP2 and secretion of interleukin-8 (IL-8) (Kegelman et al., 2014). *mda-9*-driven proliferation via p38 MAPK activation and MMP2 upregulation was similarly demonstrated in hepatocellular carcinoma (Liu, Zhang, Lv, Xiang, & Shi, 2014) and small cell lung cancer (Kim et al., 2014) cells. *mda-9* is also overexpressed in metastatic (Qian et al., 2013), as well as ER-negative (Koo et al., 2002) breast cancer. *mda-9*'s regulation of cell migration was also demonstrated in colorectal cancer (Lee et al., 2011), a cancer type in which poor clinical outcome is associated with elevated expression of this gene (Hao et al., 2010). Another recent finding from our group is *mda-9*'s regulation of urothelial cell proliferation through its modulation of EGFR signaling (Dasgupta et al., 2013).

Despite the wealth of *mda-9*-related knowledge listed above, there is still much to learn regarding *mda-9* and its involvement in cancer progression. One area that has not yet been investigated is how genetic and epigenetic factors contribute to its elevated expression during cancer progression. For any gene, two factors that can lead to elevated expression would be: a gain or amplification of copy number, and a reduced methylation at the appropriate CpG site on its promoter region. These are, by no means, the only factors which can influence the transcription of a gene. Activation of transcription factors (such as NF-κB, shown in previous experiments) and the upstream signaling pathways are also necessary for the gene's transcriptional activation. The activities of DNA methyltransferases (DNMTs), which initially methylate the CpG sites, may also factor in the gene's dysregulation. The elucidation of genetic and epigenetic factors associated with *mda-9*'s transformation into a cancer- or metastasis-promoting state may now be conducted *in silico*. This is due to the availability of comprehensive and publicly available genomic datasets (such as genome-wide expression, methylation, and copy number data) for various types of cancer. Integrated analyses of these datasets may help define the effects of *mda-9* copy number and CpG methylation on its transcription. Analysis of expression datasets may also lead to the identification of various genes and molecular pathways associated with *mda-9* dysregulation, which can then be validated experimentally.

2. PUBLICLY AVAILABLE CANCER GENOMIC DATASETS

In the past decade, cancer-related data generated using various genome-wide molecular profiling tools have been made publicly available. Currently, there are tens of thousands of cancer genomic datasets available to the public for analysis. Two very extensive repositories are NCBI's Gene Expression Omnibus (GEO) (http://www.ncbi.nlm.nih.gov/geo/) and EMBL's Array Express (www.ebi.ac.uk/arrayexpress/). A huge proportion of these datasets are genome-wide expression profiles (but also include genome-wide methylation and copy number data) of cancer tissue samples, cell lines, and mouse models. However, what is perhaps the most comprehensive and organized cancer genomic repository is The Cancer Genome Atlas (TCGA) (https://tcga-data.nci.nih.gov/tcga/). Launched in 2005 (Kaiser, 2005), TCGA has now examined the genome-wide expression (mRNA, miRNA, exon, limited protein), copy number variations, methylation status, and mutations in more than 20 adult cancer types (a total of more than 6000 tissue samples). The primary advantages of TCGA datasets are: (a) each patient sample is accompanied by very comprehensive clinicopathological data (e.g., follow-up survival records, TNM staging, treatment records), (b) a huge portion of the samples have integrated molecular profiles (i.e., same sample being profiled for expression, copy number, sequence, methylation), (c) many of the tumor samples have matched normals, and (d) the data are generated using the latest and widely considered standards in molecular profiling technology. These include Illumina HiSeq 2000 for expression profiling, Illumina Infinium 450K BeadChip for methylation analysis, Affymetrix SNP 6 array for copy number analysis, and various Next Gen Sequencing platforms for mutational profiling. Also available to the public is the Human Protein Atlas Database (http://www.proteinatlas.org/) (Uhlen et al., 2010). This database (although not as quantitative as the mRNA levels reported using RNASeq or array chips) is a genome-wide immunohistochemical staining analyses portal of a large number of human tissues, cancers, and cell lines. Another data repository, which is very useful for cancer research is ENCODE (ENCyclopedia of DNA Elements) (https://genome.ucsc.edu/ENCODE/) (ENCODE Project Consortium, 2004), a project which aims to build a comprehensive list of functional elements in the human genome, and further our understanding of gene regulations. In ENCODE, numerous experimental tools, such as ChIP Seq technology, were employed to examine how proteins (e.g., transcription factors, histones) recognize and bind to genomic sequences such as promoter regions.

3. SPECIFIC DATASETS EXAMINED FOR *MDA-9* ANALYSIS AND THE ANALYTICAL APPROACHES EMPLOYED

3.1 Publicly Available Genomic Datasets

This review is a foray into *mda-9/Syntenin*'s genetic and epigenetic regulation through a careful and rigorous examination of various public genomic datasets (see Fig. 1 for the scheme). Most datasets originated from TCGA, and some were downloaded from NCBI-GEO (described in Table 1). Specifically, we analyzed the Illumina HiSeq 2000 (expression), Illumina Infinium 450K BeadChip (CpG Methylation), and Affymetrix SNP 6-derived GISTIC2 copy number datasets for TCGA Glioma (combined Glioblastoma Multiforme and Lower Grade Glioma; GBM and LGG, respectively), Skin Cutaneous Melanoma (SKCM), Liver Hepatocellular Cancer (LIHC), Prostate Adenocarcinoma (PRAD), Colon

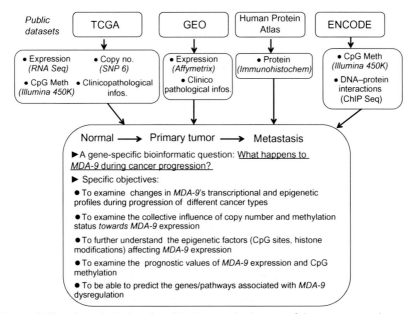

Figure 1 Flowchart depicting the objectives and schemes of the present study.

Table 1 The List of Publicly Available Genomic Datasets Analyzed for the Study

Dataset ID	Cancer Type	Platform	No. Cancer	No. Normal
A. Genome-wide expression				
TCGA_GBM[a]	Glioma (Glioblastoma Multiforme)	Illumina Hi Seq 2000	168	0
TCGA_LGG	Glioma (Lower Grade Glioma)	Illumina Hi Seq 2000	205	0
TCGA_LIHC	Liver Hepatocellular Carcinoma	Illumina Hi Seq 2000	69	36
TCGA_PRAD	Prostate Adenocarcinoma	Illumina Hi Seq 2000	142	37
TCGA_KIRP	Kidney Renal Papillary Cell Carcinoma	Illumina Hi Seq 2000	76	25
TCGA_PANCAN[b]	Pan Cancer (Combination of 22 Cancer Types)	Illumina Hi Seq 2000	5453	587
GSE4290b[c]	Glioma (all grades)	Affymetrix HG U133 Plus 2 Array	153	23
GSE3189[c]	Melanoma	Affymetrix HG U133A Array	45	7
GDS2545[c,d]	Prostate Cancer	Affymetrix HG U95A Array	90	81
B. Genome-wide copy number				
TCGA_GBM	Glioma (Glioblastoma Multiforme)	Affymetrix SNP 6 Array	544	N/A
TCGA_LGG	Glioma (Lower Grade Glioma)	Affymetrix SNP 6 Array	206	N/A
TCGA_LIHC	Liver Hepatocellular Carcinoma	Affymetrix SNP 6 Array	97	N/A
TCGA_COAD	Colon Adenocarcinoma	Affymetrix SNP 6 Array	413	N/A

Continued

Table 1 The List of Publicly Available Genomic Datasets Analyzed for the Study—cont'd

Dataset ID	Cancer Type	Platform	No. Cancer	No. Normal
TCGA_PRAD	Prostate Adenocarcinoma	Affymetrix SNP 6 Array	187	N/A
TCGA_KIRP	Kidney Renal Papillary Cell Carcinoma	Affymetrix SNP 6 Array	127	N/A
TCGA_SKCM	Skin Cutaneous Melanoma	Affymetrix SNP 6 Array	260	N/A
C. Genome-wide CpG methylation				
TCGA_GBM	Glioma (Glioblastoma Multiforme)	Illumina Methylation 450K Beadchip Array	121	0
TCGA_LGG	Glioma (Lower Grade Glioma)	Illumina Methylation 450K Beadchip Array	204	0
TCGA_LIHC	Liver Hepatocellular Carcinoma	Illumina Methylation 450K Beadchip Array	98	50
TCGA_COAD	Colon Adenocarcinoma	Illumina Methylation 450K Beadchip Array	258	38
TCGA_PRAD	Prostate Adenocarcinoma	Illumina Methylation 450K Beadchip Array	192	49
TCGA_KIRP	Kidney Renal Papillary Cell Carcinoma	Illumina Methylation 450K Beadchip Array	111	45
TCGA_SKCM	Skin Cutaneous Melanoma	Illumina Methylation 450K Beadchip Array	338[d]	1

[a] All of TCGA datasets and accompanying clinicopathological information were downloaded as preprocessed data, through the UCSC Cancer Genomics Browser (https://genome-cancer.ucsc.edu/).
[b] The TCGA Pan Cancer (PANCAN) dataset is mean-normalized, consisting of 22 TCGA datasets. Normals were not analyzed for this study.
[c] These datasets were downloaded from the NCBI Gene Expression Omnibus (GEO) site (http://www.ncbi.nlm.nih.gov/geo/).
[d] Tumor samples include 267 metastatic and 71 primary tumors.

Adenocarcinoma (COAD) and Kidney Renal Papillary Cell Carcinoma (KIRP). In addition, we also examined the TCGA Pan Cancer (PANCAN) dataset, which is the merging of all the 22 unique TCGA expression datasets (Chang et al., 2013). At level 3, The TCGA datasets coming from TCGA Genome Characterization (and Data Coordination) Centers consist of single text files representing genome-wide expression, copy number, or methylation values for individual samples (see https://tcga-data.nci.nih.gov/tcga/tcgaDataType.jsp). These datasets were then merged, further processed (e.g., normalization for expression data) and made available for download by the UCSC Cancer Genomics Brower team (https://genome-cancer.ucsc.edu/) (Goldman et al., 2013; Zhu et al., 2009). For the purposes of this report, the processed datasets along with relevant clinical information were downloaded for further analyses. Also, analyzed were tissue expression datasets from GEO: GSE4290 (glioma) (Sun et al., 2006), GSE3189 (melanoma) (Talantov et al., 2005), and GDS2545 dataset (prostate cancer) (Chandran et al., 2007). These datasets were generated using Affymetrix HG U133 Plus 2, U133A, and U95A arrays, respectively. The MDA-9 protein expression levels were also assessed in the Human Protein Atlas Database (http://www.proteinatlas.org/). Immunohistochemical images and other data were downloaded directly from the website. CpG methylation data (Illumina 450K) for cell lines included in the ENCODE project, generated from Hudson Alpha Inst. (R.M. Myers lab; Hunstsville, AL) were downloaded through the UCSC Genome Browser. The UCSC Genome Browser was also used to access information from the ENCODE project, such as the ChIP Seq-derived quantification of DNA-bound modified histones (B.E. Bernstein lab, Broad Inst., Cambridge, MA).

3.2 Analytical Tools

A number of genomic and statistical tools were employed for this study. Initial manipulations of the downloaded datasets were performed using Gene-E (Broad Institute, Cambridge, MA). Statistical analyses were performed using JMP Pro 10 software (SAS, Cary, NC), Gene-E, and Gene Set Enrichment Analysis (Subramanian et al., 2005) (GSEA; Broad Inst., Cambridge, MA). Also used during the later stages of analyses were Cytoscape (Shannon et al., 2003) and the Cytoscape Enrichment Map Plug-in (Merico, Isserlin, Stueker, Emili, & Bader, 2010). Crucial to the analyses was information gathered from the UCSC Genome Browser, UCSC Cancer Genomics Browser, Human Protein Atlas, MsigDB (www.broadinstitute.org/msigdb) (Liberzon et al., 2011), Reactome (www.reactome.org) (Joshi-Tope et al., 2005), KEGG pathway database (http://www.genome.jp/kegg/), and Biocarta (www.biocarta.com). Accessed through the UCSC Genome Browser were results from ENCODE Chromatin State Segmentation using Hidden Markov Modeling (Ernst & Kellis, 2010; Ernst et al., 2011). Necessary gene and probeset annotations were downloaded from Human Genome Organisation Gene Nomenclature Committee (HGNC) (http://www.genenames.org/) and NCBI-GEO websites.

4. THE PATTERNS OF *MDA-9* EXPRESSION DURING CANCER PROGRESSION AND THEIR CLINICAL IMPLICATIONS IN DIFFERENT CANCER TYPES

4.1 Glioma

In our recent report, we demonstrated how *mda-9* is upregulated as tumor grade progresses. The analysis of expression datasets might provide a more quantitative assessment about *mda-9*'s expression pattern in glioma. The first dataset we examined was GSE4290 (Sun et al., 2006), generated using the Affymetrix platform HG U133 Plus 2. As shown in Fig. 2A, *mda-9* expression levels increase as sample progresses from lower toward higher tumor grades, as compared to normal tissues (ANOVA, $p < 0.001$). The same trend can be

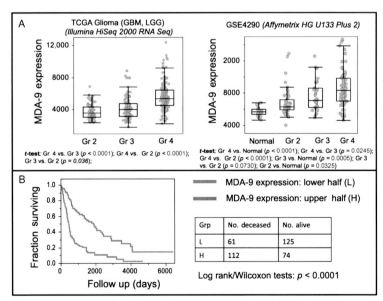

Figure 2 (A) The progressive increase in the expression level of *mda-9/syntenin* in glioma as observed in the TCGA Glioma dataset (left) and GEO dataset GSE4290 (right). (B) The overall survival of glioma patients, grouped according to *mda-9* expression level. (See the color plate.)

observed upon analysis of the combined TCGA GBM and LGG datasets, which were generated using an entirely different platform (i.e., Illumina Hi Seq 2000) (ANOVA, $p<0.001$). However, unlike the GSE4290, the TCGA glioma dataset did not include normal tissue samples. As we have previously pointed out, the expression level of *mda-9* can also be a marker of prognosis in glioma (Kegelman et al., 2014). In the Kaplan–Meir graph shown in Fig. 2B, the TCGA glioma samples were divided into two subgroups according to *mda-9* expression levels: below median (L) and above median (H). Results show that the H subgroup had much worse overall survival rate compared to the L subgroup. Unfortunately, the GSE4290 was not annotated with follow-up records, thus could not be subjected to survival analysis.

4.2 Melanoma, Prostate Cancer, Liver Hepatocellular Carcinoma, and Kidney Renal Papillary Carcinoma

The upregulation (relative to normals) of *mda-9* is also evident in other tumor types, as shown in Fig. 3. The distribution of *mda-9* expression levels

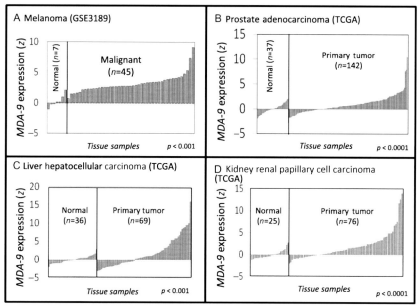

Figure 3 The expression of *mda-9* (relative to normal samples) in melanoma (A), prostate cancer (B), liver cancer (C) and renal cancer (D). The expression values were derived from public genome-wide expression datasets. The relative expression (z) is calculated as $(z = I_n - \text{Ave. } I_{norm})/\text{sd}_{norm}$, where n refers to every sample (including tumors), while norm refers to normal samples only. Inset is the resulting *p*-values for *t*-tests comparing the normal and tumor sample groups. (See the color plate.)

in tumor and normal tissue groups for the four cancer types are shown in order of increasing z score (relative expression), calculated as (I_n Average I_{norm})/standard dev$_{norm}$, where n refers to every sample (including tumors), while norm refers to normal samples only. Student t-test indicates that *mda-9* expression levels in tumors are significantly higher compared to normals. The *mda-9* levels for prostate cancer, liver hepatocellular carcinoma, and kidney renal papillary carcinoma were all extracted from the TCGA RNA Seq datasets. The TCGA dataset for melanoma (SKCM) was not used for this particular part of the analysis because the dataset did not include normal samples. Most of the TCGA SKCM samples were actually classified as metastasis (which will be discussed as a component of the TCGA Pan Cancer dataset). Instead, the dataset GSE3189 (Talantov et al., 2005) was analyzed to differentiate the expression levels of malignant melanoma ($n=45$) to a limited number of normal skin ($n=7$) samples. The original GSE3189 dataset includes nevi samples, which were not included in this analysis.

4.3 *mda-9* Expression in Metastasis Samples

The data described above were those of *mda-9* transcription levels in primary tumors. Given that recent experimental results point to *mda-9*'s role in invasion, it would be of interest to examine its expression pattern in a cohort, which includes metastasis samples. Shown in Fig. 4 is data taken from the GDS2545 dataset (Chandran et al., 2007) generated using Affymetrix U95A array. As the graph indicates, there *mda-9* level is significantly higher in metastasis samples compared to primary tumors and normal samples.

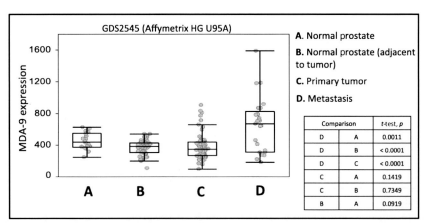

Figure 4 Comparative *mda-9* expression profiles of normal prostate, primary prostate tumors, and metastatic prostate cancer. (See the color plate.)

5. MDA-9 EXPRESSION CORRELATES WITH BOTH GENE COPY NUMBER AND THE METHYLATION STATUS OF CG1719774

5.1 mda-9 RNA Expression Generated Illumina HiSeq Data

The total RNA for each TCGA tissue sample was quantitated using the Illumina HiSeq 2000 RNA Sequencing platform (performed at a TCGA Genome Characterization Center). The matrix form dataset (which is the merging of individual level 3 processed dataset) was downloaded from the UCSC Cancer Genomics Browser and the samples annotated with the accompanying clinical data. Each dataset included the expression levels for 20,501 genes. From these datasets, the transcript levels of *mda-9* were derived and further analyzed.

5.2 mda-9 Copy Number as Determined by Affymetrix SNP Array

The genome-wide copy number data for TCGA samples were also generated in TCGA Genome Characterization Centers, using Affymetrix Genome-wide Human SNP 6.0 Array, consisting of probes for more than 906,600 SNPs, in addition to more than 946,000 probes for the detection of copy number variations. The experimental protocol described in the user manual can be downloaded from the company website (www.affymetrix.com). The TCGA FIREHOSE pipeline (Broad Institute, Cambridge, MA) employed the GISTIC2 algorithm (Mermel et al., 2011) to generate copy number estimates (CN) for the genes mapped in the genome. The GISTIC2 estimate for *mda-9* was then extracted from the dataset, and converted to copy number estimate, calculated using the formula: $CN = 2^{(GISTIC2 + 1)}$.

5.3 *mda-9* CpG Sites Interrogated in Illumina 450K Array

The genome-wide CpG methylation data for TCGA samples were generated using the Illumina Infinium Human Methylation 450K array, which is designed to interrogate more than 480,000 CpG sites in the entire human genome (Sandoval et al., 2011). Bead Studio software was then employed to generate beta values, which range from 0 (fully unmethylated) to 1 (fully methylated). For this report, we used the UCSC offset value in which -0.5

was subtracted from the original scale, resulting in beta values ranging from −0.5 (fully unmethylated) to 0 (50% methylated) to 0.5 (fully methylated). A total of 13 *mda-9* locus CpG sites covered in the Illumina 450K array registered beta values in the TCGA datasets (see Table 2). Ten of the probes are part of the CpG island group located in the promoter region, with the first five located within the transcription start sites (TSS1500, TSS200), and the next five as part of the 5′ UTR exon (see Fig. 5). The next 5 CpG sites are located in the intervening introns and the last one within the 3′ UTR exon. The distributions of the beta values, corresponding to each of the 13 CpG sites for each of the 6 TCGA cancer datasets, are shown in Fig. 6. Except for two (cg17197774 and cg10129404), the CpG sites were mostly unmethylated across all the cancer datasets. This suggests that only the variation in the beta values of the two CpG sites translate to variation in *mda-9* expression levels. The CpG site cg17197774 is exactly 1105 bases from the 3′ edge of the CpG island (Gardiner-Garden & Frommer, 1987), while cg10129404 is part of the 3′ UTR, making it less likely for the latter to be a factor in transcription of *mda-9* (this will be clarified in the next subsection). For glioma, methylation at the CpG sites decreases as the tumor grade progresses. For liver cancer, colon adenocarcinoma, and kidney papillary carcinoma, methylation at these sites decreases during the transformation from solid normals to primary tumor. There are only a few normal samples in the SKCM dataset, but it is very clear that primary and metastatic SKCM samples had the lowest levels of methylation (compared to the other five datasets). There was only a minimal change in methylation in primary prostate cancer relative to solid normals.

5.4 Which Among the Two CpG Sites Influence *mda-9* Expression?

As mentioned above, it is unlikely that the 3′ UTR CpG site cg10129404 influences *mda-9* transcript levels. A simple analysis was conducted by plotting *mda-9* expression versus the beta value for cg17197774 or cg10129404 for two select TCGA datasets (glioma and KIRP) (Fig. 7). In glioma, it is clear that the methylation at cg17197774 may influence *mda-9* expression ($R^2 = 0.38$; linear regression). In contrast, cg10129404 appears to have a negligible effect on *mda-9* expression ($R^2 = 0.051$). Similar analysis was conducted for the TCGA KIRP dataset. Linear regression analyses indicate that the R^2 value for the plot of *mda-9* expression versus cg17197774 methylation

Table 2 List of the *mda-9/Syntenin* CpG Sites Interrogated in the Illumina Infinium 450K Beadchip Array

Probe ID	Chr	Strand	Coordinate (Build 37)	Relation_to_UCSC_CpG_Island	UCSC_RefGene_Group	Forward_Sequence
cg20052227	8	R	59465520	Island	TSS1500	GTGGGTGGCA[CG]GGGCCCGCGG
cg02896624	8	R	59465528	Island	TSS1500	GCACGGGGCC[CG]CGGGCACGAA
cg17984783	8	R	59465536	Island	TSS200	CCCGCGGGCA[CG]AACAGCCGAA
cg26656684	8	F	59465596	Island	TSS200	CAGCGGACAG[CG]GGGCGCATGA
cg11550426	8	F	59465608	Island	TSS200	CGGCATGAAC[CG]CCCCACTTTG
cg27280034	8	F	59465624	Island	TSS200	CCCACTTTGC[CG]GATACCTGGA
cg06294637	8	F	59465768	Island	1stExon;5'UTR	GCCTCGGGGG[CG]GTCCTCGGGC
cg07798892	8	F	59465776	Island	1stExon;5'UTR	GGTCCTCGGG[CG]CGCACCGCTC
cg20848390	8	F	59465780	Island	1stExon;5'UTR	TCCTCGGGCG[CG]CACCGCTCTC
cg02103294	8	R	59465972	Island	5'UTR	GCATCCTGGT[CG]CAGCCGTTTT
cg12046629	8	R	59466300	S_Shore	5'UTR	TCCCAGTGCT[CG]GCGTTTCTAG
cg17197774	8	F	59467208	S_Shore	5'UTR	TAATGGTTGC[CG]GTTAAATGTA
cg10129404	8	R	59494304		3'UTR	TTAAAATTCA[CG]GCACCATGA

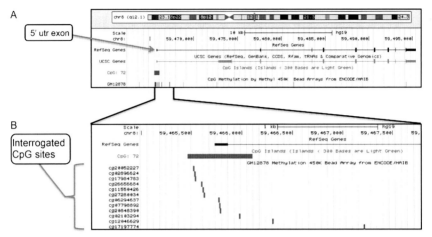

Figure 5 (A) The locus map of *mda-9* with the nine identified exons in RefSeq. UCSC Genes prediction identified a 10th exon, after the 5′ UTR exon. (B) Higher resolution of the promoter region, indicating the locations of the interrogated CpG sites. Coordinates were derived from GRCh37/hg19 assembly. (See the color plate.)

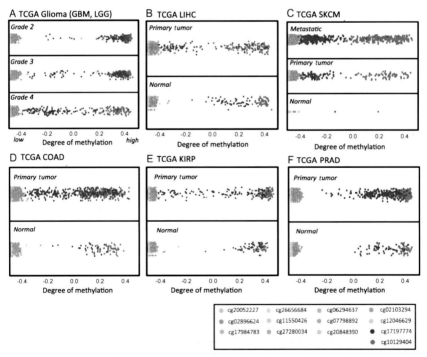

Figure 6 The degree of methylation, ranging from totally unmethylated (−0.5) to totally methylated (0.5), at the 13 CpG sites interrogated in Illumina 450K array, for tissue samples belonging to 6 TCGA cancer groups. Of the 13 sites, the most differentially methylated is cg179774 (dark blue), followed by cg10129404 (dark red). (See the color plate.)

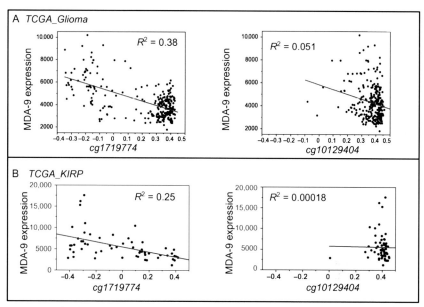

Figure 7 In both TCGA glioma (A) and KIRP (B), *mda-9* expression is clearly influenced by the methylation status of cg1719774 (the CpG site close to 5' UTR) and not by cg10129404 (the CpG site at 3' UTR). (See the color plate.)

is 0.25, while that of *mda-9* expression versus cg10129404 is almost zero. Overall, these analyses indicate that cg10129404's influence on *mda-9* expression is unlikely.

5.5 Glioma

Having established that *mda-9* expression correlates with methylation at cg17197774, the next step was to analyze the dual contribution of both methylation and copy number in *mda-9* expression for each of the six datasets included in this study. The relationships between *mda-9* expression, copy number, and the methylation status of cg17197774 are illustrated in Fig. 8. For glioma (whose *mda-9* copy number is mostly neutral), we can see the effect of both copy number and methylation status. Copy number is a likely factor ($R=0.26$; expression vs. copy number). However, it is also clear that among the samples with only two copies of *mda-9*, those with low degree of methylation at the cg1719774 tend to have a higher *mda-9* expression level ($R=-0.61$; expression vs. cg1719774 beta value). By itself, it appears that cg1719774 methylation status may be a reliable marker of survival in glioma (Fig. 9).

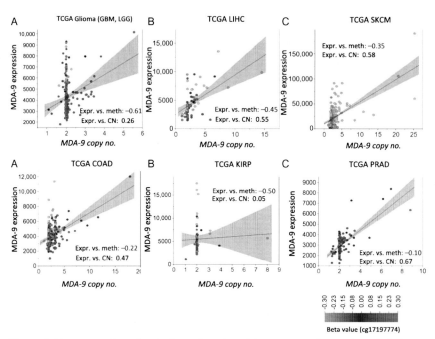

Figure 8 The expression of *mda-9* as a function of its copy number and methylation at cg1719774, among six TCGA datasets. (See the color plate.)

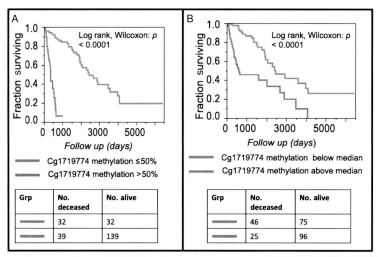

Figure 9 Methylation at cg1719774 is a marker of clinical outcome in glioma. Patients were grouped into two, according to % methylation (A), or methylation value (beta value) relative to median (B). (See the color plate.)

5.6 Melanoma, Prostate Cancer, Liver Hepatocellular Carcinoma, and Kidney Renal Papillary Carcinoma

On average, SKCM samples have the lowest beta values for cg1719774, at -0.295 (Table 3). This may also explain why among the six TCGA cancer datasets, SKCM tumors have the highest expression levels for *mda-9* (at more than 20,000 units; inv log 2 scale). Nonetheless, the influence of both factors is evident in Fig. 8C (correlation coefficients of -0.35 and 0.58 for expression vs. methylation and expression vs. copy number, respectively). The tumor sample with the highest overall *mda-9* expression level (at more than 190,000 units) has an *mda-9* copy number estimate of 25 and cg1719774 beta value of -0.45. For the TCGA COAD dataset, both copy number and cg1719774 methylation appear to factor in *mda-9* expression, with the former ($R=0.47$) having greater influence than the latter (-0.22). A great majority of KIRP samples have a neutral copy number ($CN=2$) at the *mda-9* locus. Not surprisingly, the elevated *mda-9* expression is primarily due to hypomethylation at cg1719774 (*mda-9* expression vs. cg1719774 methylation correlation coefficient $=-0.5$). In contrast to KIRP, the mode of *mda-9* dysregulation in prostate cancer samples is primarily through copy number gain ($R=0.67$), with cg1719774 methylation apparently lacking any effect on the gene's expression level. Among LIHC primary tumors, it is clear that both copy number ($R=0.55$) and cg1719774 methylation ($R=-0.45$) are factors influencing *mda-9* RNA levels.

Table 3 Comparison of *mda-9/Syntenin* Copy Number, CpG Methylation (at cg1719774), RNA Level, and Protein Level Among Six TCGA Cohorts

Tumor Cohort (TCGA)	No. of Tumor Samples	Copy Number Estimate (Lowest, Highest)	CpG Methylation at cg1719774 (-0.5 to 0.5; Low to High)	Expression Level
SKCM	241	2.80 (1.15, 24.2)	-0.295	23062 ± 20569
PRAD	175	2.18 (1.20, 8.82)	0.195	3039 ± 943
COAD	188	2.82 (1.11, 17.93)	-0.019	4172 ± 1514
GBM/LGG	257	2.09 (1.09, 5.56)	0.209	4308 ± 1500
LIHC	68	2.92 (1.09, 14.55)	-0.088	4623 ± 2505
KIRP	60	2.16 (0.84)	-0.014	5553 ± 3268

Included in the analyses are tumor samples having all the three molecular data (copy number estimate, expression, and methylation).

5.7 Combination of the Six Cancer Datasets

For a unified view on the effects of both copy number and cg1719774 methylation on *mda-9* expression, we used the PANCAN-normalized expression value for *mda-9*. As explained earlier, the PANCAN dataset is the merging of all the 22 TCGA RNASeq datasets (Chang et al., 2013). According to the information provided by the UCSC Cancer Genomics Browser (from which the dataset was downloaded), the original level 3 RNASeq V2 datasets were downloaded from TCGA, log (base 2) transformed, then mean-normalized across all the cohorts. The normalization across all the cohorts provides more reliable relative values for gene expression. Figure 10A and B shows the same Cartesian plot (i.e., PANCAN-normalized *mda-9* expression value vs. methylation at cg1719774) with varying information for each data point. As these figures indicate, there is a clear inverse exponential relationship between the two variables. As discussed previously, the highest *mda-9* expression levels were those of SKCM samples, owing to the low beta values for cg1719774. The effect of copy number on *mda-9*

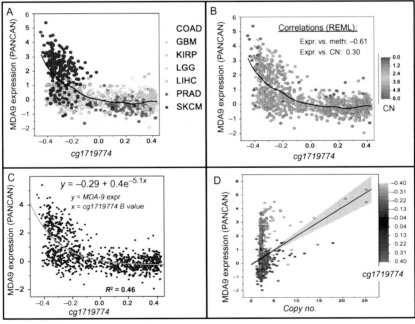

Figure 10 *mda-9* expression (PANCAN-normalized) versus cg1719774 beta values with data points labeled according to TCGA cohort (A), and copy number (B). (C) The same dataset with the exponential 3P regression plot. (D) *mda-9* expression versus copy number, with data points marked according to cg1719774 Beta value. (See the color plate.)

expression is illustrated in Fig. 10B, wherein solid red circles (those samples whose *mda-9* copy number is 6 or higher) mostly appear in the upper edges of the graph. This indicates that copy number can elevate *mda-9* expression levels irrespective of the methylation status of cg1719774. Overall, the correlation coefficients (Multivariate REML statistics) for expression vs. methylation and expression vs. copy number are -0.61 and 0.30, respectively. This shows that the methylation status of cg1719774 has greater influence (compared to *mda-9* copy number) toward the gene's expression level. Figure 10C shows a superimposed exponential regression model (JMP Pro 10) relating *mda-9* expression and cg1719774. Lastly, Fig. 10D illustrates the relationship between *mda-9* expression and copy number (essentially a merging of Fig. 8). At this point, we already know that tumor samples with neutral copy number for *mda-9* can have an elevated expression of the gene if it is hypomethylated at cg1719774.

6. *MDA-9* IS HIGHLY EXPRESSED IN MELANOMA

6.1 mda-9 RNA Expression Across all Cancer Types (PANCAN Dataset)

The results illustrated above point to *mda-9* being most highly expressed in melanoma (among the 6 cancer types analyzed). We then proceeded to examine *mda-9* expression in all of the TCGA cohorts through analysis of the TCGA PANCAN dataset. Covering 22 cancer types, the dataset includes 6040 samples (4982 primary tumors, 271 metastatic, 27 recurrent tumors, 173 peripheral blood, 587 solid tissue normals). First, the 6040 samples were grouped according to TCGA-defined sample types. On average, the highest *mda-9* levels were those of metastasis samples, which were largely the melanoma (SKCM) samples (Fig. 11a). When *mda-9* expression was divided according to cohort (without the normals), it is clear that *mda-9* is most elevated in melanoma samples (Fig. 11b).

6.2 MDA-9 Protein Expression Across all Cancer Types (The Human Protein Atlas)

TCGA also has datasets for protein expression (using Reverse Phase Protein Array). However, it only covers a few hundred select proteins (the most widely studied cancer-related genes) and does not include MDA-9. However, a very comprehensive source of genome-wide protein expression is the Human Protein Atlas (http://www.proteinatlas.org/), where proteins were immunohistochemically analyzed in different cancer types (*note*: Human

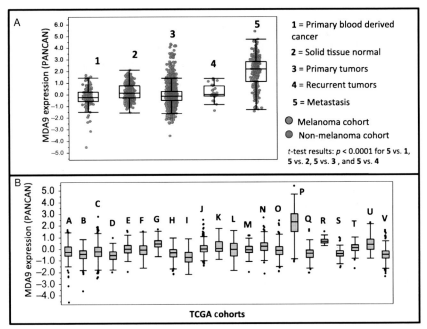

Figure 11 The expression of *mda-9* (PANCAN-normalized) with samples grouped according to sample type (a) and TCGA cohort (solid normals not included (b)). The cohorts are the following: (A) acute myeloid leukemia (LAML), (B) bladder urothelial carcinoma (BLCA), (C) breast invasive carcinoma (BRCA), (D) cervical squamous cell carcinoma and endocervical adenocarcinoma (Francescone et al., 2011), (E) colon adenocarcinoma (COAD), (F) uterine corpus endometrial carcinoma (UCEC), (G) glioblastoma multiforme (GBM), (H) head and neck squamous cell carcinoma (HNSC), (I) kidney chromophobe (KICH), (J) kidney renal clear cell carcinoma (KIRC), (K) kidney renal papillary cell carcinoma (KIRP), (L) liver hepatocellular carcinoma (LIHC), (M) brain lower grade glioma (LGG), (N) lung adenocarcinoma (LUAD), (O) lung squamous cell carcinoma (LUSC), (P) skin cutaneous melanoma (SKCM), (Q) ovarian serous cystadenocarcinoma (OV), (R) pancreatic adenocarcinoma (PAAD), (S) prostate adenocarcinoma (PRAD), (T) rectum adenocarcinoma (READ), (U) sarcoma (SARC), (V) thyroid carcinoma (THCA). (See the color plate.)

Protein Atlas samples are independent of the TCGA cohorts). For MDA-9, two different antibodies were used: HPA023840 from Sigma–Aldrich and CAB012245 from Abcam. Consistent with results from the PANCAN analysis, MDA-9 according to Human Protein Atlas, is most highly expressed in melanoma (see Fig. 12). As shown in the figure (bottom right), the HPA023840 signal was much stronger in melanoma compared to glioma. The staining results were evaluated by a trained pathologist, who assigned intensity (negative, weak, moderate, or strong), fraction of stained cells (rare,

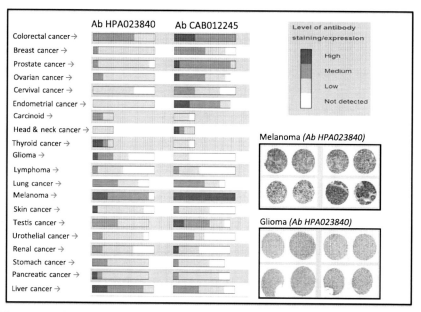

Figure 12 MDA-9 protein levels, as assessed by immunohistochemistry (semiquantitative) using two different antibodies. The bars indicate that melanoma has the highest MDA-9 expression among different cancer types. Insets are the immunohistochemical images of melanoma and glioma sections, indicative of the former's stronger MDA-9 signal. All of the images above were taken from the Protein Atlas (http://www.proteinatlas.org). (See the color plate.)

<25%, 25–75%, or >75%), as well as subcellular localization (nuclear and/or cytoplasmic/membranous).

7. HYPOMETHYLATION AT CG1719774 IS ASSOCIATED WITH HISTONE MODIFICATIONS INDICATIVE OF HIGHER TRANSCRIPTIONAL ACTIVITY

7.1 Information from the ENCODE Project

The ENCODE project was launched several years ago (through the National Human Genome Research Institute), with the primary aim of identifying all functional elements in the human genome sequence (ENCODE Project Consortium, 2011). It is a consortium involving a number of laboratories mostly in the United States. The investigators examined regulatory elements using various tools such as DNA methylation assays, DNA hypersensitivity assays, RNA sequencing, and chromatin immunoprecipitation (ChIP) of proteins interacting with DNA (e.g., histones, transcription factors), followed by sequencing (ChIP Seq). In a typical ChIP Seq experiment, proteins are cross-linked to cancer cell line DNA, followed by shearing of the DNA strands. An antibody to the protein of interest (e.g., TF, histones), attached to a bead will then be used to immunoprecipitate the target protein, which is linked to the DNA sequences it recognized. The protein is then unlinked from the DNA, which is then sequenced (using Next Gen sequencing) and mapped to the genome. The binding of a protein to a DNA (or RNA) sequence is directly proportional to the number of readouts for that particular sequence. In essence, a ChIP Seq experiment is supposed to: (a) identify the sequence the protein binds to and (b) measure the extent of this binding. Results from numerous ENCODE projects are now publicly available through the UCSC Genome Browser, for either on-site analysis or data download for further analysis. There are various types of analyses, which can be done on these data.

7.2 How CpG Methylation is Associated with Differential Binding of Modified Histones

Since we have already established (from TCGA analysis) that the methylation at the CpG site cg1719774 correlates with decreased *mda-9* expression, it would be of interest to investigate how this CpG site may also affect the constitution of chromatin covering the *mda-9* locus. Take for example, the cell lines H1-hESC (human embryonic stem cell) and HUVEC (human umbilical vein endothelial cell). Both cell lines have available genome-wide CpG methylation data (Illumina 450K methylation array), as well as ChIP Seq data for various modified histones (ENCODE data available for download and analysis through UCSC Genome Browser). As indicated in Fig. 13A, the MDA-9 promoter region of HUVEC (hypomethylated at cg1719774; beta value $= -0.353$) exhibits greater affinity (darker bands) for Histone H3 mono-, di-, or trimethylated at K4 (H3K4me1, H3K4me2, H3K4me3), and acetylated at K9 (H3K9ac), compared to that of H1-hESC (highly methylated at cg1719774; beta value $= 0.230$). Those histone modifications, just like hypomethylation at cg1719774, are associated with more active transcription (Campbell & Turner, 2013; Ernst et al., 2011; Ghirlando et al., 2012). Four other markers (H4K20me1 which is associated with transcriptional activity; H3K36me3 which is associated with RNAPII elongation; H3K27me3 which is associated with promoter silencing; and the transcriptional repressor CTCF) were not significantly different between the two cell lines. Ernst, Kellis, and colleagues devised an analytical scheme, which summarizes the collective transcriptional

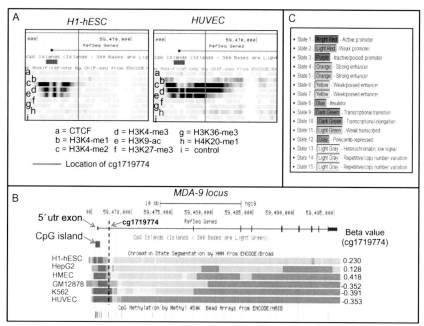

Figure 13 (A) The "affinity" of modified histones and CTCF toward the *mda-9* promoter region of the HUVEC and H1-hESC cell lines. (B) Chromatin state segmentation at the *mda-9* locus, for six cell lines (using the method of Ernst & Kellis, 2010). Cell lines with low beta values at cg1719774 tend to have a more active promoter (dark red) and transcription (dark green). (C) Color codes for the different chromosomal states (copied from UCSC Genome Browser; http://genome.ucsc.edu). (See the color plate.)

influences of the aforementioned histone modifications (and CTCF) in a given segment of the chromosome (Ernst & Kellis, 2010; Ernst et al., 2011). A Hidden Markov Model was used to predict the state of a chromosomal segment with respect to the activity of the promoter, strength of enhancer sequences, and openness for transcription, based on the ChIP Seq data specific for these markers. Of the nine cell lines examined using this approach (thus far), only six have corresponding Illumina 450K methylation data. In addition to H1-hESC and HUVEC, the other cell lines were: the hepatocellular carcinoma cell line HepG2, the microvascular endothelial cell line HMEC, the lymphoblastoid cell line GM12878, and the chronic myelogenous leukemia cell line K562. As shown in Fig. 13B, H1-hESC, HepG2, and HMEC, which are all highly methylated at cg1719774, are predicted to exhibit less active MDA-9 promoter and transcriptional activity (interpretations for segmental colors are listed in Fig. 13C). In contrast,

MDA-9 expression is expected to be higher for cell lines GM12878, K562, and HUVEC, which are all weakly methylated at cg1719774.

8. PREDICTING GENES AND PATHWAYS ASSOCIATED WITH *MDA-9* DYSREGULATION IN GLIOMA

8.1 Comparing *mda-9*-High and *mda-9*-Low Subtypes of Glioma

Genome-wide expression datasets may also be useful in predicting genes or pathways associated with *mda-9* dysregulation. One approach (see Fig. 14) is to start with the isolation of two subsets (out of the complete cohort of tumor samples) at the opposite tail ends of distribution according to *mda-9* expression: (a) *mda-9-high*, i.e., those whose *mda-9* expression levels are at least 1 standard deviation higher than the average and (b) *mda-9-low*, i.e.,

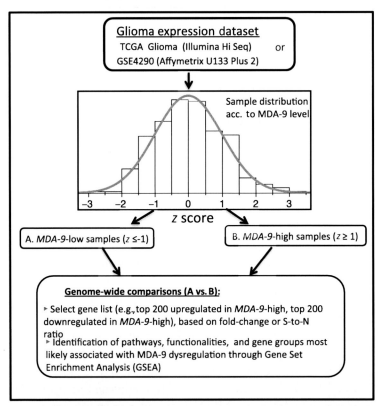

Figure 14 *Virtual Gene Overexpression or Repression (VIGOR).* An analytical scheme to identify genes and pathways associated with *mda-9* dysregulation in glioma.

those cases whose *mda-9* expression levels are at least 1 standard deviation lower than the average. The comparative genome-wide expression of these two subsets will then be examined to identify the genes, functionalities and molecular pathways, which are most likely associated with *mda-9* upregulation or *mda-9* downregulation. This comparison between the *mda-9*-high and *mda-9*-low subtypes is akin to a gene being ectopically over-expressed or repressed (with RNAi) in a cancer cell line (an experimental approach employed in one of our recent publications; Sokhi et al., 2013) with the objective of examining the molecular changes, which occur after the process. For that reason, this approach is being referred to as *Virtual Gene Overexpression or Repression (VIGOR)*. Genes will then be ranked according to a comparative statistic (either signal-to-noise ratio or fold-change) between the two groups. The *VIGOR* approach was applied to the two glioma datasets (TCGA and GSE4290) and the results were highly concordant. As shown in Fig. 15, the resulting signal-to-noise ratios (for all genes) in TCGA glioma dataset directly correlate with those of the GSE4290 dataset. Another statistical tool employed is the GSEA available through the Broad Institute (www.broadinstitute.org/gsea/) (Subramanian et al., 2005). GSEA analysis starts with the recognition that genes are associated with particular

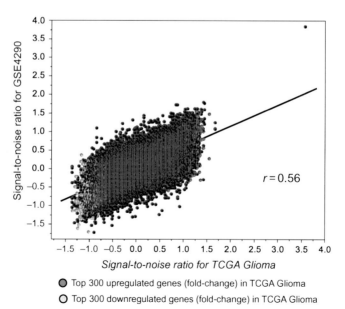

Figure 15 There is a great degree of similarity between TCGA Glioma and GSE4290 datasets, in terms of signal-to-noise ratio (*mda-9*-high vs. *mda-9*-low) for every gene. (See the color plate.)

Table 4 The Top 200 Upregulated and Top 50 Downregulated Genes in *mda-9* High Glioma—cont'd

ID	Gene Title	Chromosome	Fold Change (TCGA)	S-to-N Ratio (TCGA)	p-Value (TCGA)	Affymetrix ID	Fold Change (GSE4290)
COL3A1	Collagen, type III, alpha 1	2q32.2	30.99	0.99	2.0×10^{-3}	211161_s_at	16.7
COL1A1	Collagen, type I, alpha 1	17q21.33	30.08	0.99	2.0×10^{-3}	202310_s_at	17.27
TREM1	Triggering receptor expressed on myeloid cells 1	6p21.1	30.06	1.14	2.0×10^{-3}	219434_at	5.12
HOXC10	Homeobox C10	12q13.13	28.58	0.9	2.0×10^{-3}	218959_at	2.38
CHI3L2	Chitinase 3-like 2	1p13.3	27.18	1.07	2.0×10^{-3}	213060_s_at	7.83
TIMP1	TIMP metallopeptidase inhibitor 1	Xp11.3-p11.23	27.17	1.29	2.0×10^{-3}	201666_at	10.89
CD163	CD163 molecule	12p13	26.87	1.3	2.0×10^{-3}	215049_x_at	10.62
SHOX2	Short stature homeobox 2	3q25.32	26.33	0.96	2.0×10^{-3}	210135_s_at	7.65
FCGR2B	Fc fragment of IgG, low affinity IIb, receptor (CD32)	1q23	25.18	1.25	2.0×10^{-3}	210889_s_at	3.49
PI3	Peptidase inhibitor 3, skin-derived	20q13.12	24.76	0.84	2.0×10^{-3}	203691_at	3.8
SRPX2	Sushi-repeat-containing protein, X-linked 2	Xq21.33-q23	23.25	1.29	2.0×10^{-3}	205499_at	11.53
GDF15	Growth differentiation factor 15	19p13.11	23.08	1.23	2.0×10^{-3}	221577_x_at	6.27

Gene	Description	Location			p-value	Probe	
MEOX2	Mesenchyme homeobox 2	7p22.1-p21.3	22.45	0.84	2.0×10^{-3}	206201_s_at	9.29
SERPINA5	Serpin peptidase inhibitor, clade A (alpha-1 antiproteinase, antitrypsin), member 5	14q32.1	22.32	1.33	2.0×10^{-3}	209443_at	3.67
HOXA10	Homeobox A10	7p15.2	21.96	0.88	2.0×10^{-3}	213150_at	6.32
HOXD13	Homeobox D13	2q31.1	21.57	0.86	2.0×10^{-3}	207397_s_at	3.73
TNFSF12-TNFSF13	TNFSF12-TNFSF13 readthrough	17p13.1	21.56	1.15	2.0×10^{-3}	209499_x_at	1.60
HOXB3	Homeobox B3	17q21.32	21.52	0.9	2.0×10^{-3}	228904_at	4.18
PDPN	Podoplanin	1p36.21	21.39	1.2	2.0×10^{-3}	204879_at	27.01
LIF	Leukemia inhibitory factor (cholinergic differentiation factor)	22q12.2	21.38	1.05	2.0×10^{-3}	205266_at	3.82
SERPINE1	Serpin peptidase inhibitor, clade E (nexin, plasminogen activator inhibitor type 1), member 1	7q22.1	21.1	1.17	2.0×10^{-3}	202627_s_at	8.69
IL2RA	Interleukin 2 receptor, alpha	10p15-p14	20.86	1.12	2.0×10^{-3}	206341_at	1.49
HOXD11	Homeobox D11	2q31.1	20.3	0.93	2.0×10^{-3}	214604_at	3.85
ESM1	Endothelial cell-specific molecule 1	5q11	19.95	1.02	2.0×10^{-3}	208394_x_at	3.89
ANXA1	Annexin A1	9q21.13	19.6	1.55	2.0×10^{-3}	201012_at	10.05

Continued

Table 4 The Top 200 Upregulated and Top 50 Downregulated Genes in *mda-9* High Glioma—cont'd

ID	Gene Title	Chromosome	Fold Change (TCGA)	S-to-N Ratio (TCGA)	*p*-Value (TCGA)	Affymetrix ID	Fold Change (GSE4290)
HOXA3	Homeobox A3	7p15.2	19.2	0.93	2.0×10^{-3}	235521_at	4.21
LOX	Lysyl oxidase	5q23.3-q31.2	19.07	1.25	2.0×10^{-3}	215446_s_at	14.4
EMP3	Epithelial membrane protein 3	19q13.3	18.95	1.26	2.0×10^{-3}	203729_at	9.7
IL8	Interleukin 8	4q13-q21	18.93	0.79	2.0×10^{-3}	211506_s_at	7.86
HOXD10	Homeobox D10	2q31.1	18.87	0.89	2.0×10^{-3}	229400_at	4.74
SPOCD1	SPOC domain containing 1	1p35.1	18.77	0.97	2.0×10^{-3}	235417_at	14.27
GPX8	Glutathione peroxidase 8 (putative)	5q11.2	18.76	1.23	2.0×10^{-3}	228141_at	7.38
HOXA4	Homeobox A4	7p15.2	18.33	0.82	2.0×10^{-3}	206289_at	2.21
CXCL10	Chemokine (C-X-C motif) ligand 10	4q21	18.28	1.01	2.0×10^{-3}	204533_at	2.73
HOXA7	Homeobox A7	7p15.2	18.2	0.7	2.0×10^{-3}	235753_at	2.63
HOXA5	Homeobox A5	7p15.2	17.67	0.95	2.0×10^{-3}	213844_at	3.83
GALNT5	UDP-N-acetyl-alpha-D-galactosamine:polypeptide N-acetylgalactosaminyltransferase 5 (GalNAc-T5)	2q24.1	17.34	1.14	2.0×10^{-3}	236129_at	1.76

CA3	Carbonic anhydrase III, muscle specific	8q21.2	17.26	0.92	2.0×10^{-3}	204865_at	6.67
STC1	Stanniocalcin 1	8p22-p12	17.13	1.08	2.0×10^{-3}	204597_x_at	10.02
PTX3	Pentraxin 3, long	3q25	16.81	0.98	2.0×10^{-3}	206157_at	9.7
CSTA	Cystatin A (stefin A)	3q21	16.67	1.37	2.0×10^{-3}	204971_at	3.71
METTL7B	Methyltransferase like 7B	12q13.2	16.52	1.14	2.0×10^{-3}	227055_at	4.37
S100A4	S100 calcium-binding protein A4	1q12-q22	16.37	1.19	2.0×10^{-3}	203186_s_at	6.94
TRPM8	Transient receptor potential cation channel, subfamily M, member 8	2q37	16.14	1.11	2.0×10^{-3}	243483_at	3.94
IGFBP2	Insulin-like growth factor binding protein 2, 36 kDa	2q33-q34	16.03	0.98	2.0×10^{-3}	202718_at	19.33
FMOD	Fibromodulin	1q32	15.94	1.06	2.0×10^{-3}	202709_at	3.63
H19	H19, imprinted maternally expressed transcript (nonprotein coding)	11p15.5	15.94	0.62	2.0×10^{-3}	224646_x_at	4.4
IL13RA2	Interleukin 13 receptor, alpha 2	Xq13.1-q28	15.87	0.71	2.0×10^{-3}	206172_at	6.65
HPD	4-Hydroxyphenylpyruvate dioxygenase	12q24.31	15.45	1.08	2.0×10^{-3}	206024_at	1.53
SERPINA3	Serpin peptidase inhibitor, clade A (alpha-1 antiproteinase, antitrypsin), member 3	14q32.1	15.08	1.04	2.0×10^{-3}	202376_at	6.22

Continued

Table 4 The Top 200 Upregulated and Top 50 Downregulated Genes in mda-9 High Glioma—cont'd

ID	Gene Title	Chromosome	Fold Change (TCGA)	S-to-N Ratio (TCGA)	p-Value (TCGA)	Affymetrix ID	Fold Change (GSE4290)
COL6A3	Collagen, type VI, alpha 3	2q37	14.99	0.89	2.0×10^{-3}	201438_at	8.45
SLC47A2	Solute carrier family 47, member 2	17p11.2	14.93	0.95	2.0×10^{-3}	231068_at	3.24
WISP1	WNT1 inducible signaling pathway protein 1	8q24.22	14.67	1.04	2.0×10^{-3}	206796_at	3.64
ANXA2P2	Annexin A2 pseudogene 2	9p13.3	14.52	1.3	2.0×10^{-3}	208816_x_at	4.43
CHRNA9	Cholinergic receptor, nicotinic, alpha 9	4p14	14.42	0.89	2.0×10^{-3}	221107_at	2.97
F13A1	Coagulation factor XIII, A1 polypeptide	6p24.2-p23	14.38	0.93	2.0×10^{-3}	203305_at	3.38
TNFRSF12A	Tumor necrosis factor receptor superfamily, member 12A	16p13.3	14.35	1.17	2.0×10^{-3}	218368_s_at	7.47
ADM	Adrenomedullin	11p15.4	14.32	1.01	2.0×10^{-3}	202912_at	7.57
COL8A1	Collagen, type VIII, alpha 1	3q11.1-q13.2	14.29	1	2.0×10^{-3}	226237_at	9.88
MOXD1	Monooxygenase, DBH-like 1	6q23.2	14.24	0.83	2.0×10^{-3}	1554474_a_at	3.6
COL6A2	Collagen, type VI, alpha 2	21q22.3	13.96	1	2.0×10^{-3}	209156_s_at	9.88
SOCS3	Suppressor of cytokine signaling 3	17q25.3	13.77	1.08	2.0×10^{-3}	227697_at	10.04

Gene	Description	Location					
EMR1	egf-like module containing, mucin-like, hormone receptor-like 1	19p13.3	13.7	1.08	2.0×10^{-3}	207111_at	1.2
TCTEX1D1	Tctex1 domain containing 1	1p31.2	13.69	0.91	2.0×10^{-3}	1553635_s_at	2.27
NKX2-5	NK2 transcription factor related, locus 5 (Drosophila)	5q34	13.66	0.77	2.0×10^{-3}	206578_at	3
COL5A1	Collagen, type V, alpha 1	9q34.2-q34.3	13.22	0.87	2.0×10^{-3}	203325_s_at	5.38
PLEK2	Pleckstrin 2	14q23.3	13.11	1.17	2.0×10^{-3}	218644_at	1.24
HOXB4	Homeobox B4	17q21.32	13.06	0.87	2.0×10^{-3}	231767_at	1.69
CLCF1	Cardiotrophin-like cytokine factor 1	11q13.3	12.92	1.12	2.0×10^{-3}	219500_at	2.52
CA9	Carbonic anhydrase IX	9p13.3	12.85	0.67	2.0×10^{-3}	205199_at	1.4
LYZ	Lysozyme	12q15	12.75	1.01	2.0×10^{-3}	213975_s_at	8.16
HOTAIR	Hox transcript antisense RNA (nonprotein coding)	12q13.13	12.71	0.88	2.0×10^{-3}	239153_at	4.45
MSR1	Macrophage scavenger receptor 1	8p22	12.66	1.35	2.0×10^{-3}	214770_at	4.12
GPRC5A	G protein-coupled receptor, family C, group 5, member A	12p13-p12.3	12.59	1.07	2.0×10^{-3}	203108_at	3.36
PLAU	Plasminogen activator, urokinase	10q24	12.59	1.08	2.0×10^{-3}	205479_s_at	5.57
EN1	Engrailed homeobox 1	2q14.2	12.44	0.72	2.0×10^{-3}	220559_at	3.92

Continued

Table 4 The Top 200 Upregulated and Top 50 Downregulated Genes in *mda-9* High Glioma—cont'd

ID	Gene Title	Chromosome	Fold Change (TCGA)	S-to-N Ratio (TCGA)	p-Value (TCGA)	Affymetrix ID	Fold Change (GSE4290)
FCGBP	Fc fragment of IgG-binding protein	19q13.2	12.41	1.04	2.0×10^{-3}	203240_at	6.56
ADAM12	ADAM metallopeptidase domain 12	10q26	12.39	1.06	2.0×10^{-3}	202952_s_at	5.39
IFI30	Interferon, gamma-inducible protein 30	19p13.1	12.33	1.21	2.0×10^{-3}	201422_at	4.83
PAX3	Paired box 3	2q36.1	12.23	0.78	2.0×10^{-3}	231666_at	2.35
DDIT4L	DNA-damage-inducible transcript 4-like	4q23	12.23	0.91	2.0×10^{-3}	228057_at	2.58
COL4A1	Collagen, type IV, alpha 1	13q34	12.02	1.12	2.0×10^{-3}	211981_at	15.36
LUM	Lumican	12q21.3-q22	11.92	0.9	2.0×10^{-3}	201744_s_at	7.11
FCGR2C	Fc fragment of IgG, low affinity IIc, receptor for (CD32) (gene/pseudogene)	1q23	11.84	1.26	2.0×10^{-3}	210992_x_at	11.95
S100A9	S100 calcium-binding protein A9	1q21	11.82	0.83	2.0×10^{-3}	203535_at	6.04
CCL20	Chemokine (C–C motif) ligand 20	2q36.3	11.8	1	2.0×10^{-3}	205476_at	3.21
HOXB2	Homeobox B2	17q21.32	11.73	0.8	2.0×10^{-3}	205453_at	3.17
NCRNA00152	Nonprotein coding RNA 152		11.73	1.02	2.0×10^{-3}	1552258_at	1.53

HOXA2	Homeobox A2	7p15.2	11.68	0.78	2.0×10^{-3}	214457_at	3.04
ADAMDEC1	ADAM-like, decysin 1	8p12	11.66	0.76	2.0×10^{-3}	206134_at	2.41
CCL2	Chemokine (C–C motif) ligand 2	17q11.2-q21.1	11.51	0.96	2.0×10^{-3}	216598_s_at	8.73
G0S2	G0/G1switch 2	1q32.2-q41	11.49	0.92	2.0×10^{-3}	213524_s_at	1.9
HOXA1	Homeobox A1	7p15.2	11.49	1.16	2.0×10^{-3}	214639_s_at	5.7
HOXC6	Homeobox C6	12q13.13	11.39	0.88	2.0×10^{-3}	206858_s_at	3.08
SH2D4A	SH2 domain containing 4A	8p21	11.38	1.35	2.0×10^{-3}	219749_at	1.35
COL1A2	Collagen, type I, alpha 2	7q21.3	11.3	0.94	2.0×10^{-3}	202404_s_at	11.33
DPEP1	Dipeptidase 1 (renal)	16q24	11.2	0.75	2.0×10^{-3}	205983_at	1.08
MMP7	Matrix metallopeptidase 7 (matrilysin, uterine)	11q21-q22	11.2	0.66	2.0×10^{-3}	204259_at	1.8
S100A8	S100 calcium-binding protein A8	1q12-q22	11.08	0.81	2.0×10^{-3}	202917_s_at	5.21
PTPN22	Protein tyrosine phosphatase, nonreceptor type 22 (lymphoid)	1p13.2	11.04	1.29	2.0×10^{-3}	236539_at	2.91
MARCO	Macrophage receptor with collagenous structure	2q14.2	10.95	0.77	2.0×10^{-3}	205819_at	3.02
HLA-DQA1	Major histocompatibility complex, class II, DQ alpha 1	6p21.3	10.77	0.87	2.0×10^{-3}	203290_at	3.12

Continued

Table 4 The Top 200 Upregulated and Top 50 Downregulated Genes in *mda-9* High Glioma—cont'd

ID	Gene Title	Chromosome	Fold Change (TCGA)	S-to-N Ratio (TCGA)	p-Value (TCGA)	Affymetrix ID	Fold Change (GSE4290)
STEAP3	STEAP family member 3	2q14.2	10.74	1.28	2.0×10^{-3}	1554830_a_at	9.76
IGF2BP3	Insulin-like growth factor 2 mRNA-binding protein 3	7p15.3	10.71	0.88	2.0×10^{-3}	203819_s_at	6.63
CTHRC1	Collagen triple helix repeat containing 1	8q22.3	10.66	1.02	2.0×10^{-3}	225681_at	4.52
C7orf57	Chromosome 7 open-reading frame 57	7p12.3	10.64	0.81	2.0×10^{-3}	1557636_a_at	6.67
TGFBI	Transforming growth factor, beta-induced, 68 kDa	5q31	10.63	0.98	2.0×10^{-3}	201506_at	6.66
MYBPH	Myosin-binding protein H	1q32.1	10.61	0.86	2.0×10^{-3}	206304_at	1.42
HOXB7	Homeobox B7	17q21.32	10.59	0.87	2.0×10^{-3}	204778_x_at	2.95
HOXC13	Homeobox C13	12q13.13	10.58	0.71	2.0×10^{-3}	219832_s_at	1.53
TNFRSF11B	Tumor necrosis factor receptor superfamily, member 11b	8q24	10.56	1.03	2.0×10^{-3}	204933_s_at	12.3
FPR3	Formyl peptide receptor 3	19q13.3–q13.4	10.35	0.92	2.0×10^{-3}	230422_at	3.26
MGP	Matrix Gla protein	12p12.3	10.31	0.95	2.0×10^{-3}	202291_s_at	5.36

IGF2BP2	Insulin-like growth factor 2 mRNA-binding protein 2	3q27.2	10.3	0.86	2.0×10^{-3}	218847_at	1.53
CCDC109B	Coiled-coil domain containing 109B	4q25	10.29	1.38	2.0×10^{-3}	218802_at	3.55
TUBA1C	Tubulin, alpha 1c	12q13.12	10.27	1.12	2.0×10^{-3}	209251_x_at	1.02
MS4A6A	Membrane-spanning 4-domains, subfamily A, member 6A	11q12.1	10.27	1.06	2.0×10^{-3}	224356_x_at	4.92
GPR65	G protein-coupled receptor 65	14q31-q32.1	10.26	1.24	2.0×10^{-3}	214467_at	3.96
FAM20A	Family with sequence similarity 20, member A	17q24.3	10.06	1.03	2.0×10^{-3}	226804_at	3.31
GPR82	G protein-coupled receptor 82	Xp11.4	10.03	1.1	2.0×10^{-3}	244434_at	3.85
ANXA2	Annexin A2	15q22.2	10.02	1.28	2.0×10^{-3}	213503_x_at	6.09
C15orf48	Chromosome 15 open-reading frame 48	15q21.1	10.02	0.94	2.0×10^{-3}	223484_at	6.09
HAMP	Hepcidin antimicrobial peptide	19q13.1	9.96	0.85	2.0×10^{-3}	220491_at	2.03
MYO1G	Myosin IG	7p13-p11.2	9.93	1.09	2.0×10^{-3}	227799_at	1.22
SERPINA1	Serpin peptidase inhibitor, clade A (alpha-1 antiproteinase, antitrypsin), member 1	14q32.1	9.84	1.19	2.0×10^{-3}	202833_s_at	4.61
AEBP1	AE binding protein 1	7p	9.81	1.04	2.0×10^{-3}	201792_at	5.23

Continued

Table 4 The Top 200 Upregulated and Top 50 Downregulated Genes in *mda-9* High Glioma—cont'd

ID	Gene Title	Chromosome	Fold Change (TCGA)	S-to-N Ratio (TCGA)	p-Value (TCGA)	Affymetrix ID	Fold Change (GSE4290)
ADAM6	ADAM metallopeptidase domain 6 (pseudogene)	14q32.33	9.81	0.68	2.0×10^{-3}	237909_at	−2.17
PLAUR	Plasminogen activator, urokinase receptor	19q13	9.79	1.23	2.0×10^{-3}	211924_s_at	2.82
TNFAIP6	Tumor necrosis factor, alpha-induced protein 6	2q23.3	9.77	0.91	2.0×10^{-3}	206025_s_at	3.38
MAP1LC3C	Microtubule-associated protein 1 light chain 3 gamma	1q43	9.73	1.17	2.0×10^{-3}	221697_at	2.64
LOC541471	Hypothetical LOC541471		9.65	0.96	2.0×10^{-3}	1562876_s_at	3.06
HMGA2	High mobility group AT-hook 2	12q15	9.63	0.74	2.0×10^{-3}	1558682_at	−1.67
RARRES2	Retinoic acid receptor responder (tazarotene induced) 2	7q36.1	9.6	0.82	2.0×10^{-3}	209496_at	4.01
GPR1	G protein-coupled receptor 1	2q33.3	9.53	0.9	2.0×10^{-3}	214605_x_at	1.6
DPYD	Dihydropyrimidine dehydrogenase	1p22	9.44	1.42	2.0×10^{-3}	1554536_at	4.89
RNASE2	Ribonuclease, RNase A family, 2 (liver, eosinophil-derived neurotoxin)	14q24-q31	9.36	1.07	2.0×10^{-3}	206111_at	2.52
CD48	CD48 molecule	1q21.3-q22	9.26	0.99	2.0×10^{-3}	204118_at	2

Gene	Description	Location			p-value	Probe	Fold
IGFBP3	Insulin-like growth factor binding protein 3	7p13-p12	9.25	0.87	2.0×10^{-3}	210095_s_at	6.64
PDLIM1	PDZ and LIM domain 1	10q23.1	9.24	1.14	2.0×10^{-3}	208690_s_at	4.53
DES	Desmin	2q35	9.23	0.71	2.0×10^{-3}	202222_s_at	1.72
GPNMB	Glycoprotein (transmembrane) nmb	7p	9.2	0.97	2.0×10^{-3}	1554018_at	5.2
LRRN4CL	LRRN4 C-terminal like	11q12.3	9.19	1.07	2.0×10^{-3}	1556427_s_at	2.61
THBS1	Thrombospondin 1	15q15	9.09	0.88	2.0×10^{-3}	201109_s_at	8.69
IDO1	Indoleamine 2,3-dioxygenase 1	8p12-p11	9.04	0.88	2.0×10^{-3}	210029_at	1.48
CLEC12A	C-type lectin domain family 12, member A	12p13.31	9.03	0.97	2.0×10^{-3}	1552398_a_at	1.82
OSR2	Odd-skipped related 2 (Drosophila)	8q22.2	9.02	0.87	2.0×10^{-3}	213568_at	2.12
DMRTA2	DMRT-like family A2	1p33	9.01	0.65	2.0×10^{-3}	1558856_at	2.21
IL1RN	Interleukin 1 receptor antagonist	2q14.2	9.01	0.89	2.0×10^{-3}	212657_s_at	2.12
CCR2	Chemokine (C–C motif) receptor 2	3p21	8.97	1.07	2.0×10^{-3}	206978_at	1.43
HOXD9	Homeobox D9	2q31.1	8.97	0.64	2.0×10^{-3}	205604_at	1.95
KDELR3	KDEL (Lys–Asp–Glu–Leu) endoplasmic reticulum protein retention receptor 3	22q13	8.97	1.15	2.0×10^{-3}	204017_at	2.33
IL1R2	Interleukin 1 receptor, type II	2q12	8.96	0.85	2.0×10^{-3}	205403_at	2.62

Continued

Table 4 The Top 200 Upregulated and Top 50 Downregulated Genes in *mda-9* High Glioma—cont'd

ID	Gene Title	Chromosome	Fold Change (TCGA)	S-to-N Ratio (TCGA)	p-Value (TCGA)	Affymetrix ID	Fold Change (GSE4290)
STAC	SH3 and cysteine rich domain	3p22.3	8.95	0.72	2.0×10^{-3}	205743_at	1.55
AREG	Amphiregulin	4q13.3	8.87	0.84	2.0×10^{-3}	205239_at	1.9
ACTG2	Actin, gamma 2, smooth muscle, enteric	2p13.1	8.84	0.75	2.0×10^{-3}	202274_at	1.91
ANXA2P1	Annexin A2 pseudogene 1	4q31.3	8.81	1.13	2.0×10^{-3}	210876_at	2.65
SPAG4	Sperm associated antigen 4	20q11.2	8.8	0.98	2.0×10^{-3}	219888_at	3.73
PITX1	Paired-like homeodomain 1	5q31.1	8.75	0.71	2.0×10^{-3}	208502_s_at	3.39
CLIC1	Chloride intracellular channel 1	6p21.3	8.72	1.34	2.0×10^{-3}	208659_at	4.19
C21orf7	Chromosome 21 open-reading frame 7		8.69	1.2	2.0×10^{-3}	221211_s_at	2.8
LSP1	Lymphocyte-specific protein 1	11p15.5	8.66	1.09	2.0×10^{-3}	205523_at	5.02
HOXC11	Homeobox C11	12q13.13	8.65	0.73	2.0×10^{-3}	206745_at	1.44
PLBD1	Phospholipase B domain containing 1	12p13.1	8.63	1.33	2.0×10^{-3}	218454_at	2.21
VDR	Vitamin D (1,25-dihydroxyvitamin D3) receptor	12q12-q14	8.61	1.1	2.0×10^{-3}	204254_s_at	1.88
HOXC9	Homeobox C9	12q13.13	8.61	0.77	2.0×10^{-3}	231936_at	1.5

CP	Ceruloplasmin (ferroxidase)	3q23-q25	8.6	0.85	2.0×10^{-3}	1558034_s_at	6.76
BCL2A1	BCL2-related protein A1	15q24.3	8.58	0.9	2.0×10^{-3}	205681_at	3.23
TYMP	Thymidine phosphorylase	22q13	8.58	1.06	2.0×10^{-3}	204858_s_at	2.47
SFRP4	Secreted frizzled-related protein 4	7p14.1	8.57	0.98	2.0×10^{-3}	204051_s_at	3.64
HLA-DRA	Major histocompatibility complex, class II, DR alpha	6p21.3	8.56	1.12	2.0×10^{-3}	210982_s_at	4.69
SLC11A1	Solute carrier family 11 (proton-coupled divalent metal ion transporters), member 1	2q35	8.54	1.1	2.0×10^{-3}	217507_at	3.23
PTPN7	Protein tyrosine phosphatase, nonreceptor type 7	1q32.1	8.51	1.15	2.0×10^{-3}	204852_s_at	1.23
C1R	Complement component 1, r subcomponent	12p13.31	8.45	1.22	2.0×10^{-3}	212067_s_at	4.36
IL6	Interleukin 6 (interferon, beta 2)	7p21-p15	8.44	0.73	2.0×10^{-3}	205207_at	2.15
PDLIM4	PDZ and LIM domain 4	5q31.1	8.43	0.83	2.0×10^{-3}	211564_s_at	6.01
CASP4	Caspase 4, apoptosis-related cysteine peptidase	11q22.2-q22.3	8.42	1.38	2.0×10^{-3}	209310_s_at	3.59
GZMA	Granzyme A (granzyme 1, cytotoxic T-lymphocyte-associated serine esterase 3)	5q11-q12	8.38	1.01	2.0×10^{-3}	205488_at	1.25

Continued

Table 4 The Top 200 Upregulated and Top 50 Downregulated Genes in *mda-9* High Glioma—cont'd

ID	Gene Title	Chromosome	Fold Change (TCGA)	S-to-N Ratio (TCGA)	p-Value (TCGA)	Affymetrix ID	Fold Change (GSE4290)
FPR2	Formyl peptide receptor 2	19q13.3-q13.4	8.36	0.93	2.0×10^{-3}	210772_at	2.61
GBP5	Guanylate binding protein 5	1p22.2	8.36	1.03	2.0×10^{-3}	229625_at	4.48
F2RL2	Coagulation factor II (thrombin) receptor-like 2	5q13	8.32	0.9	2.0×10^{-3}	230147_at	7.55
PCOLCE	Procollagen C-endopeptidase enhancer	7q22	8.29	0.86	2.0×10^{-3}	202465_at	5.24
TMEM71	Transmembrane protein 71	8q24.22	8.29	1.08	2.0×10^{-3}	238429_at	10.76
SPP1	Secreted phosphoprotein 1	4q22.1	8.29	0.89	2.0×10^{-3}	1568574_x_at	2.21
KYNU	Kynureninase (L-kynurenine hydrolase)	2q22.2	8.29	1.17	2.0×10^{-3}	210663_s_at	3.45
PDCD1LG2	Programmed cell death 1 ligand 2	9p24.2	8.28	1.18	2.0×10^{-3}	220049_s_at	1.36
FAP	Fibroblast activation protein, alpha	2q23	8.24	1.01	2.0×10^{-3}	209955_s_at	2.45
OR51E1	Olfactory receptor, family 51, subfamily E, member 1	11p15.4	8.22	0.86	2.0×10^{-3}	229768_at	1.08
RBP1	Retinol binding protein 1, cellular	3q21-q23	8.21	0.89	2.0×10^{-3}	203423_at	2.71
DSG2	Desmoglein 2	18q12.1	8.21	0.64	2.0×10^{-3}	217901_at	2.18

OTP	Orthopedia homeobox	5q14.1	8.19	0.66	2.0×10^{-3}	223835_x_at	1.85
NOX4	NADPH oxidase 4	11q14.2-q21	8.18	0.96	2.0×10^{-3}	219773_at	2.9
KRT75	Keratin 75	12q13.13	8.17	0.74	2.0×10^{-3}	207065_at	1.59
B. Genes highly downregulated in MDA-9-high glioma							
GRIN1	Glutamate receptor, ionotropic, N-methyl D-aspartate 1	9q34.3	−47.19	−1.03	2.0×10^{-3}	210781_x_at	−5.19
CUX2	Cut-like homeobox 2	12q24.12	−30.25	−1.23	2.0×10^{-3}	213920_at	−6.58
INA	Internexin neuronal intermediate filament protein, alpha	10q24	−28.94	−1.11	2.0×10^{-3}	204465_s_at	−6.42
SVOP	SV2-related protein homolog (rat)	12q24.11	−27.91	−1.02	2.0×10^{-3}	229818_at	−7.29
TNR	Tenascin R (restrictin, janusin)	1q24	−27.59	−0.96	2.0×10^{-3}	1564897_at	−1.87
CHGA	Chromogranin A (parathyroid secretory protein 1)	14q32	−26.84	−0.99	2.0×10^{-3}	204697_s_at	−7.24
LRTM2	Leucine-rich repeats and transmembrane domains 2	12p13.33	−26.48	−1.15	2.0×10^{-3}	1558530_at	−2.86
SLC1A6	Solute carrier family 1 (high affinity aspartate/glutamate transporter), member 6	19p13.12	−26.13	−1.18	2.0×10^{-3}	1554593_s_at	−7.03
JPH3	Junctophilin 3	16q24.3	−25.52	−1.27	2.0×10^{-3}	229294_at	−10.33

Continued

Table 4 The Top 200 Upregulated and Top 50 Downregulated Genes in *mda-9* High Glioma—cont'd

ID	Gene Title	Chromosome	Fold Change (TCGA)	S-to-N Ratio (TCGA)	p-Value (TCGA)	Affymetrix ID	Fold Change (GSE4290)
C2orf85	Chromosome 2 open reading frame 85		−24.44	−1.29	2.0×10^{-3}	1560035_at	−1.69
CACNG2	Calcium channel, voltage-dependent, gamma subunit 2	22q13.1	−24.06	−1.19	2.0×10^{-3}	214495_at	−7.69
FAM123C	Family with sequence similarity 123C		−23.91	−1.15	2.0×10^{-3}	244709_at	−5.02
PRLHR	Prolactin releasing hormone receptor	10q25.3-q26	−23.89	−1.01	2.0×10^{-3}	231805_at	−2.34
CSMD3	CUB and Sushi multiple domains 3	8q23.3	−23.02	−0.96	2.0×10^{-3}	240228_at	−5.32
RIMS2	Regulating synaptic membrane exocytosis 2	8q22.3	−22.68	−1.09	2.0×10^{-3}	229823_at	−7.59
CPLX2	Complexin 2	5q35.2	−22.32	−0.95	2.0×10^{-3}	225815_at	−8.52
CACNA1B	Calcium channel, voltage-dependent, N type, alpha 1B subunit	9q34	−22.31	−1.03	2.0×10^{-3}	235781_at	−4.37
MYT1L	Myelin transcription factor 1 like	2p25.3	−22.01	−0.91	2.0×10^{-3}	210016_at	−9.13
SSTR1	Somatostatin receptor 1	14q13	−21.55	−0.99	2.0×10^{-3}	235591_at	−7.38
PTPRT	Protein tyrosine phosphatase, receptor type, T	20q12-q13	−21.4	−1.19	2.0×10^{-3}	205948_at	−8.59
PCDH15	Protocadherin-related 15	10q21.1	−21.24	−0.91	2.0×10^{-3}	1553344_at	−7.92

SCRT1	Scratch homolog 1, zinc finger protein (Drosophila)	8q24.3	−21.23	−1.16	2.0×10^{-3}	228761_at	−5.7
CBLN1	Cerebellin 1 precursor	16q12.1	−20.32	−1.25	2.0×10^{-3}	205747_at	−2.33
VSTM2A	V-set and transmembrane domain containing 2A	7p11.2	−20.15	−0.94	2.0×10^{-3}	1554530_at	−8.52
HRH3	Histamine receptor H3	20q13.33	−20.07	−1.13	2.0×10^{-3}	220447_at	−6.95
ST8SIA3	ST8 alpha-N-acetyl-neuraminide alpha-2,8-sialyltransferase 3	18q21.31	−20.05	−0.99	2.0×10^{-3}	230262_at	−4.69
GABRG2	Gamma-aminobutyric acid (GABA) A receptor, gamma 2	5q34	−19.84	−0.8	2.0×10^{-3}	1568612_at	−9.37
SNAP91	Synaptosomal-associated protein, 91 kDa homolog (mouse)	6q15	−19.6	−1.05	2.0×10^{-3}	204953_at	−10.7
SPHKAP	SPHK1 interactor, AKAP domain containing	2q36.3	−19.43	−0.94	2.0×10^{-3}	228509_at	−3.95
UNC13C	unc-13 homolog C (*C. elegans*)	15q21.3	−19.11	−0.91	2.0×10^{-3}	1556095_at	−11.56
PCDH11X	Protocadherin 11 X-linked	Xq21.3	−18.92	−1.05	2.0×10^{-3}	208366_at	−1.03
CHRNA4	Cholinergic receptor, nicotinic, alpha 4	20q13.33	−18.91	−1.12	2.0×10^{-3}	206736_x_at	−1.63
HMP19	HMP19 protein		−18.87	−1.04	2.0×10^{-3}	218623_at	−7.22

Continued

Table 4 The Top 200 Upregulated and Top 50 Downregulated Genes in *mda-9* High Glioma—cont'd

ID	Gene Title	Chromosome	Fold Change (TCGA)	S-to-N Ratio (TCGA)	p-Value (TCGA)	Affymetrix ID	Fold Change (GSE4290)
KSR2	Kinase suppressor of ras 2	12q24.22-q24.23	−18.45	−0.95	2.0×10^{-3}	230551_at	−4.61
ELFN2	Extracellular leucine-rich repeat and fibronectin type III domain containing 2	22q13.1	−18.03	−1.09	2.0×10^{-3}	1559072_a_at	−5.83
GABRA1	Gamma-aminobutyric acid (GABA) A receptor, alpha 1	5q34	−17.92	−0.71	2.0×10^{-3}	244118_at	−17.68
PAK7	p21 protein (Cdc42/Rac)-activated kinase 7	20p12	−17.89	−1.11	2.0×10^{-3}	210721_s_at	−8.97
WSCD2	WSC domain containing 2	12q23.3	−17.87	−1.01	2.0×10^{-3}	229032_at	−4.8
SLC8A2	Solute carrier family 8 (sodium/calcium exchanger), member 2	19q13.32	−17.81	−1.06	2.0×10^{-3}	215267_s_at	−7.58
ACTL6B	Actin-like 6B	7q22	−17.65	−1	2.0×10^{-3}	206013_s_at	−11.81
PCSK2	Proprotein convertase subtilisin/kexin type 2	20p11.2	−17.5	−0.85	2.0×10^{-3}	204870_s_at	−10.56
GLRA3	Glycine receptor, alpha 3	4q34.1	−17.2	−1.06	2.0×10^{-3}	207928_s_at	−2.7
USH1C	Usher syndrome 1C (autosomal recessive, severe)	11p14.3	−17.12	−0.87	2.0×10^{-3}	205137_x_at	−3.81

HPSE2	Heparanase 2	10q23-q24	−17.07	−0.93	2.0×10^{-3}	220927_s_at	−2.73
CHRM1	Cholinergic receptor, muscarinic 1	11q12-q13	−17.04	−0.95	2.0×10^{-3}	231783_at	−9.4
KCNT1	Potassium channel, subfamily T, member 1	9q34.3	−16.93	−0.95	2.0×10^{-3}	1563608_a_at	−2.52
OPALIN	Oligodendrocytic myelin paranodal and inner loop protein	10q23-q24	−16.92	−0.66	2.0×10^{-3}	239575_at	−3.82
CHD5	Chromodomain helicase DNA binding protein 5	1p36.3	−16.68	−0.93	2.0×10^{-3}	213965_s_at	−4.45
CACNG3	Calcium channel, voltage-dependent, gamma subunit 3	16p12.1	−16.05	−0.78	2.0×10^{-3}	206384_at	−9.57
SHANK2	SH3 and multiple ankyrin repeat domains 2	11q13.2	−15.85	−1.15	2.0×10^{-3}	243681_at	−6.99
ZDHHC22	Zinc finger, DHHC-type containing 22	14q24.3	−15.52	−1.23	2.0×10^{-3}	229875_at	−15.29

intermediate filament protein, alpha), *SVOP* (SV2-related protein homolog), and *CHGA* (*chromogranin A (parathyroid secretory protein 1*)). The *MMPs*, *TIMPs*, *COLs*, and *IBSP* are all components of ECM processes crucial to cell invasion and metastasis; thus, these genes' connections to *mda-9* are quite clear (to be discussed further in subsequent sections). *ILs* and *ILRs* are components of various immuno-related pathways and are known to play important roles in glioma progression (Yeung, McDonald, Grewal, & Munoz, 2013). *SAA1* codes for an acute phase protein and is a potential serum-based cancer biomarker. Recently, it was found to be a transcriptional target of the metastasis-inducing S100A (also highly upregulated in *mda-9*-high glioma) (Hansen et al., 2015). HOX proteins are oftentimes transcription factors and many of them have been linked to cancer progression (Shah & Sukumar, 2010). PLA2G2A, which may induce astrocytoma proliferation (Hernandez, Martin, Garcia-Cubillas, Maeso-Hernandez, & Nieto, 2010), is an enzyme, which hydrolyzes phosphatidylinositol (PI) during lipid metabolism and membrane assembly (Singer et al., 2002). MDA-9 may be closely associated with this particular enzyme, given that MDA-9 has been shown to bind to phosphatidylinositol 4,5-bisphosphate (and may assist in transporting proteins to membranes) (Wawrzyniak, Kashyap, & Zimmermann, 2013). *ABCC3* belongs to the ABC subfamily of multidrug resistance proteins, known to be a major detriment to glioma therapy (Decleves, Amiel, Delattre, & Scherrmann, 2006). The glycoprotein CHI3L1 has previously been linked to glioma progression through its upregulation of VEGF (and angiogenesis) (Francescone et al., 2011). The stromal protein POSTN has also been linked to glioma invasion and its poor clinical outcome (Mikheev et al., 2015; Tian, Zhang, & Zhang, 2014; Wang, Wang, & Jiang, 2013). Its possible association with *mda-9* is yet to be investigated. How the iron-binding and multifunctional protein LTF (Gibbons, Kanwar, & Kanwar, 2011; Ward, Paz, & Conneely, 2005) may be positively associated with glioma progression or *mda-9* functionality is not yet clear. Many of the very highly downregulated genes in *mda-9*-high glioma appear to have neuronal functions. *GRIN1*, which codes for a critical part of the NMDA (*N*-methyl-D-aspartate) receptor (Gibb, 2004), is among the most highly downregulated genes in *mda-9*-high glioma. At this point, it is not exactly clear how the downregulation of this gene may be connected to glioma pathogenesis or *mda-9* expression. *CUX2* is a neuronal transcription factor crucial to regulation of dendritic branching, as well as synapses of the upper cortical neurons (Cubelos et al., 2010). It is also not clear how this gene may be involved in glioma progression. *CHGA* is the precursor to vasostatin, pancreastatin, and parastatin, which are negative modulators of the neuroendocrine system. Its role in pathogenesis of glioma has not been established, but was found to be inhibitory to angiogenesis (Loh, Cheng, Mahata, Corti, & Tota, 2012). *INA* or *alpha internexin* is a Type IV intermediate filament involved in neuronal morphogenesis (Lepinoux-Chambaud & Eyer, 2013). Its downregulation among *mda-9*-high glioma is consistent with recent observations that relatively low expression of the gene correlates with the absence of 1p/19q codeletion (which are established markers of good prognosis) (Ducray et al., 2009).

8.3 Pathways and Gene Groups Identified Through GSEA

Instead of inspecting the likely *mda-9*-associated genes one at a time (like what was accomplished in the previous section), GSEA analysis was able to identify the groups of genes representing molecular pathways and related functionalities. The first six rows of Table 5 are the top sets of genes (i.e., the highest ES values) directly associated with *mda-9*-high glioma. The last row includes the few gene sets found to be likely associated with *mda-9* downregulation. Some of the notable gene sets are discussed below.

8.3.1 Association with ECM, Cell Adhesion, and Migration
8.3.1.1 The Metallopepties and Associated Proteins

Among the gene sets exhibiting the highest ES values are those related to the ECM. These include gene groups labeled as Reactome's *degradation of ECM* (ES=0.79), GO Molecular Function's *metallopeptidase activity* (ES=0.84) and GO Cellular Component's *ECM* (ES=0.77) (Fig. 16). All of these gene sets essentially refer to the same group of genes, which include the MMPs and its activator proteins (*TIMPs*). MMPs are zinc and calcium dependent proteases, which upon secretion can degrade the ECM (VanMeter et al., 2001), an essential process for the cells to accomplish invasion and metastasis. A closer look at the GSEA plot for the Reactome gene set *degradation of ECM* indicates 12 genes (out of 28 members) considered as part of core enrichment (Fig. 16A). These include *MMP9* and *TIMP1*, which were upregulated (*mda-9*-high vs. *mda-9*-low) 44-fold and 27-fold, respectively, as well as *MMP13*, *MMP1*, *TIMP1*, *MMP11*, *MMP9*, and *TPSAB1* (*tryptase alpha/beta 1*). Shown in Fig. 16B are specific pathways involving the aforementioned genes (taken from www.reatome.org). These are mostly associated with degradation of the glycoproteins fibrillin (left), fibronectin (center), and decorin (right). Experimentally, our group has already demonstrated that overexpression of *mda-9* in melanoma and glioma cell lines leads

Collagen fibers provide the network of roads which cells travel on during migration (Egeblad, Rasch, & Weaver, 2010). There has never been any report implicating *mda-9* as a regulator of collagen expression (or the reverse case).

8.3.1.3 Integrins and Focal Adhesion Genes

Other ECM-related gene sets highly enriched in *mda-9*-high glioma belong to the following categories: Reactome's *integrin cell surface interactions* and KEGG's *focal adhesion*. Among the components of these sets are genes coding for various collagen molecules (*COL1A2, COL1A1, COL4A1, COL4A2*), integrins (*ITGA4, ITGA5, ITGB3, ITGB2*), *IBSP* (integrin-binding sialoprotein), the adhesive glycoprotein *THBS1* (thrombospondin 1), the cell surface glycoprotein *ICAM1* (*intercellular adhesion molecule 1*), laminins (*LAMB1, LAMC1*), and fibronectin (*FN1*). There were recent reports indicating that *mda-9* is indeed involved in the activation of focal adhesion kinase (FAK). MDA-9's role in the activation of FAK has been demonstrated in breast cancer cells (through the crosstalk between Protein Kinase C α and MDA-9) (Hwangbo, Kim, Lee, & Lee, 2010), dendritic cells (Cho, Kim, Lee, Hong, & Choe, 2013), glioma (through FAK-JNK and FAK-AKT signaling) (Zhong et al., 2012), and small cell lung cancer cells (Kim et al., 2014). MDA-9 was also shown to positively regulate the scaffold function of ILK (integrin-linked kinase) (Hwangbo, Park, & Lee, 2011), part of KEGG's *focal adhesion* gene set, which is also upregulated in *mda-9*-high glioma.

8.3.1.4 Neurite Outgrowth

The Reactome gene set centered around *NCAM (neural cell adhesion molecule), important for neurite outgrowth* (ES = 0.76), is also enriched among *mda-9*-high glioma samples. NCAM, a member of the immunoglobulin (Ig) superfamily, is a mediator of the neurite outgrowth process (Hansen, Berezin, & Bock, 2008). In glioma cells, it was shown that neurite outgrowth can be promoted through a cadherin-dependent adhesion (Cifarelli, Titus, & Yeoh, 2011). The computational prediction of *MDA-9*'s association with the neurite outgrowth process was primarily driven by the overexpression of various collagen genes (e.g., *COL6A3, COL3A1, COL1A2, COL1A1, COL6A2, COL5A1*) among *mda-9*-high glioma. Chen and colleagues have demonstrated that 3D collagen scaffolds can enhance neurite outgrowth in neuron cancer stem cells (Chen et al., 2012). Another possibly important gene in the neurite outgrowth process is *ST8SIA4* (*ST8 alpha-N-acetyl-neuraminide alpha-2,8-sialyltransferase 4*), an enzyme necessary for synthesis of polysialic acid, which modulates NCAM1's adhesive properties (Takahashi et al., 2012). The gene is also highly upregulated in *mda-9*-high tumors. The connection between *mda-9* and neurite outgrowth stemmed from earlier reports of MDA-9 physically interacting with two molecules implicated in axon (a neurite) formation: NFASC (*neurofascin*) (Koroll, Rathjen, & Volkmer, 2001) and ULK1 (*unc-51-like autophagy activating kinase 1*) (Tomoda et al., 2004).

8.3.2 Association with Immuno Signaling Pathways

8.3.2.1 Interleukins

Among the most highly upregulated genes in *mda-9*-high gliomas are, *IL8* (*interleukin 8*) (19-fold) and *IL2RA* (*interleukin 2 receptor, alpha*) (21-fold). Indeed, GSEA analysis identifies a number of IL-related gene sets directly correlating with *mda-9* upregulation. These include the following: Biocarta gene sets belonging to the *IL17 pathway* (ES = 0.85), *Inflammation pathway* (ES = 0.78), *IL2 pathway* (ES = 0.74), *IL7 pathway* (ES = 0.73), and *IL12 pathway* (ES = 0.70); GO Molecular Function gene sets involving *interleukin receptor activity* (0.83) (Fig. 17A) and *interleukin binding* (0.79). The upregulation of many IL-related genes and pathways is likely related to the glioma cells trying to maintain their immunosuppressive state as much as possible (Parney, 2012). The involvement of *mda-9* in these processes has been verified in a number of reports. Geijsen and colleagues showed that MDA-9 binds to the cytoplasmic tail of IL5RA, which is necessary for IL5-mediated SOX4 activation (Geijsen et al., 2001). Another interleukin (*IL16*), which contains 4 PDZ domains, in theory may interact with MDA-9, as predicted in Struc2Net (http://groups.csail.mit.edu/cb/struct2net/webserver/) (Singh, Park, Xu, Hosur, & Berger, 2010). A recent report from our laboratory indicated that MDA-9 positively regulates the transcription of IL8 (Kegelman et al., 2014). An earlier report by Stier and colleagues suggested that tumor necrosis factor-alpha (TNFα) is capable of upregulating the expression of both IL8 and *mda-9* in endothelial cells (Stier et al., 2000).

8.3.2.2 Interferon

The direct correlation between IFNs and *mda-9* can be traced back to the observation that IFN-γ was found to induce *mda-9* expression in HO-1 human melanoma cells (Lin et al., 1998). Many genes involved in the IFN-γ pathway were found to be upregulated in *mda-9*-high gliomas as

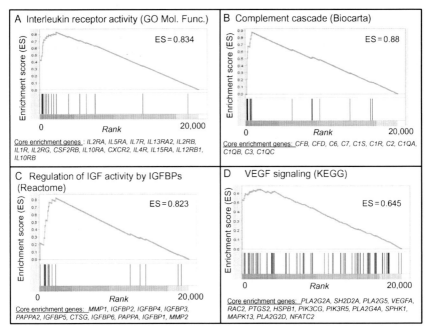

Figure 17 Four representative pathways and processes, which exhibited gene enrichment in GSEA analysis. Indicated are the core enrichment genes in order of decreasing rank. (See the color plate.)

shown in the GSEA plot for Reactome's *IFN-γ pathway* gene set (ES = 0.73). These include the following: *major histocompatibility complex class II genes* (*HLA-DQA1, HLA-DQA2, HLA-DRB1, HLA-DRB5*), *guanylate-binding proteins* (*GBP5, GBP1, GBP2*), CD44, the nuclear antigen *SP100*, suppressor of cytokine signaling (*SOCS3, SOCS1*), and *interferon gamma* (*IFNG*). The top gene on the list (most highly upregulated in *mda-9*-high glioma) is *GBP5*, proven to be upregulated by IFN-γ induction (Kitaya, Yasuo, Yamaguchi, Fushiki, & Honjo, 2007).

8.3.2.3 CTLA4 Pathway

Also noticeable is the apparent activation of the CTLA4 pathway among *mda-9*-high glioma. CTLA4 (fold-change ~3) is a receptor on the surface of Helper T cells, which can downregulate the immune system. Other *mda-9*-high upregulated genes belonging to Biocarta's *CTLA4 pathway* gene set (ES = 0.86) are genes coding for proteins that form a complex with T cell receptors (*CD3D, CD3E*), lymphocyte-specific protein tyrosine kinase (*LCK*), proteins forming the MHC Class II complex (*HLA-D3A*,

HLA-DRB1), IL2-inducible T cell kinase (*ITK*), inducible T-cell costimulator (*ICOS*), and membrane-bound proteins necessary for T cell activation (*CD80*, *CD86*). The identification of these genes is consistent with the knowledge of the presence of infiltrating cells in the glioma environment (Parney, 2012). Whether or not *mda-9* participates in the CTLA4 pathway requires experimental confirmation.

8.3.2.4 Complement Cascade
The complement cascade is another pathway found to be heavily dysregulated among glioma samples with elevated MDA-9. As shown in GSEA plots for complement cascade-related gene sets in both Reactome (ES=0.84) and Biocarta (ES=0.88) (Fig. 17B), the complement component genes *CFB, CFD, C6, C7, C1S, C1R, C2, C1QA, CR1, C4BPA* are indeed among the most highly upregulated genes in the *MDA-9*-high glioma subset. The possible contribution of the complement proteins to cancer proliferation is a relatively new concept. Nonetheless, recent evidence points to the roles of these proteins in the promotion of mitogenic signaling pathways, angiogenesis, migration, resistance to apoptosis, and immunosurveillance (Rutkowski, Sughrue, Kane, Mills, & Parsa, 2010). For now, the assumption is that glioma cells themselves are producing these proteins. It would be interesting to investigate how MDA-9 possibly interacts with these proteins.

8.3.3 Association with Transcriptional Activation and Other Signaling Pathways

8.3.3.1 IGF Signaling
As our GSEA analysis indicates, there is a significant dysregulation of the Reactome gene set "*regulation of insulin like growth factor IGF activity by insulin like growth factor binding proteins IGFBPs*," among the *MDA-9*-high glioma group (ES=0.82) (Fig. 17C). The most highly ranked genes in this gene set include *IGF binding proteins (IGFBPs (1–5))*, *matrix metallopeptidases (MMP1, MMP2)*, *cathepsin (CTSG)*, *pappalysin 2 (PAPPA2)*, and *vascular endothelial growth factor A (VEGFA)*. IGFBPs serve as modulators of IGF1/2, whose downstream transcriptional targets are proinvasion genes such as *MMP2* and *VEGFA*. On the other hand, PAPPA2 and CTSG are proteases, which regulate IGFBPs. IGFBP2 in particular proved to promote glioma progression (Dunlap et al., 2007). A recent report published by our group demonstrated the connection between MDA-9 and IGFBP2 (Das et al., 2013). As described in that report, overexpression of MDA-9 in

melanoma can lead to Src and FAK activation, phosphorylation of Akt, induction of HIF-1α (hypoxia-inducible factor 1-α), eventually leading to elevated IGFBP2 transcription, VEGFA production, and angiogenesis.

8.3.3.2 VEGF Pathway

As mentioned above, VEGFA is highly upregulated in the glioma subset with elevated *MDA-9* expression. The *VEGF signaling pathway* (KEGG) gene set is indeed enriched among *MDA-9*-high gliomas, according to results from the GSEA analysis (ES=0.65) (Fig. 17D). Aside from VEGFA, other highly ranked genes belonging to the gene set are several of its downstream targets including: the phospholipase genes *PLA2G2A*, *PLG2G5*, and *PLG2G4A*; the adaptor protein *SH2D2A* (*SH2 domain containing 2A*, or *VRAP*); the RAS-related gene *RAC2*; *prostaglandin-endoperoxide synthase 2* (*PTGS2* or *COX-2*); and the phosphatidylinositol-4,5-bisphosphate 3-kinase genes *PIK3CG* (catalytic) and *PIK3R5* (regulatory). As mentioned above, we recently demonstrated that VEGFA is a downstream target of MDA-9 (Das et al., 2013).

8.3.3.3 NF-κB Activation

Our group has shown that overexpression of MDA-9 eventually leads to NF-κB activation and the transcriptional activation of MMP2 and IL8 (Boukerche et al., 2010; Boukerche, Su, Emdad, Sarkar, & Fisher, 2007; Kegelman et al., 2014). Results from this bioinformatic analysis are consistent with those experimental observations. Three NF-κB-specific gene sets appear to be enriched in the MDA-9-high subset. These are Reactome's "*TAK1 activates NF-κB by phosphorylation and activation of IKKS complex*" (ES=0.76), "*NF-κB activation via DAI*" (ES=0.78), and "*TRAF6-mediated NF-κB activation*" (ES=0.78). Common to those three pathways are two proteins (the serum amyloid protein SAA1 and the calcium-binding protein S100A12), both very highly upregulated in *MDA-9*-high gliomas. SAA1 is a previously identified transcriptional target of NF-κB (Li & Liao, 1991). On the other hand, S100A12 is a known mediator of NF-κB signaling (Hofmann et al., 1999; Srikrishna et al., 2010).

8.3.4 Association with Phospholipid Metabolism

Zimmerman and colleagues were responsible for elucidating how MDA-9's PDZ domains bind to phosphatidylinositol 4,5-bisphosphate (PIP$_2$), a process believed to be crucial to MDA-9's transport to the cell membrane, signal transduction, as well as nuclear assembly (Wawrzyniak et al.,

2013; Zimmermann, 2006; Zimmermann et al., 2005). *mda-9*'s role in phospholipid metabolism is also predicted in this bioinformatic analysis. GSEA results indicate a number of enriched gene sets, which refer to the five Reactome gene sets which refer to the acyl chain remodeling of PI, phosphatidylglycerol, phosphatidylserine, phosphatidylcholine, and phosphatidylethanolamine. A closer examination of these gene sets points to the phospholipases *PLA2G2A* and *PLA2G5* as the two highest ranked genes common to the gene sets. Another phospholipase (*PLBD1*) is specific to PI. *PLA2G2A* is among the most highly upregulated genes in *mda-9*-high gliomas (at close to 100-fold). The enzyme is involved in the hydrolysis of PI so it can be reacylated by acyltransferases (Singer et al., 2002); a process that diversifies molecules distributed in the membrane. Although not exactly clear at this point, it would be interesting to investigate the possibility of MDA-9 interacting with PLA2G2A.

8.3.5 Pathways Directly Associated with mda-9 Downregulation

For the *mda-9*-low subgroup, only two gene sets exhibited ES and FDR q values within the acceptable range. One is the Reactome gene set *GABA-A receptor activation*, which has ES and FDR q values of -0.88 and 0.005, respectively. The other one is the Reactome gene set *GABA synthesis release uptake and degradation* (ES $= -0.64$, FDR $q = 0.084$). For the first gene set, the core enrichment genes (i.e., those which are highly downregulated in *mda-9*-high glioma) include various GABA-A receptor genes (*GABRG3*, *GABRB1*, *GABRA6*, *GABRA2*, *GABRA3*, *GABRA4*, *GABRB3*, *GABRA5*, *GABRG2*, *GABRA1*, and *GABRB2*). For the other gene set, the genes most likely to influence the pathway are *SNAP25* (*synaptosomal-associated protein, 25 kDa*), *SYT1* (*synaptotagmin I*), *SLC32A1* [*solute carrier family 32 (GABA vesicular transporter), member 1*], glutamate decarboxylase genes (*GAD2*, *GAD1*), *RIMS* (*regulating synaptic membrane exocytosis 1*), and *STX1A* [*syntaxin 1A (brain)*]. These results may be related to several studies describing downregulation of GABA-A receptor during glioma progression (Desrues et al., 2012; D'Urso et al., 2012; Smits et al., 2012).

8.4 A Network of GSEA-Identified Gene Sets

The GSEA-identified gene sets (partially shown in Table 5) can be quite difficult to interpret given the volume of information. However, many of the Reactome, Biocarta, KEGG, and GO gene sets are overlapping in terms of component genes. One way to simplify the results and identify gene sets (e.g., pathways, functionalities), which form clusters, is through network analysis. Useful for this purpose is Cytoscape (Shannon et al., 2003) and its Enrichment Map plug-in (Merico et al., 2010). Data generated from GSEA were analyzed directly using these applications. In order to find clusters, a filter was set so that only gene sets registering $p \leq 0.005$ and FDR $q \leq 0.015$. This resulted in the narrowing down of the very complex interactions into several prominent clusters including the *Extracellular Matrix Cluster* (Fig. 18A) where we can see the overlap and interactions of genes defined by peptidase activity, roles in degradation of the ECM matrix, neurite outgrowth, as well as collagen formation. Another interesting cluster identified through this approach is the *Autoimmunity Disease Cluster* (Fig. 18B), even though the dataset used in the analyses came from glioma. This suggests that the same sets of *mda-9*-associated genes may also be important in certain autoimmune diseases including asthma (something that the authors are already aware of) and type I diabetes. This approach also identified a cytokine-related cluster (figure not shown) connecting GO's

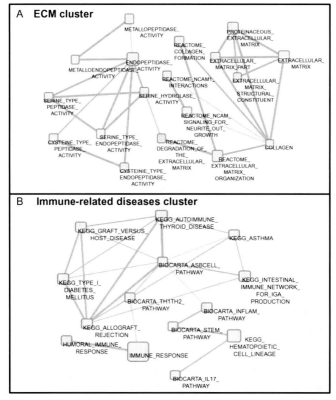

Figure 18 Two of the prominent clusters of GSEA-identified gene sets. (See the color plate.)

leukocyte migration, leukocyte chemotaxis, locomotory behavior, chemokine activity, GPCR binding, chemokine receptor binding, and Reactome's chemokine receptor binds chemokines. As indicated in Table 5, these gene sets are all enriched in the *mda-9*-high glioma subset.

9. SUMMARY, CONCLUSION, AND FUTURE DIRECTIONS

Through the integrated bioinformatic analyses of publicly available cancer genomic datasets, we were able to comprehensively analyze the various epigenetic and molecular factors associated with the dysregulation of *mda-9* in various types of cancer. The following is a list of the most important conclusions obtained from this exercise:

(a) *The elevation of mda-9 expression during cancer progression correlates with both a copy number increase and reduced methylation at a key CpG site (cg1719774) located at the intron, more than 1000 bases 3′ from the CpG island.* These observations derive from analyses of genome-wide expression, CpG methylation, and copy number for TCGA Glioma (Glioblastoma Multiforme, Lower Grade Glioma), SKCM (Skin Cutaneous Melanoma), PRAD (Prostate Adenocarcinoma), COAD (colon adenocarcinoma), LIHC (Liver Hepatocellular Carcinoma), and KIRP (Kidney Renal Papillary Cell Carcinoma) datasets. The degree of influence of each molecular factor (copy number, methylation) toward *mda-9* expression varies among cancer types.

(b) In glioma, both *mda-9* expression level and methylation at cg1719774 can provide prognostic markers, with elevated expression and hypomethylation, respectively, correlating with poor survival.

(c) Among the cancer types included in the PANCAN dataset, melanoma exhibits the highest level of *mda-9* (which is consistent with information downloaded from the Human Protein Atlas), which may be explained by its generally low methylation level at the cg1719774 CpG site (which has the lowest among the 6 TCGA datasets analyzed for CpG methylation).

(d) Hypomethylation at cg1719774 is associated with higher intensity of modified histones such as H3K4 di- and trimethylation and H3K9 acetylation in the promoter area. H3K4 methylations are characteristics of active (or poised to be activated) promoters, while H3K9 acetylation describes a likely active transcription. Data from the ENCODE project (ChIP Seq plus CpG Methylation) were used for this particular set of observations.

(e) An analytical scheme (the *VIGOR*) was designed to predict the groups of genes, functionalities, and molecular pathways associated with *mda-9* dysregulation. The scheme started with the formation of two subsets of glioma patients, according to *mda-9* expression: (1) "*mda-9*-high" group consists of samples whose *mda-9* expression is at least 1 sd higher than the average ($z \geq 1$; upper 15.8%) and (2) "*mda-9*-low" group consists of samples whose *mda-9* expression is at most 1 sd lower than the average ($z \leq -1$; lower 15.8%). The genome-wide expressions of the two subgroups were compared statistically. GSEA (Subramanian et al., 2005) was then employed to identify gene sets likely associated with *mda-9* dysregulation. Among the top gene groups and molecular pathways identified using the scheme described above are those previously linked with both *mda-9* dysregulation and glioma progression (e.g., MMPs, interleukins, IGFBP2 signaling, NF-κB activation). Other interesting immune-related pathways such as that of CTLA-4 have been associated with glioma progression but their connections to *mda-9* functionality are yet to be defined.

The driving forces behind a gene's dysregulated expression in cancer (either upregulation or downregulation) involve a multitude of factors including gene copy number variations, epigenetic regulation, and transcription factors. As for *mda-9*, the gene is located in chromosome 8, which in many cancer types exhibits a gain of the entire or a greater part of the *q* arm (Knuutila et al., 1998). A deeper reason may be a resident oncogene in the 8q arm (*MYC*), whose expression often gets elevated in various cancer types (Prochownik, 2008). The 8p loss/8q gain aberration commonly occurs in colorectal cancer (Diep et al., 2006), liver cancer (van Malenstein, van Pelt, & Verslype, 2011), prostate cancer (Brothman, 2002), and melanoma (Carlson et al., 2005). In glioma and renal cancer, this specific type of chromosomal aberration is far less frequent (Gardina, Lo, Lee, Cowell, & Turpaz, 2008; Kim et al., 2009). This explains why the *R* values (REML analysis) for *mda-9* expression versus *mda-9* copy number are lower in these two cancer types (compared to the other four) (Fig. 8). In contrast, the copy number influence is most pronounced in PRAD samples, which are mostly highly methylated at cg1719774. The influence of cg1719774 methylation toward *mda-9* levels is readily observed in the five other tumor types.

Our identification of cg1719774 as the most differentially methylated (and most highly associated with *mda-9* transcript levels) is consistent with recent observations by Irizarry et al. (2009). In their genome-wide methylation analysis of colon cancer samples, they made the general observation that differentially methylated CpG sites tend to be located outside the

CpG island group, as described by Gardiner-Garden and Frommer (1987). Instead, many of the differentially methylated CpG sites are a few kilobases 5′ (referred to as N Shore) or 3′ (referred to as S Shore) from the CpG island cluster. The cg1719774 is actually 1105 bases from the 3′ end of the CpG island site.

Another epigenetic factor, which can influence *mda-9* expression are the type of modifications which nucleosomes associated with a gene locus might undergo. For example, an H3 histone di- or trimethylated at K4 is indicative of a poised or active promoter. A mono-methylated H3 histone is often associated with enhancer downstream of a start site. Transcriptional activity is characterized by the presence of H3 histones whose modifications may be acetylation at K9 or K27, or trimethylation at K36. Active transcription may also be characterized by an H4 histone methylated at K20. In contrast to the aforementioned nucleosomal markers, the locus binding of H3 histone trimethylated at K27, as well as the transcription factor CTCF are both indicative of transcriptional silencing (Campbell & Turner, 2013; Ernst et al., 2011; Ghirlando et al., 2012). Those were the possible scenarios considered in the chromosomal segmentation analysis developed by Ernst, Kellis, and coworkers, meant to measure activity of transcription at every gene locus in the entire genome (Ernst & Kellis, 2010; Ernst et al., 2011). We then scrutinized the ChIP Seq-generated ENCODE data accessible through the UCSC Genome Browser. What we observed is that cell lines weakly methylated at cg1719774 (such as GM12878, K562, and HUVEC) tend to have more intense binding by H3 methylated at K4 or acetylated at K9, and H4 methylated at K20. Results from chromosomal segmentation analysis indicate that those three cell lines have more active transcription of *mda-9*. For now, we can just assume that the methylation at cg1719774 may be detrimental to the binding affinity of these modified histones. However, this is a readily testable hypothesis.

The GSEA analyses of two glioma subsets at the opposite ends of distribution (*mda-9*-high vs. *mda-9*-low), apparently was capable of correctly predicting pathways and functionalities that have proven associations with *mda-9*. The top gene sets listed in Table 5 were predicted out of 217 Biocarta, 186 KEGG, 674 Reactome, 825 GO Biological Process, 233 GO Cellular Component, and 396 Molecular Function gene sets. This suggests that other predicted pathways without any known link to *mda-9* may be worthy of further investigations.

Many of the gene sets identified by GSEA as likely associated with *mda-9* upregulation refer to ECM processes and functionalities. These results are consistent with the metastasis-promoting property of *mda-9*. Nonetheless, a closer look at those gene sets leads to questions pertaining to the exact role *mda-9* might play in these processes. For example, it would be interesting to define the details connecting *mda-9* with the various collagen molecules found to be highly upregulated in *mda-9*-high glioma. GSEA analysis also identified gene sets related to signaling pathways involving IGFBP2, NF-κB, and VEGF. Previous reports have linked *mda-9* to these proteins. Another very interesting observation is the elevated expression of the phospholipase PLA2G2A, an enzyme involved in the acyl chain remodeling of PI. The possible connection between *mda-9* and PLAG2A and how it may relate to glioma pathogenesis is an interesting topic to pursue, since MDA-9 is known to bind to phosphoinositides. PLA2G2A is involved in the hydrolysis of phospholipids, a process crucial to the asymmetric distribution of molecules in cell membranes.

The upregulation of various immune-related genes and pathways (such as interleukins, interferons, and complement cascade) pertaining to the *mda-9*-high glioma datasets is not surprising since glioma is known to express many immunosuppressive and immunomodulatory factors (Parney, 2012), such as IL-6, IL-10, and VEGFA, all of which are overexpressed in *mda-9*-high glioma. As we have already proven, overexpression of *mda-9* in glioma cells can indeed stimulate the expression of IL-8 (Kegelman et al., 2014). However, it is important to point out that infiltrating cells such as microglia and monocytes may also contribute to the expression of these cytokines (Parney, 2012). Of additional interest is the predicted association of the CTLA-4 pathway with *mda-9*-high glioma. CTLA4 (expressed in Helper T cells and a negative regulator of T lymphocytes) (Daga, Bottino, Castriconi, Gangemi, & Ferrini, 2011) increased ~3-fold in *mda-9*-high relative to *mda-9*-low glioma. At this point, there is still very limited knowledge relative to *mda-9*'s role in glioma immunosuppression.

There are additional interesting observations, which came out of this database analysis. These include the observation that lactotransferrin is the most overexpressed gene in *mda-9*-high glioma. Its role in glioma pathogenesis and certainly its possible physical interaction with MDA-9 needs to be investigated. It is also of interest to investigate how *mda-9* contributes to the regulatory protein S100A4's already proven induction of SAA1 (serum amyloid A1) and subsequently IL-8, MMPs, and metastasis (Bettum et al., 2014; Hansen et al., 2015; Zhang, Liu, & Wang, 2014). All of the aforementioned genes are very highly upregulated in *mda-9*-high gliomas.

What we just demonstrated in this report is the potential of bioinformatic analyses to provide potential insights into our understanding of the biology of the metastasis-promoting gene *mda-9* (its regulation and associated pathways), which could be readily applied to other genes of interest in regulating defined cellular phenotypes. It is now possible to conduct these types of analyses using integrated genomic datasets because of the availability of genomic datasets for various types of molecular profiles for a wide range of caner types, which continues to expand. Many of the datasets (TCGA in particular) are integrated, which means that a single tumor case has been profiled for genome-wide expression, copy number, mutation, and methylation. The approaches employed including *VIGOR* may also be applicable to other genes. We are not implying that the *VIGOR* approach is universally applicable (which will require additional analyses to support the universality of this approach). However, it has proven to be a very useful tool (along with GSEA) in predicting *mda-9*-associated genes and pathways. Overall, this bioinformatic exercise yielded a plethora of information regarding *mda-9*'s functionality (cancer-related or not), which is not surprising given the presence of its promiscuously interacting PDZ domains.

ACKNOWLEDGMENTS

The present work was supported in part by National Institutes of Health Grants R01 CA134721, the National Foundation for Cancer Research (NFCR), the Samuel Waxman Cancer Research Foundation (SWCRF), and cancer research development funds from the VCU Massey Cancer Center. D.S. is a Harrison and Blick Scholar in the VCU Massey Cancer Center and VCU School of Medicine. P.B.F. holds the Thelma Newmeyer Corman Chair in Cancer Research in the VCU Massey Cancer Center. We sincerely thank Mary Goldman (UCSC) and Joshua Gould (Broad Inst.) for their very helpful advice regarding the use of UCSC Cancer Genomics Browser and Gene-E, respectively.

REFERENCES

Beekman, J. M., & Coffer, P. J. (2008). The ins and outs of syntenin, a multifunctional intracellular adaptor protein. *Journal of Cell Science*, *121*(Pt 9), 1349–1355.

Bettum, I. J., Vasiliauskaite, K., Nygaard, V., Clancy, T., Pettersen, S. J., Tenstad, E., et al. (2014). Metastasis-associated protein S100A4 induces a network of inflammatory cytokines that activate stromal cells to acquire pro-tumorigenic properties. *Cancer Letters*, *344*(1), 28–39.

Boukerche, H., Aissaoui, H., Prevost, C., Hirbec, H., Das, S. K., Su, Z. Z., et al. (2010). Src kinase activation is mandatory for MDA-9/syntenin-mediated activation of nuclear factor-kappaB. *Oncogene*, *29*(21), 3054–3066.

Boukerche, H., Su, Z. Z., Emdad, L., Baril, P., Balme, B., Thomas, L., et al. (2005). mda-9/Syntenin: A positive regulator of melanoma metastasis. *Cancer Research*, *65*(23), 10901–10911.

Boukerche, H., Su, Z. Z., Emdad, L., Sarkar, D., & Fisher, P. B. (2007). mda-9/Syntenin regulates the metastatic phenotype in human melanoma cells by activating nuclear factor-kappaB. *Cancer Research*, *67*(4), 1812–1822.

Boukerche, H., Su, Z. Z., Prevot, C., Sarkar, D., & Fisher, P. B. (2008). mda-9/Syntenin promotes metastasis in human melanoma cells by activating c-Src. *Proceedings of the National Academy of Sciences of the United States of America, 105*(41), 15914–15919.

Brothman, A. R. (2002). Cytogenetics and molecular genetics of cancer of the prostate. *American Journal of Medical Genetics, 115*(3), 150–156.

Campbell, M. J., & Turner, B. M. (2013). Altered histone modifications in cancer. *Advances in Experimental Medicine and Biology, 754*, 81–107.

Carlson, J. A., Ross, J. S., Slominski, A., Linette, G., Mysliborski, J., Hill, J., et al. (2005). Molecular diagnostics in melanoma. *Journal of the American Academy of Dermatology, 52*(5), 743–775, quiz 775–748.

Chandran, U. R., Ma, C., Dhir, R., Bisceglia, M., Lyons-Weiler, M., Liang, W., et al. (2007). Gene expression profiles of prostate cancer reveal involvement of multiple molecular pathways in the metastatic process. *BMC Cancer, 7*, 64.

Chang, K., Creighton, C. J., Davis, C., Donehower, L., Drummond, J., Wheeler, D., et al. (2013). The Cancer Genome Atlas Pan-Cancer analysis project. *Nature Genetics, 45*(10), 1113–1120.

Chen, C. H., Kuo, S. M., Liu, G. S., Chen, W. N., Chuang, C. W., & Liu, L. F. (2012). Enhancement of neurite outgrowth in neuron cancer stem cells by growth on 3-D collagen scaffolds. *Biochemical and Biophysical Research Communications, 428*(1), 68–73.

Cho, W., Kim, H., Lee, J. H., Hong, S. H., & Choe, J. (2013). Syntenin is expressed in human follicular dendritic cells and involved in the activation of focal adhesion kinase. *Immune Network, 13*(5), 199–204.

Cifarelli, C. P., Titus, B., & Yeoh, H. K. (2011). Cadherin-dependent adhesion of human U373MG glioblastoma cells promotes neurite outgrowth and increases migratory capacity. Laboratory investigation. *Journal of Neurosurgery, 114*(3), 663–669.

Cubelos, B., Sebastian-Serrano, A., Beccari, L., Calcagnotto, M. E., Cisneros, E., Kim, S., et al. (2010). Cux1 and Cux2 regulate dendritic branching, spine morphology, and synapses of the upper layer neurons of the cortex. *Neuron, 66*(4), 523–535.

Daga, A., Bottino, C., Castriconi, R., Gangemi, R., & Ferrini, S. (2011). New perspectives in glioma immunotherapy. *Current Pharmaceutical Design, 17*(23), 2439–2467.

Das, S. K., Bhutia, S. K., Azab, B., Kegelman, T. P., Peachy, L., Santhekadur, P. K., et al. (2013). MDA-9/syntenin and IGFBP-2 promote angiogenesis in human melanoma. *Cancer Research, 73*(2), 844–854.

Das, S. K., Bhutia, S. K., Kegelman, T. P., Peachy, L., Oyesanya, R. A., Dasgupta, S., et al. (2012). MDA-9/syntenin: A positive gatekeeper of melanoma metastasis. *Frontiers in Bioscience: A Journal and Virtual Library, 17*, 1–15.

Dasgupta, S., Menezes, M. E., Das, S. K., Emdad, L., Janjic, A., Bhatia, S., et al. (2013). Novel role of MDA-9/syntenin in regulating urothelial cell proliferation by modulating EGFR signaling. *Clinical Cancer Research, 19*(17), 4621–4633.

Decleves, X., Amiel, A., Delattre, J. Y., & Scherrmann, J. M. (2006). Role of ABC transporters in the chemoresistance of human gliomas. *Current Cancer Drug Targets, 6*(5), 433–445.

Desrues, L., Lefebvre, T., Lecointre, C., Schouft, M. T., Leprince, J., Compere, V., et al. (2012). Down-regulation of GABA(A) receptor via promiscuity with the vasoactive peptide urotensin II receptor. Potential involvement in astrocyte plasticity. *PloS One, 7*(5), e36319.

Diep, C. B., Kleivi, K., Ribeiro, F. R., Teixeira, M. R., Lindgjaerde, O. C., & Lothe, R. A. (2006). The order of genetic events associated with colorectal cancer progression inferred from meta-analysis of copy number changes. *Genes, Chromosomes & Cancer, 45*(1), 31–41.

Ducray, F., Criniere, E., Idbaih, A., Mokhtari, K., Marie, Y., Paris, S., et al. (2009). alpha-Internexin expression identifies 1p19q codeleted gliomas. *Neurology, 72*(2), 156–161.

Dunlap, S. M., Celestino, J., Wang, H., Jiang, R., Holland, E. C., Fuller, G. N., et al. (2007). Insulin-like growth factor binding protein 2 promotes glioma development and progression. *Proceedings of the National Academy of Sciences of the United States of America, 104*(28), 11736–11741.

D'Urso, P. I., D'Urso, O. F., Storelli, C., Mallardo, M., Gianfreda, C. D., Montinaro, A., et al. (2012). miR-155 is up-regulated in primary and secondary glioblastoma and promotes tumour growth by inhibiting GABA receptors. *International Journal of Oncology, 41*(1), 228–234.

Egeblad, M., Rasch, M. G., & Weaver, V. M. (2010). Dynamic interplay between the collagen scaffold and tumor evolution. *Current Opinion in Cell Biology, 22*(5), 697–706.

ENCODE Project Consortium. (2004). The ENCODE (ENCyclopedia of DNA elements) project. *Science, 306*(5696), 636–640.

ENCODE Project Consortium. (2011). A user's guide to the encyclopedia of DNA elements (ENCODE). *PLoS Biology, 9*(4), e1001046.

Ernst, J., & Kellis, M. (2010). Discovery and characterization of chromatin states for systematic annotation of the human genome. *Nature Biotechnology, 28*(8), 817–825.

Ernst, J., Kheradpour, P., Mikkelsen, T. S., Shoresh, N., Ward, L. D., Epstein, C. B., et al. (2011). Mapping and analysis of chromatin state dynamics in nine human cell types. *Nature, 473*(7345), 43–49.

Fernandez-Larrea, J., Merlos-Suarez, A., Urena, J. M., Baselga, J., & Arribas, J. (1999). A role for a PDZ protein in the early secretory pathway for the targeting of proTGF-alpha to the cell surface. *Molecular Cell, 3*(4), 423–433.

Fisher, P. B., Prignoli, D. R., Hermo, H., Jr., Weinstein, I. B., & Pestka, S. (1985). Effects of combined treatment with interferon and mezerein on melanogenesis and growth in human melanoma cells. *Journal of Interferon Research, 5*(1), 11–22.

Francescone, R. A., Scully, S., Faibish, M., Taylor, S. L., Oh, D., Moral, L., et al. (2011). Role of YKL-40 in the angiogenesis, radioresistance, and progression of glioblastoma. *The Journal of Biological Chemistry, 286*(17), 15332–15343.

Gardina, P. J., Lo, K. C., Lee, W., Cowell, J. K., & Turpaz, Y. (2008). Ploidy status and copy number aberrations in primary glioblastomas defined by integrated analysis of allelic ratios, signal ratios and loss of heterozygosity using 500K SNP Mapping Arrays. *BMC Genomics, 9*, 489.

Gardiner-Garden, M., & Frommer, M. (1987). CpG islands in vertebrate genomes. *Journal of Molecular Biology, 196*(2), 261–282.

Geijsen, N., Uings, I. J., Pals, C., Armstrong, J., McKinnon, M., Raaijmakers, J. A., et al. (2001). Cytokine-specific transcriptional regulation through an IL-5Ralpha interacting protein. *Science, 293*(5532), 1136–1138.

Ghirlando, R., Giles, K., Gowher, H., Xiao, T., Xu, Z., Yao, H., et al. (2012). Chromatin domains, insulators, and the regulation of gene expression. *Biochimica et Biophysica Acta, 1819*(7), 644–651.

Gibb, A. J. (2004). NMDA receptor subunit gating—Uncovered. *Trends in Neurosciences, 27*(1), 7–10, discussion 10.

Gibbons, J. A., Kanwar, R. K., & Kanwar, J. R. (2011). Lactoferrin and cancer in different cancer models. *Frontiers in Bioscience (Scholar Edition), 3*, 1080–1088.

Goldman, M., Craft, B., Swatloski, T., Ellrott, K., Cline, M., Diekhans, M., et al. (2013). The UCSC Cancer Genomics Browser: Update 2013. *Nucleic Acids Research, 41*(Database issue), D949–D954.

Gordon-Alonso, M., Rocha-Perugini, V., Alvarez, S., Moreno-Gonzalo, O., Ursa, A., Lopez-Martin, S., et al. (2012). The PDZ-adaptor protein syntenin-1 regulates HIV-1 entry. *Molecular Biology of the Cell, 23*(12), 2253–2263.

Hansen, S. M., Berezin, V., & Bock, E. (2008). Signaling mechanisms of neurite outgrowth induced by the cell adhesion molecules NCAM and N-cadherin. *Cellular and Molecular Life Sciences, 65*(23), 3809–3821.

Hansen, M. T., Forst, B., Cremers, N., Quagliata, L., Ambartsumian, N., Grum-Schwensen, B., et al. (2015). A link between inflammation and metastasis: Serum amyloid A1 and A3 induce metastasis, and are targets of metastasis-inducing S100A4. *Oncogene, 34*, 424–435.

Hao, J. M., Chen, J. Z., Sui, H. M., Si-Ma, X. Q., Li, G. Q., Liu, C., et al. (2010). A five-gene signature as a potential predictor of metastasis and survival in colorectal cancer. *Journal of Pathology, 220*(4), 475–489.

Hernandez, M., Martin, R., Garcia-Cubillas, M. D., Maeso-Hernandez, P., & Nieto, M. L. (2010). Secreted PLA2 induces proliferation in astrocytoma through the EGF receptor: Another inflammation-cancer link. *Neuro-Oncology, 12*(10), 1014–1023.

Hirbec, H., Martin, S., & Henley, J. M. (2005). Syntenin is involved in the developmental regulation of neuronal membrane architecture. *Molecular and Cellular Neuroscience, 28*(4), 737–746.

Hofmann, M. A., Drury, S., Fu, C., Qu, W., Taguchi, A., Lu, Y., et al. (1999). RAGE mediates a novel proinflammatory axis: A central cell surface receptor for S100/calgranulin polypeptides. *Cell, 97*(7), 889–901.

Hwangbo, C., Kim, J., Lee, J. J., & Lee, J. H. (2010). Activation of the integrin effector kinase focal adhesion kinase in cancer cells is regulated by crosstalk between protein kinase Calpha and the PDZ adapter protein mda-9/Syntenin. *Cancer Research, 70*(4), 1645–1655.

Hwangbo, C., Park, J., & Lee, J. H. (2011). mda-9/Syntenin protein positively regulates the activation of Akt protein by facilitating integrin-linked kinase adaptor function during adhesion to type I collagen. *The Journal of Biological Chemistry, 286*(38), 33601–33612.

Irizarry, R. A., Ladd-Acosta, C., Wen, B., Wu, Z., Montano, C., Onyango, P., et al. (2009). The human colon cancer methylome shows similar hypo- and hypermethylation at conserved tissue-specific CpG island shores. *Nature Genetics, 41*(2), 178–186.

Ivarsson, Y. (2012). Plasticity of PDZ domains in ligand recognition and signaling. *FEBS Letters, 586*(17), 2638–2647.

Jiang, H., Lin, J. J., Su, Z. Z., Goldstein, N. I., & Fisher, P. B. (1995). Subtraction hybridization identifies a novel melanoma differentiation associated gene, mda-7, modulated during human melanoma differentiation, growth and progression. *Oncogene, 11*(12), 2477–2486.

Joshi-Tope, G., Gillespie, M., Vastrik, I., D'Eustachio, P., Schmidt, E., de Bono, B., et al. (2005). Reactome: A knowledgebase of biological pathways. *Nucleic Acids Research, 33*(Database issue), D428–D432.

Kaiser, J. (2005). National Institutes of Health. NCI gears up for cancer genome project. *Science, 307*(5713), 1182.

Kegelman, T. P., Das, S. K., Emdad, L., Hu, B., Menezes, M. E., Bhoopathi, P., et al. (2015). Targeting tumor invasion: The roles of MDA-9/Syntenin. *Expert Opinion on Therapeutic Targets, 19*(1), 97–112.

Kegelman, T. P., Das, S. K., Hu, B., Bacolod, M. D., Fuller, C. E., Menezes, M. E., et al. (2014). MDA-9/syntenin is a key regulator of glioma pathogenesis. *Neuro-Oncology, 16*(1), 50–61.

Kim, W. Y., Jang, J. Y., Jeon, Y. K., Chung, D. H., Kim, Y. G., & Kim, C. W. (2014). Syntenin increases the invasiveness of small cell lung cancer cells by activating p38, AKT, focal adhesion kinase and SP1. *Experimental and Molecular Medicine, 46*, e90.

Kim, H. J., Shen, S. S., Ayala, A. G., Ro, J. Y., Truong, L. D., Alvarez, K., et al. (2009). Virtual-karyotyping with SNP microarrays in morphologically challenging renal cell neoplasms: A practical and useful diagnostic modality. *American Journal of Surgical Pathology, 33*(9), 1276–1286.

Kitaya, K., Yasuo, T., Yamaguchi, T., Fushiki, S., & Honjo, H. (2007). Genes regulated by interferon-gamma in human uterine microvascular endothelial cells. *International Journal of Molecular Medicine, 20*(5), 689–697.

Knuutila, S., Bjorkqvist, A. M., Autio, K., Tarkkanen, M., Wolf, M., Monni, O., et al. (1998). DNA copy number amplifications in human neoplasms: Review of comparative genomic hybridization studies. *American Journal of Pathology, 152*(5), 1107–1123.

Koo, T. H., Lee, J. J., Kim, E. M., Kim, K. W., Kim, H. D., & Lee, J. H. (2002). Syntenin is overexpressed and promotes cell migration in metastatic human breast and gastric cancer cell lines. *Oncogene, 21*(26), 4080–4088.

Koroll, M., Rathjen, F. G., & Volkmer, H. (2001). The neural cell recognition molecule neurofascin interacts with syntenin-1 but not with syntenin-2, both of which reveal self-associating activity. *The Journal of Biological Chemistry, 276*(14), 10646–10654.

Lee, H., Kim, Y., Choi, Y., Choi, S., Hong, E., & Oh, E. S. (2011). Syndecan-2 cytoplasmic domain regulates colon cancer cell migration via interaction with syntenin-1. *Biochemical and Biophysical Research Communications, 409*(1), 148–153.

Lepinoux-Chambaud, C., & Eyer, J. (2013). Review on intermediate filaments of the nervous system and their pathological alterations. *Histochemistry and Cell Biology, 140*(1), 13–22.

Li, X. X., & Liao, W. S. (1991). Expression of rat serum amyloid A1 gene involves both C/EBP-like and NF kappa B-like transcription factors. *The Journal of Biological Chemistry, 266*(23), 15192–15201.

Liberzon, A., Subramanian, A., Pinchback, R., Thorvaldsdottir, H., Tamayo, P., & Mesirov, J. P. (2011). Molecular signatures database (MSigDB) 3.0. *Bioinformatics, 27*(12), 1739–1740.

Lin, J. J., Jiang, H., & Fisher, P. B. (1998). Melanoma differentiation associated gene-9, mda-9, is a human gamma interferon responsive gene. *Gene, 207*(2), 105–110.

Liu, X., Zhang, X., Lv, Y., Xiang, J., & Shi, J. (2014). Overexpression of syntenin enhances hepatoma cell proliferation and invasion: Potential roles in human hepatoma. *Oncology Reports, 32*(6), 2810–2816.

Loh, Y. P., Cheng, Y., Mahata, S. K., Corti, A., & Tota, B. (2012). Chromogranin A and derived peptides in health and disease. *Journal of Molecular Neuroscience, 48*(2), 347–356.

Merico, D., Isserlin, R., Stueker, O., Emili, A., & Bader, G. D. (2010). Enrichment map: A network-based method for gene-set enrichment visualization and interpretation. *PloS One, 5*(11), e13984.

Mermel, C. H., Schumacher, S. E., Hill, B., Meyerson, M. L., Beroukhim, R., & Getz, G. (2011). GISTIC2.0 facilitates sensitive and confident localization of the targets of focal somatic copy-number alteration in human cancers. *Genome Biology, 12*(4), R41.

Mikheev, A. M., Mikheeva, S. A., Trister, A. D., Tokita, M. J., Emerson, S. N., Parada, C. A., et al. (2015). Periostin is a novel therapeutic target that predicts and regulates glioma malignancy. *Neuro-Oncology, 17*(3), 372–382.

Parney, I. F. (2012). Basic concepts in glioma immunology. *Advances in Experimental Medicine and Biology, 746,* 42–52.

Prochownik, E. V. (2008). c-Myc: Linking transformation and genomic instability. *Current Molecular Medicine, 8*(6), 446–458.

Qian, X. L., Li, Y. Q., Yu, B., Gu, F., Liu, F. F., Li, W. D., et al. (2013). Syndecan binding protein (SDCBP) is overexpressed in estrogen receptor negative breast cancers, and is a potential promoter for tumor proliferation. *PloS One, 8*(3), e60046.

Rutkowski, M. J., Sughrue, M. E., Kane, A. J., Mills, S. A., & Parsa, A. T. (2010). Cancer and the complement cascade. *Molecular Cancer Research, 8*(11), 1453–1465.

Sala-Valdes, M., Gordon-Alonso, M., Tejera, E., Ibanez, A., Cabrero, J. R., Ursa, A., et al. (2012). Association of syntenin-1 with M-RIP polarizes Rac-1 activation during chemotaxis and immune interactions. *Journal of Cell Science, 125*(Pt 5), 1235–1246.

Sandoval, J., Heyn, H., Moran, S., Serra-Musach, J., Pujana, M. A., Bibikova, M., et al. (2011). Validation of a DNA methylation microarray for 450,000 CpG sites in the human genome. *Epigenetics, 6*(6), 692–702.

Sarkar, D., Boukerche, H., Su, Z. Z., & Fisher, P. B. (2008). mda-9/Syntenin: More than just a simple adapter protein when it comes to cancer metastasis. *Cancer Research, 68*(9), 3087–3093.

Shah, N., & Sukumar, S. (2010). The Hox genes and their roles in oncogenesis. *Nature Reviews Cancer, 10*(5), 361–371.

Shannon, P., Markiel, A., Ozier, O., Baliga, N. S., Wang, J. T., Ramage, D., et al. (2003). Cytoscape: A software environment for integrated models of biomolecular interaction networks. *Genome Research, 13*(11), 2498–2504.

Singer, A. G., Ghomashchi, F., Le Calvez, C., Bollinger, J., Bezzine, S., Rouault, M., et al. (2002). Interfacial kinetic and binding properties of the complete set of human and mouse groups I, II, V, X, and XII secreted phospholipases A2. *The Journal of Biological Chemistry, 277*(50), 48535–48549.

Singh, R., Park, D., Xu, J., Hosur, R., & Berger, B. (2010). Struct2Net: A web service to predict protein-protein interactions using a structure-based approach. *Nucleic Acids Research, 38*(Web Server issue), W508–W515.

Smits, A., Jin, Z., Elsir, T., Pedder, H., Nister, M., Alafuzoff, I., et al. (2012). GABA-A channel subunit expression in human glioma correlates with tumor histology and clinical outcome. *PloS One, 7*(5), e37041.

Sokhi, U. K., Bacolod, M. D., Dasgupta, S., Emdad, L., Das, S. K., Dumur, C. I., et al. (2013). Identification of genes potentially regulated by human polynucleotide phosphorylase (hPNPase old-35) using melanoma as a model. *PloS One, 8*(10), e76284.

Srikrishna, G., Nayak, J., Weigle, B., Temme, A., Foell, D., Hazelwood, L., et al. (2010). Carboxylated N-glycans on RAGE promote S100A12 binding and signaling. *Journal of Cellular Biochemistry, 110*(3), 645–659.

Stier, S., Totzke, G., Grunewald, E., Neuhaus, T., Fronhoffs, S., Sachinidis, A., et al. (2000). Identification of syntenin and other TNF-inducible genes in human umbilical arterial endothelial cells by suppression subtractive hybridization. *FEBS Letters, 467*(2–3), 299–304.

Subramanian, A., Tamayo, P., Mootha, V. K., Mukherjee, S., Ebert, B. L., Gillette, M. A., et al. (2005). Gene set enrichment analysis: A knowledge-based approach for interpreting genome-wide expression profiles. *Proceedings of the National Academy of Sciences of the United States of America, 102*(43), 15545–15550.

Sun, L., Hui, A. M., Su, Q., Vortmeyer, A., Kotliarov, Y., Pastorino, S., et al. (2006). Neuronal and glioma-derived stem cell factor induces angiogenesis within the brain. *Cancer Cell, 9*(4), 287–300.

Takahashi, K., Mitoma, J., Hosono, M., Shiozaki, K., Sato, C., Yamaguchi, K., et al. (2012). Sialidase NEU4 hydrolyzes polysialic acids of neural cell adhesion molecules and negatively regulates neurite formation by hippocampal neurons. *The Journal of Biological Chemistry, 287*(18), 14816–14826.

Talantov, D., Mazumder, A., Yu, J. X., Briggs, T., Jiang, Y., Backus, J., et al. (2005). Novel genes associated with malignant melanoma but not benign melanocytic lesions. *Clinical Cancer Research, 11*(20), 7234–7242.

Tian, B., Zhang, Y., & Zhang, J. (2014). Periostin is a new potential prognostic biomarker for glioma. *Tumour Biology, 35*(6), 5877–5883.

Tomoda, T., Kim, J. H., Zhan, C., & Hatten, M. E. (2004). Role of Unc51.1 and its binding partners in CNS axon outgrowth. *Genes and Development, 18*(5), 541–558.

Uhlen, M., Oksvold, P., Fagerberg, L., Lundberg, E., Jonasson, K., Forsberg, M., et al. (2010). Towards a knowledge-based Human Protein Atlas. *Nature Biotechnology, 28*(12), 1248–1250.

van Malenstein, H., van Pelt, J., & Verslype, C. (2011). Molecular classification of hepatocellular carcinoma anno 2011. *European Journal of Cancer, 47*(12), 1789–1797.

VanMeter, T. E., Rooprai, H. K., Kibble, M. M., Fillmore, H. L., Broaddus, W. C., & Pilkington, G. J. (2001). The role of matrix metalloproteinase genes in glioma invasion: Co-dependent and interactive proteolysis. *Journal of Neuro-Oncology, 53*(2), 213–235.

Wang, H., Wang, Y., & Jiang, C. (2013). Stromal protein periostin identified as a progression associated and prognostic biomarker in glioma via inducing an invasive and proliferative phenotype. *International Journal of Oncology, 42*(5), 1716–1724.

Ward, P. P., Paz, E., & Conneely, O. M. (2005). Multifunctional roles of lactoferrin: A critical overview. *Cellular and Molecular Life Sciences, 62*(22), 2540–2548.

Wawrzyniak, A. M., Kashyap, R., & Zimmermann, P. (2013). Phosphoinositides and PDZ domain scaffolds. *Advances in Experimental Medicine and Biology, 991*, 41–57.

Yeung, Y. T., McDonald, K. L., Grewal, T., & Munoz, L. (2013). Interleukins in glioblastoma pathophysiology: Implications for therapy. *British Journal of Pharmacology, 168*(3), 591–606.

Zhang, W., Liu, Y., & Wang, C. W. (2014). S100A4 promotes squamous cell laryngeal cancer Hep-2 cell invasion via NF-kB/MMP-9 signal. *European Review for Medical and Pharmacological Sciences, 18*(9), 1361–1367.

Zhong, D., Ran, J. H., Tang, W. Y., Zhang, X. D., Tan, Y., Chen, G. J., et al. (2012). Mda-9/syntenin promotes human brain glioma migration through focal adhesion kinase (FAK)-JNK and FAK-AKT signaling. *Asian Pacific Journal of Cancer Prevention, 13*(6), 2897–2901.

Zhu, J., Sanborn, J. Z., Benz, S., Szeto, C., Hsu, F., Kuhn, R. M., et al. (2009). The UCSC Cancer Genomics Browser. *Nature Methods, 6*(4), 239–240.

Zimmermann, P. (2006). The prevalence and significance of PDZ domain-phosphoinositide interactions. *Biochimica et Biophysica Acta, 1761*(8), 947–956.

Zimmermann, P., Tomatis, D., Rosas, M., Grootjans, J., Leenaerts, I., Degeest, G., et al. (2001). Characterization of syntenin, a syndecan-binding PDZ protein, as a component of cell adhesion sites and microfilaments. *Molecular Biology of the Cell, 12*(2), 339–350.

Zimmermann, P., Zhang, Z., Degeest, G., Mortier, E., Leenaerts, I., Coomans, C., et al. (2005). Syndecan recycling [corrected] is controlled by syntenin-PIP2 interaction and Arf6. *Developmental Cell, 9*(3), 377–388.

CHAPTER THREE

Exploitation of the Androgen Receptor to Overcome Taxane Resistance in Advanced Prostate Cancer

Sarah K. Martin[*], Natasha Kyprianou[*,†,‡,§,1]

[*]Department of Molecular and Cellular Biochemistry, University of Kentucky College of Medicine, Lexington, Kentucky, USA
[†]Department of Urology, University of Kentucky College of Medicine, Lexington, Kentucky, USA
[‡]Department of Pathology and Toxicology, University of Kentucky College of Medicine, Lexington, Kentucky, USA
[§]Markey Cancer Center, University of Kentucky College of Medicine, Lexington, Kentucky, USA
[1]Corresponding author: e-mail address: nkypr2@email.uky.edu

Contents

1. Introduction — 125
2. AR Signaling Finds Its Intracellular "Zip-Code" — 126
3. Can ADT Overcome AR Addiction in Prostate Tumors? — 128
   3.1 Abiraterone: Targeting Alternative Androgen Synthesis — 128
   3.2 Enzalutamide (MDV3100): Targeting AR Activation — 130
   3.3 EPI-002: Targeting the NTD — 130
4. Taxane Action in Prostate Cancer Cells: Up, Close, and "Personal" with Microtubules — 131
5. Taxane "Promiscuous" Actions Beyond Microtubule Stabilization — 133
   5.1 Blocking AR Translocation — 133
6. Motor Proteins: An Intracellular Accomplice — 134
7. Mechanisms of Therapeutic Resistance — 139
   7.1 "Pumping" Issues — 139
   7.2 Tubulin Mutations — 140
   7.3 Microtubule Binding Dynamics — 140
   7.4 Impairing Centrosome Clustering — 142
   7.5 The EMT Landscape — 142
8. Overcoming Taxane Resistance: The Translational Challenge — 143
   8.1 Intermittent Chemotherapy — 143
9. Chemotherapy in Combination: "The Golden Fleece" of Cancer Treatment — 144
   9.1 Platinum-Based Therapy — 144
   9.2 Abiraterone Acetate — 144
   9.3 Enzalutamide — 145
   9.4 Clusterin Targeting (OGX-011) — 146

9.5 Combinations of Promise and Therapeutic Value	146
9.6 The Taxane "Sisterhood" Finds Solace in Microtubules	147
10. Conclusions	147
References	148

Abstract

Prostate cancer is a tumor addicted to androgen receptor (AR) signaling, even in its castration resistant state, and recently developed antiandrogen therapies including Abiraterone acetate and enzalutamide effectively target the androgen signaling axis, but there is ultimately recurrence to lethal disease. Development of advanced castration-resistant prostate cancer (CRPC) is a biological consequence of lack of an apoptotic response of prostate tumor cells to androgen ablation. Taxanes represent the major clinically relevant chemotherapy for the treatment of patients with metastatic CRPC; unfortunately, they do not deliver a cure but an extension of overall survival. First-generation taxane chemotherapies, Docetaxel (Taxotere), effectively target the cytoskeleton by stabilizing the interaction of β-tubulin subunits of microtubules preventing depolymerization, inducing G2M arrest and apoptosis. Shifting the current paradigm is a growing evidence to indicate that Docetaxel can effectively target the AR signaling axis by blocking its nuclear translocation and transcriptional activity in androgen-sensitive and castration-resistant prostate cancer cells, implicating a new mechanism of cross-resistance between microtubule-targeting chemotherapy and antiandrogen therapies. More recently, Cabazitaxel has emerged as a second-line taxane chemotherapy capable of conferring additional survival benefit to patients with CRPC previously treated with Docetaxel or in combination with antiandrogens. Similar to Docetaxel, Cabazitaxel induces apoptosis and G2M arrest; in contrast to Docetaxel, it sustains AR nuclear accumulation although it reduces the overall AR levels and FOXO1 expression. Cabazitaxel treatment also leads to downregulation of the microtubule-depolymerizing mitotic kinesins, MCAK, and HSET, preventing their ability to depolymerize microtubules and thus enhancing sensitivity to taxane treatment. The molecular mechanisms underlying taxane resistance involve mutational alterations in the tubulin subunits, microtubule dynamics, phenotyping programming of the epithelial-to-mesenchymal transition landscape, and the status of AR activity. This chapter discusses the mechanisms driving the therapeutic resistance of taxanes and antiandrogen therapies in CRPC, and the role of AR in potential interventions toward overcoming such resistance in patients with advanced metastatic disease.

ABBREVIATIONS

AA Abiraterone acetate
ADT androgen deprivation therapy
AR androgen receptor
ARE androgen responsive elements
ARG androgen responsive genes

CBZ Cabazitaxel
CRPC castration-resistant prostate cancer
DBD DNA-binding domain
DHT dihydrotestosterone
EMT epithelial-to-mesenchymal transition
FDA Food and Drug Administration
FOXO1 Forkhead box protein 1
HSET human kinesin-14
LBD ligand-binding domain
LHRH luteinizing hormone response hormone
MAP microtubule-associated protein
MCAK mitotic centromere-associated kinesin
MDR multidrug resistance protein
NLS nuclear localization signal
NTD N-terminal domain
P-gp P-glycoprotein
PIN prostatic intraepithelial neoplasia
PSA prostate-specific antigen
SHBP sex-hormone-binding protein
STLC S-trityl-L-cysteine
TGF-β transforming growth factor beta
TRAMP/DNTβRII transgenic adenocarcinoma of the mouse prostate/dominant-negative TGF-β receptor II mouse model

1. INTRODUCTION

Over 70 years ago, male steroid hormones were causally implicated in prostate cancer by the pioneering work of Huggins and Hodges (1941), with androgen withdrawal shown to suppress tumorigenic growth of endocrine-dependent cancers (Huggins & Hodges, 1941). This cornerstone of our clinical treatment paradigm stands to this day. Androgen deprivation therapy (ADT) can be achieved by a variety of surgical and/or pharmacological methods, but ultimately fails to effectively cure patients with prostate cancer. Castration-resistant prostate cancer (CRPC) emerges as the growth kinetics of prostate tumor cells result in uncontrolled proliferation and consistent apoptosis evasion, independently of the presence of androgens, leading to advanced, aggressive disease (Debes & Tindall, 2004). Localized prostate cancer is highly curable with radical prostatectomy or radiation therapy, but among men who progress to CRPC, 90% of them will develop bone metastasis with clinical progression to metastatic CRPC (mCRPC) being associated with survival of less than 2 years. Biochemical recurrence can be monitored in patients by sequential evaluation of serum prostate-specific antigen (PSA); nearly 70,000 American men develop this biochemical recurrence per year (Freedland & Moul, 2007; Hu, Denmeade, & Luo, 2010), as assessed by PSA screening facilitating risk identification in patients progressing to mCRPC. Progression to CRPC is characterized by increased androgen receptor (AR) expression in prostate tumors and perpetually active AR signaling despite physiologically castrate levels of androgens (Chen et al., 2004; Feldman & Feldman, 2001). Thus, the clinical challenge presenting itself is to effectively treat patients with CRPC by increasing survival and enhancing quality of life toward a complete cure.

For the last decade, taxanes prevailed as the only class of Food and Drug Administration (FDA)-approved chemotherapeutic agents to confer additional survival and palliative benefit to CRPC patients (Tannock et al., 2004). Taxanes are cytotoxic chemotherapeutic agents which bind to the β-tubulin subunit of the protofilament of the microtubule, stabilizing the structure of the cellular cytoskeleton. This stabilization prevents cells from dividing, inducing mitotic arrest and apoptosis. In view, however, of the slow growth kinetics that characterizes prostate tumor cells, one may consider additional effects by taxanes, transcending the catastrophic consequences on mitotic events. In this chapter, we discuss the recent evidence on paradigm-shift action of taxane treatment beyond the antimitotic effects in preclinical models of prostate tumor progression and in clinical mCRPC, and the identification of a lead role for AR and its splice variants as critical contributors to therapeutic cross-resistance to taxanes and antiandrogens in mCRPC.

2. AR SIGNALING FINDS ITS INTRACELLULAR "ZIP-CODE"

The AR is a member of the steroid–thyroid–retinoid nuclear receptor superfamily found on the X chromosome (Xq11-12), spanning approximately 180 kb of DNA with eight exons (Gelmann, 2002). In normal AR signaling, testosterone synthesized in the testis or adrenal gland is sequestered by sex-hormone-binding protein (SHBP) in the circulation. Testosterone dissociates from SHBP and diffuses across the plasma membrane, bringing testosterone into close proximity with 5α-reductase (SRD5A1, SRD5A2; cytochrome p450 enzyme) producing the cognate ligand of AR:dihydrotestosterone (DHT) (Lonergan & Tindall, 2011; Schmidt & Tindall, 2011; Wilson, 2001). Upon AR binding to its cognate ligand

DHT, there is rearrangement of AR protein domains that facilitate conformational change within the heat-shock protein 90 (Hsp90) super complex, and subsequent nuclear translocation and transcriptional activation. DHT-bound AR undergoes homodimerization and phosphorylation by the protein kinase A signaling pathway resulting in activation (Brinkmann et al., 1999; Nazareth & Weigel, 1996). The AR homodimers translocate to the cell nucleus and bind androgen responsive genes (ARG) at specific palindromic DNA sequences known as androgen responsive elements (ARE) (Feng, Zheng, Wennuan, Isaacs, & Xu, 2011). Upon binding to ARE, AR dimers can act as a scaffold toward the recruitment of accessory proteins to assemble an active transcription complex (Feng et al., 2011; Heinlein & Chang, 2002; Roy, Lavrosky, & Song, 1999).

Nuclear translocation of AR is inhibited by its sequestration in the Hsp90 super complex, tethering it to the cytoskeleton. The nuclear localization signal (NLS1) of AR is bipartite and spans the DNA-binding domain (DBD) and hinge regions of the protein with exons 3 and 4 represented (Zhou, Sar, Simental, Lane, & Wilson, 1994). Nuclear translocation of the full-length AR is protected by the bipartite nature of the NLS; this feature safeguards that cooperation between the domains occurs before nuclear transport may proceed. Binding of AR to its cognate ligand facilitates a conformational change that brings the NLS into a functional orientation for signaling translocation (Zhou et al., 1994). After translocation, NLS allows binding of AR to the importin-α adaptor protein and importin-β carrier protein (Chan, Li, & Dehm, 2012), facilitating movement through the nuclear pore complex and Ran-dependent release into the nucleus (Black & Paschal, 2004; Brodsky & Silver, 1999; Corbett & Silver, 1997; Cutress, Whitaker, Mills, Stewart, & Neal, 2008; Gorlich, 1997; Nigg, 1997). A second NLS sequence (NLS2) exists in the ligand-binding domain (LBD), and allows AR to enter the nucleus via importin-α independent mechanism (Picard & Yamamoto, 1987; Poukka et al., 2000; Savory et al., 1999).

Various coregulatory proteins interacting with AR have been mechanistically characterized (Heemers & Tidall, 2007). Of particular interest are the intramolecular interactions within the AR protein domains and between AR subunits in a homodimeric complex, which are of utmost importance to AR activation and nuclear translocation (van Royen, van Cappellen, de Vos, Houtsmuller, & Trapman, 2012). Upon binding DHT, the D-box of the DBD interacts with the TAU-1 domain of the N-terminal domain (NTD), an N-terminal to C-terminal protein domain interaction that is initiated in the cytoplasm (Schaufele et al., 2005; van Royen et al., 2012).

The interaction between the D-box and the NTD of the AR is essential for the transition toward inter-AR molecule homodimerization (Schaufele et al., 2005), occurring in the cytosol and triggering the transport into the nucleus. Upon nuclear translocation and binding to DNA, the N-terminal/C-terminal intramolecular interaction is finished freeing these domains to interact with AR coregulators; the homodimer ultimately settles into the major groove of the DNA double helix (Van Royen et al., 2007; van Royen et al., 2012).

3. CAN ADT OVERCOME AR ADDICTION IN PROSTATE TUMORS?

CRPC is a disease addicted to AR signaling developed through the course of ADT. ADT can be achieved through impairing adrenal androgen synthesis, antiandrogen antagonists, and luteinizing hormone response hormone (LHRH) analogs. The therapeutic efficacy of ADT in prostate cancer patients is driven via apoptosis-mediated tumor regression of androgen-sensitive prostate cancer cells (Kahn, Collazo, & Kyprianou, 2014). Biochemically, ADT induces a chemical castration state in the patient characterized by serum testosterone levels <50 ng/mL (Kahn et al., 2014). AR signaling can become reactivated through a variety of mechanisms (Feldman & Feldman, 2001). AR signaling can be restored by promiscuous mutations to the AR that allow it to become activated by androgen-independent ligands such as growth factors including epidermal growth factor, overamplification of AR expression that causes hypersensitivity in the signaling pathway, and expression of constitutively activated AR splice variants (ARvs) which lack or have modified LBDs, enabling AR to become phosphorylated and activated without requirement for its cognate ligand (Dehm, Schmidt, Heemers, Vessella, & Tindall, 2008; Feldman & Feldman, 2001; Sun et al., 2010).

3.1 Abiraterone: Targeting Alternative Androgen Synthesis

Abiraterone acetate (AA) is a novel antiandrogen therapy designed to target the adrenal androgen-mediated signaling axis by blocking the synthesis of adrenal products which serve as precursors for testosterone and DHT synthesis (Fig. 1; Di Lorenzo, Buonerba, De Placido, & Sternberg, 2010; Sartor, 2011; Walcak & Carducci, 2007). AA acts as a pregnenolone analog, inhibiting the rate-limiting enzyme, cytochrome P450 (CYP17A1), further inhibiting androgen biosynthesis (Di Lorenzo et al., 2010; Sartor, 2011).

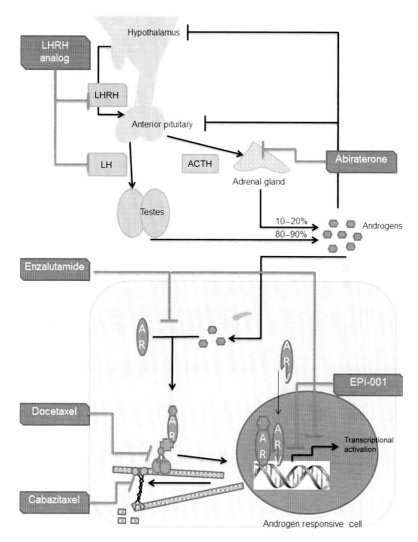

Figure 1 Therapeutic targeting of prostate cancer. Therapies employed to treat prostate cancer including: inhibition of signaling through the hypothalamic pituitary axis via LHRH analogs, inhibition of adrenal androgen synthesis with Abiraterone, direct inhibition of AR activity by treatment with AR antagonist enzalutamide, targeting of the NTD of AR with EPI-001 to prevent transcriptional activation, microtubule (MT) stabilization by Docetaxel leading to inhibition of dynein-mediated AR nuclear translocation, and Cabazitaxel-mediated MT stabilization leading to AR stranded in the nucleus and decreased expression of MT-depolymerizing kinesins. (See the color plate.)

AA inhibits both the 17α-hydroxylase and 17,20 functions of CYP17A1 (Di Lorenzo et al., 2010). The efficacy of AA was demonstrated in the COU-AA-301 trial, confirming that AA imparted additional survival benefit compared to DR-CRPC men treated with placebo and prednisone.

3.2 Enzalutamide (MDV3100): Targeting AR Activation

The emergence of the direct, AR antagonist, MDV3100, only slightly changed the landscape of prostate cancer therapy (Di Lorenzo et al., 2010; Shen & Balk, 2009). This drug is a diarylthiohydantoin member of the family of AR antagonists rationally designed from the crystal structure of the AR bound to its ligand (Shen & Balk, 2009). The therapeutic efficacy of MDV3100 is found in the context of AR overexpression, in addition to inhibiting AR nuclear translocation, preventing binding of the AR to DNA, blocking recruitment of coactivators to AR target genes, and inducing apoptosis (Fig. 1; Scher et al., 2010; Shen & Balk, 2009; Vishnu & Tan, 2010). Clinically, MDV3100 has been efficacious in improving survival in therapy-naïve patients, those previously treated with ADT, and in patients previously treated with Docetaxel and in patients treated with both (Attard, Cooper, & de Bono, 2009; Di Lorenzo et al., 2010; Nadal et al., 2014; Tran et al., 2009). MDV3100 inhibits translocation of full-length AR and prevents activation of ARvs lacking portions or all of the LBD by preventing dimerization of those ARv with full-length AR (Watson et al., 2010). Interestingly, treatment of MDV3100 and other ADT strategies has been shown to induce expression of ARvs, evidence emphasizing the need to identify effective therapeutic targets to impair CRPC outside of the "classic" AR signaling (Watson et al., 2010).

3.3 EPI-002: Targeting the NTD

Antiandrogenic targeting of the AR has primarily been focused on binding the LBD and thereby preventing activation of AR, but one AR antagonist has emerged which takes an enlightened approach. The small-molecule inhibitor EPI-001/002 series of antiandrogens bind the intrinsically disordered domain of the NTD, preventing it from initiating transcription activation functions and DNA binding (Fig. 1; Andersen et al., 2010). The EPI series was originally identified from peptides isolated from marine sponges (Sadar, 2011). This novel antiandrogen has demonstrated efficacy in androgen-dependent and CRPC models of prostate cancer, providing therapeutic promise for advanced cancer (Andersen et al., 2010). Recent evidence from this laboratory documented that EPI small-molecule inhibitors offer a unique feature to AR targeting in that they can bind and inhibit the clinically relevant AR variant isoforms found in advanced CRPC (Martin, Banuelos, Sadar, & Kyprianou, 2015; Sadar, 2011). EPI can still effectively bind and inhibit activation of several AR variants (Martin, Banuelos, et al., 2015). After progression to advanced CRPC, median survival is less than 2 years and 90% of these men will endure bone metastases (Hotte & Saad, 2010), with taxane-based chemotherapy the primary clinically relevant chemotherapeutic intervention for patients with mCRPC.

4. TAXANE ACTION IN PROSTATE CANCER CELLS: UP, CLOSE, AND "PERSONAL" WITH MICROTUBULES

Taxanes are derived from naturally occurring molecules identified in the bark of yew trees, exhibiting strong cytotoxic effects against cancer cell in diverse tumor types (Huizing et al., 1995). The classic mechanism of action driving the antitumor action of taxanes has been attributed to their inherent ability to bind and stabilize the architectural component of the cell: microtubules (Huizing et al., 1995). Taxanes bind and stabilize the interaction between two subunits of β-tubulin, preventing depolymerization of the protofilament substructure within the microtubule (Fig. 2; Kraus et al., 2003). This stabilization results in G2M arrest and apoptosis (Huizing et al., 1995; Kraus et al., 2003). Significantly enough the apoptosis suppressor, Bcl-2 is frequently overexpressed in prostate cancer, and taxanes are functionally capable of counteracting the effects of this prosurvival protein effector in apoptotic signaling in prostate cancer cells (Bruckheimer & Kyprianou, 2001, 2002; Debes & Tindall, 2004; Oliver et al., 2005).

The therapeutic value in the clinical use of taxanes in patients with CRPC was established by two lead clinical trials over a decade ago, TAX327 and SWOG (Southwest Oncology Group) 9916, which demonstrated a significant survival benefit of Docetaxel-based treatment compared to control-treated patients (Tannock et al., 2004). Docetaxel treatment conferred palliative relief and overall survival benefits (Berthold et al., 2008; Petrylak et al., 2004). Since FDA approval, taxane chemotherapy stood alone for a decade now as the only clinically relevant intervention for mCRPC patients. Unfortunately, taxane treatment ultimately fails with the majority of patients developing resistance and recurrence to lethal disease. Certain mechanisms of resistance are attributed to the adenosine triphosphate-dependent drug efflux pump: P-glycoprotein-1. Indeed, it

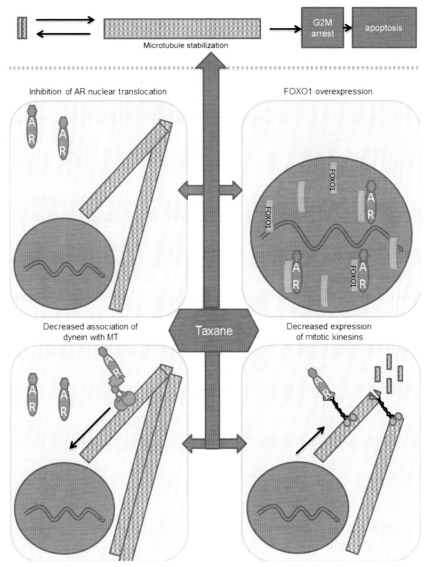

Figure 2 Taxane mechanism of action. Taxanes (Docetaxel and Paclitaxel) directly stabilize the interaction between tubulin subunits to prevent depolymerization of MTs, leading to G2M arrest and apoptosis. In addition, taxanes can inhibit AR nuclear translocation along the MT, induce overexpression of FOXO1, and thus causing sequestration of AR away from ARE; they also inhibit the association of dynein with MTs leading to diffuse AR localization. Moreover, Cabazitaxel therapy decreases the expression of mitotic kinesins in prostate cancer cells. (See the color plate.)

has been suggested that Docetaxel has a high affinity for this pump, and that an increase in expression of the efflux pump itself is observed over the course of prostate cancer progression (Abdulla & Kapoor, 2011; Attard, Greystroke, Kaye, & De Bono, 2006). Biochemical recurrence is often with metastasis to bone, brain, and lymph nodes, as well as increasing amounts of pain secondary to bone metastatic lesions common in CRPC patients (Abdulla & Kapoor, 2011).

Cabazitaxel is a next-generation semisynthetic taxane chemotherapeutic agent, approved by the FDA and shown to be effective in the Docetaxel-resistant CRPC landscape (Galsky, Dritselis, Kirkpatrick, & Oh, 2010; Sartor, 2011). Cabazitaxel is structurally very similar to Docetaxel (but in place of hydroxyl groups there are methoxyl groups in those positions; Azarenko, Smiyun, Mah, Wilson, & Jordan, 2014; Vrignaud et al., 2013). This second-generation taxane is highly cytotoxic with a low affinity for the adenosine triphosphate-dependent drug efflux pump: P-glycoprotein-1, known to confer chemotherapeutic resistance (Di Lorenzo et al., 2010). Unlike Docetaxel, Cabazitaxel can cross the blood–brain barrier, and has a 95-h terminal half-life in humans (vs. 12 h for Docetaxel) (Bruno & Sanderink, 1993; Cisternino, Bourasset, & Archimbaud, 2003; Sanofi-Aventis, 2014; Schutz, Buzaid, & Sartor, 2014). Cabazitaxel was shown in a multicenter, randomized, Phase III clinical trial (treatment of hormone-refractory metastatic prostate cancer (TROPIC)) to impart a statistically significant increase in overall survival (De Bono et al., 2010; Galsky et al., 2010). Tumor response, biochemical recurrence, and tumor progression all favored the Cabazitaxel treatment and was approved by the US FDA for use in DR-CRPC patients (De Bono et al., 2010; Galsky et al., 2010; Sartor, 2011). In addition to imparting overall survival benefits to chemotherapeutic-naïve patients, the exciting findings associated with Cabazitaxel are its ability to confer additional survival benefits in patients experiencing biochemical recurrence on ADT, Docetaxel chemotherapy, or both (Abdulla & Kapoor, 2011; Sartor, 2011).

5. TAXANE "PROMISCUOUS" ACTIONS BEYOND MICROTUBULE STABILIZATION

5.1 Blocking AR Translocation

In addition to the intramolecular ballet required for the AR to induce conformational change suitable to facilitate nuclear translocation, AR must be physically transported from the cytoplasm to the nucleus; an event stroke to "flick" tubulin subunits off the end of the structure during mitotic progression (Desai, Verma, Mitchison, & Walczak, 1999; Ogawa, Nitta, Okada, & Hirokawa, 2004; Rath & Kozielski, 2012). The mitotic kinase Aurora B controls the localization and activity of MCAK at the centromere/kinetochore, while Aurora A controls the same functions at the spindle poles (Sanhaji et al., 2010; Sanhaji, Friel, Worderman, Louwen, & Yuan, 2011; Zhang, Lan, Ems-McClung, Stukenberg, & Walczak, 2007). Polo-like kinase 1 regulates MCAK enzymatic activity, and actually controls microtubule depolymerization in cells making it of particular interest in taxane-resistant cells (Sanhaji et al., 2011). In normal cells, MCAK tracks to the plus-end tips of microtubules and utilizes microtubule-depolymerizing properties to correct improper kinetochore attachments at the centromere during mitosis (Fig. 2). During interphase, MCAK localizes to the plus end of microtubules and actively depolymerizes them (Sanhaji et al., 2011). Cancer cells of diverse origin exhibit gross overexpression of MCAK. Data mining has revealed MCAK as an important protein overexpressed in CRPC data sets that is indicative of resistance to taxane chemotherapy (Fig. 3; Sircar et al., 2012). Early attempts to target MCAK via inhibitors have been described (Aoki, Ohta, Yamazaki, Sugawara, & Sakaguchi, 2005; Rickert, 2008). Recent advances in our lab have shown that while MCAK expression may be indicative of Docetaxel resistance, MCAK expression may actually be a target of Cabazitaxel (Fig. 4; Martin, Pu, et al., 2015). Using a dominant-negative TGFβRII × TRAMP mouse model of advanced prostate cancer progression, we also demonstrated a marked increase in MCAK protein expression in response to castration-induced androgen depletion, but such an increase was abrogated by cotreatment with Cabazitaxel. Furthermore, in castration-resistant 22Rv1, and androgen-dependent VCaP, and LNCaP cell lines, MCAK expression was decreased in response to Cabazitaxel in a time-dependent fashion, regardless of the presence or absence of androgens (Martin, Pu, et al., 2015).

Human kinesin-14 (HSET) is a member of the kinesin-14 subfamily, and like the rest of this subfamily is a "minus" end-directed motor protein unlike the rest of the kinesins. HSET promotes proper bi-spindle pole formation and facilitates proper cytokinesis. Furthermore, HSET is overexpressed in Docetaxel-resistant tumors (De, Cipriano, Jackson, & Stark, 2009). Chromosomal instability (CIN) and centrosome amplification are common interrelated phenomena in tumor cells (Kwon et al., 2008; Ogden, Rida, & Aneja, 2013). CIN leads to missegregation of chromosome during mitosis,

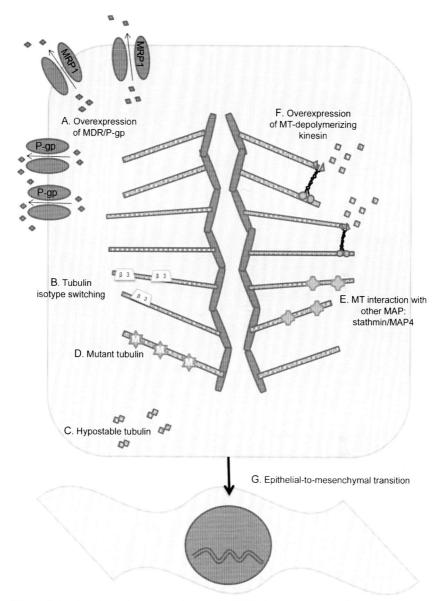

Figure 3 Mechanisms of taxane resistance in CRPC. Overexpression of MDR1/P-gp ATP-dependent drug efflux pumps can promote resistance by increasing the efficiency with which cells may pump out taxanes (A). Alterations in the tubulin subunit and its structural/spatial dynamics lead to taxane resistance via isotype switching from βI- to βIII-tubulin (B). Expression of tubulin with a hypostable binding site leads to a shift in MT binding dynamics toward dimerized tubulin and away from polymerized tubulin (C). Mutational alterations in the tubulin subunit lead to decreased binding of
(Continued)

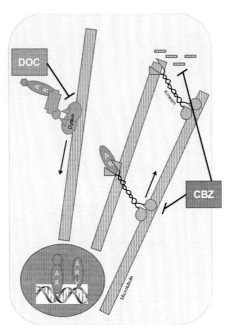

Figure 4 Effect of Docetaxel (DOC) and Cabazitaxel (CBZ) on ATP-dependent motor proteins: dynein and kinesin. Docetaxel decreases the association between AR and dynein, thereby inhibiting the translocation of AR from cytoplasm to nucleus. Cabazitaxel therapy decreases expression of MT-depolymerizing kinesins (MCAK and HSET), causing multinucleation and centrosomal amplification. (See the color plate.)

aneuploidy of daughter cells, and cell death. There is compelling evidence to suggest that centrosomal amplification and subsequent clustering has evolved to prevent cancer cells harboring aneuploidy chromosomes from undergoing apoptosis (Chandhok & Pellman, 2009; Ogden, Rida, & Aneja, 2012). The kinesin HSET is an essential regulator of centrosomal clustering; normal mitotic cells possess exactly two centrosomes and do not require HSET to coordinate the clustering of their centrosomes, while tumor cells that cannot appropriately cluster their (amplified) centrosomes will undergo mitotic arrest and apoptosis (Chandhok & Pellman, 2009). HSET remains elusive as a druggable target in advanced mCRPC; our recent studies indicate that HSET expression is downregulated at both

Figure 3—Cont'd Docetaxel to MT structure, consequently conferring resistance to taxane (D). MT interaction with other MT-associated proteins (MAPs) such as MAP4 and stathmin affect stability of MT structure (E). Overexpression of MT-depolymerizing kinesins (HSET, MCAK, and Eg5) can lead to taxane resistance by circumventing effects of stabilized MT (F). Induction of EMT also causes resistance to taxane therapy (G). (See the color plate.)

the mRNA and protein level in prostate tumors in response to Cabazitaxel treatment (Martin, Pu, et al., 2015). Significantly, enough prostate cancer cells treated with Cabazitaxel demonstrate centrosomal amplification and clustering, but do not appear to achieve mitosis caused by not only microtubule stabilization but the stark downregulation of mitotic kinesin HSET. Preclinical studies in our lab utilizing the transgenic mouse model of prostate cancer (DNTβRII × TRAMP) demonstrated that mice treated with Cabazitaxel had increased incidence of multinucleated cells in their prostate tumors, suggesting a defect in ability to undergo cytokinesis, potentially due to downregulation of HSET (Martin, Pu, et al., 2015).

7. MECHANISMS OF THERAPEUTIC RESISTANCE
7.1 "Pumping" Issues

ATP-binding cassette (ABC transporter) P-gp/multidrug resistance protein (MDR) is overexpressed in the cell membranes of tumors, and overexpression of such has been identified in prostate tumor specimens from CRPC patients (Fig. 3A; Mahon, Henshall, Sutherland, & Horvath, 2011; Siegsmund, Kreukler, & Steidler, 1997; Theyer, Schirmbock, & Thalhammer, 1993). This protein serves as a membrane-bound efflux pump physically pumping a wide range of substrates, but of particular interest, Docetaxel out of treated tumor cells decreasing efficacy at stabilizing microtubules. Variable response to P-gp can be seen in chemoresistant prostate cancer cell lines; chemoresistant PC3 cells do not overexpress P-gp, whilst chemoresistant DU145 do overexpress P-gp and that resistance can be modulated with knockdown thereof (Mahon et al., 2011; Makarovskiy, Siryaporn, Hixson, & Akerley, 2002; Takeda et al., 2007). Another member of the MDRP family, MRP1, is upregulated in chemoresistant prostate cancer cell lines which do not overexpress P-gp (Zalcberg et al., 2000). Taken together, this evidence suggests that expression of ABC transporters and subsequent effects on chemoresistance can be modulated by p53 and PIM1 kinase (Mahon et al., 2011; Sullivan et al., 2000; Xie et al., 2008). Hydroxyl groups on the structure of Docetaxel are substituted for methoxyl groups in the structure of Cabazitaxel (Vrignaud et al., 2013). In human breast cancer cells, MCF-7, Cabazitaxel uptake was higher than Docetaxel. Interestingly, washing did not impact the intracellular concentration of the drug, unlike Docetaxel-treated MCF-7 cells which demonstrated 50% concentration of drug postwashing, indicating that Cabazitaxel is better retained and thusly efficacious inside cells (Azarenko et al., 2014). Despite this improvement in intracellular retention, MDR-resistant MCF-7 variant cells still demonstrated resistance to the effects of Cabazitaxel although it was lower than that observed for Paclitaxel and Docetaxel (Duran et al., 2014).

7.2 Tubulin Mutations

Mutational alterations to taxane binding site of β-tubulin or isotype switching also confer resistance to taxane action via modulating binding of microtubule-associated proteins (MAPs) (Giannakakou et al., 2000; Huzil, Chen, Kurgan, & Tuszynski, 2007; Madan, Pal, Sartor, & Dahut, 2011). There are at least seven isotypes of β-tubulin known, and the predominant isotype expressed in normal tissue is isotype I (Luadena, 1998; Mahon et al., 2011). Increased expression of the isotype βIII-tubulin has been clinically demonstrated for lung, breast, and ovarian cancers; in addition, overexpression of the βIII-tubulin isoform has been associated with progression to mCRPC, and is predictive not only of Docetaxel efficacy (Fig. 3B; Ploussard et al., 2010; Terry et al., 2009), but also of Cabazitaxel to therapeutic resistance (Duran et al., 2014).

7.3 Microtubule Binding Dynamics

The binding site of taxane only exists on microtubules, when tubulin is polymerized and not on tubulin dimers (Orr, Verdier-Pinard, McDaid, & Band Horwitz, 2003). A shift toward preference of the dimerized form of tubulin and away from polymerized tubulin would represent a survival advantage to cells exposed to taxane (Fig. 3C). Certain taxane-resistant human cell lines have developed a requirement for exposure to taxane, suggesting that the microtubules have become "hypostable" and must be exposed to taxane to stabilize their microtubules adequately for normal function (Orr et al., 2003). Thus, the A549 cells continually exposed to taxane were selected for resistance and were developed to be between 9- and 17-fold more resistant than parental cells, becoming dependent on taxane exposure for growth, and remain blocked in G2/M phase, exhibiting increased dynamic instability toward dimerized tubulin (Goncalves et al., 2001; Orr et al., 2003).

ERG (ETS-Related Gene) is a transcription factor commonly subject to chromosomal rearrangement coming under the androgen driver of TMPRSS2 promoter elements in prostate cancer and a predominant molecular subtype of the disease (Tomlins et al., 2005). ERG overexpression in prostate cancer has been implicated in progression toward invasion and increased incidence of prostatic intraepithelial neoplasia (PIN) (Tomlins et al., 2008). This rearrangement is found in 40–80% of prostate cancers (Clark et al., 2007; Hermans et al., 2006). Recently, ERG was implicated in affecting the shift of microtubules toward instability (Galletti et al., 2014). ERG rearrangements are the most frequently recurring genetic alteration in prostate cancers and are known to undergo gene fusion with the 5′ promoters of TMPRSS2, SLC45A3, and NDRG1 (Tomlins et al., 2005). Moreover, ERG is able to bind soluble tubulin in the cytoplasm and shift tubulin binding dynamics toward catastrophe contributing to taxane resistance (Galletti et al., 2014). Mutations in the LBD of AR change the binding affinity of the transcription factor for its cytoplasmic-localized ligand. One must recognize that accumulation of mutations altering the binding kinetics of a drug to their target is a common theme in drug resistance (Fig. 3D). Giannakakou et al. originally identified two human ovarian carcinoma cell lines, resistant to the microtubule-stabilizing taxane, Taxol, but hypersensitive to the microtubule-depolymerizing agent, Vinblastine (Giannakakou et al., 1997; Orr et al., 2003). The primary β-tubulin isotype βI was identified to have a Phe_{270} to valine substitution in one cell line and an Ala_{364} to threonine substitution in the other (Giannakakou et al., 1997). These substitutions are spatially close the tubulin domain responsible for interacting with the taxane ring system, and are thus capable of disrupting Taxol binding (Orr et al., 2003).

Altered expression of MAPs is associated with progression to chemotherapeutic resistance in prostate cancer (Fig. 3E). Stathmin is a microtubule-destabilizing protein that plays an important role in mitotic spindle formation and disassembly (Belmont & Mitchison, 1996; Marklund, Larsson, Gradin, Brattsand, & Gullberg, 1996; Mistry & Atweh, 2006). Stathmin is expressed at very high levels in a variety of human cancers including prostate, and its expression has been shown to correlate with malignant phenotype of the cancer. It has been shown that stathmin may serve as a prognostic marker of prostate cancer progression (Friedrich, Grongberg, Landstrom, Gullberg, & Bergh, 1995). Inhibition of stathmin expression has been associated with shift toward epithelial-to-mesenchymal transition (EMT) in prostate cancer and the extent of expression is stage specific (Williams et al., 2012). Overexpression of stathmin may contribute to taxane resistance by shifting the equilibrium between soluble and polymerized tubulin dimers toward unpolymerized tubulin counteracting the action of tubulin polymer-stabilizing taxanes. Treatment of prostate cancer cells

(LNCaP) with stathmin interference plus taxane induces a synergistic effect on cell death induction compared to either treatment alone (Mistry & Atweh, 2006). Conversely, MAP4 is a microtubule-stabilizing protein, and phosphorylation of MAP4 causes it to dissociate from microtubules resulting in the loss of stabilizing function (Chang et al., 2001; Chien & Moasser, 2008). Increased phosphorylation of MAP4 is associated with decreased taxane sensitivity in ovarian cancer cell lines (Poruchynsky et al., 2001). Furthermore, MAP4 expression is suppressed in the presence of wild-type p53 (Murphy, Hinman, & Levine, 1996; Zhang et al., 1998).

7.4 Impairing Centrosome Clustering

The ATP-driven motor protein kinesins have emerged as significant protagonists in driving the mechanisms of taxane resistance (Fig. 3F). In breast cancer, elevated levels of KIFC3, KIFC1, and KIF5A can confer or enhance Docetaxel resistance (Liu, Gong, & Huang, 2013). In prostate cancer and other solid tumors, KIF2A (MCAK) overexpression results in enhanced Docetaxel resistance (Ganguly, Yang, & Cabral, 2011; Sircar et al., 2012). Moreover, other members of the kinesin family including KIF2B, CENPE, and HSET have also been heavily associated with resistance to microtubule-targeting chemotherapy (Schmizzi, Currie, & Rogers, 2010; Yang, Liu, Ikui, & Horwitz, 2010). Functionally, the ATP-binding domain of the kinesin is required to confer taxane resistance (Tan et al., 2012). MCAK is a plus-end tracking, microtubule-depolymerizing kinesin which induces an unnatural bend conformation to the microtubule structure promoting removal of tubulin subunits. Thus, kinesins can prominently counteract the effect of taxane by depolymerizing from the ends of stabilized microtubules. Overexpression of these kinesins allows tumor cells to "outrun" the functional consequences of taxanes in tumor cells (De et al., 2009; Liu et al., 2013; Tan et al., 2012).

7.5 The EMT Landscape

The significance of EMT in cancer emerges as tumor cells must physically detach from their immediate primary tumor, invade into the surrounding microenvironment, intravasate into the vasculature, endure the turbulence of circulation in the blood stream or lymphatics, and extravasate from the circulatory system to a secondary site. EMT has been implicated in the development, progression, and therapeutic failure in different cancers (Fig. 3G; Duran et al., 2014; Kalluri & Weinberg, 2009; Yilmaz & Christofori, 2009).

Each step toward EMT programming is navigated by a series of sophisticated molecular events (Kalluri & Weinberg, 2009; Matuszak & Kyprianou, 2011; Yilmaz & Christofori, 2009). Epithelial cells must begin their transition to a mesenchymal phenotype by disrupting their intercellular adhesive contacts (Acloque, Adams, Fishwick, Bronner-Fraser, & Nieto, 2009). This initial change in cellular behavior proceeds via formation of apical constrictions and disorganization of the basal cytoskeleton resulting in the detachment and the loss of apical-basal organization (Acloque et al., 2009; Barrallo- Gimeno & Nieto, 2005; Moreno-Bueno, Portillo, & Cano, 2008; Peindao, Olmeda, & Cano, 2007). The phenotype of the detached cell becomes spindle-like and exhibits a front–rear polarity conferring enhanced motility and invasive shape (Kalluri, 2009; Matuszak & Kyprianou, 2011; Thiery, Acloque, Huang, & Nieto, 2009; Yang & Weinberg, 2008). Furthermore, breakdown of the basal membrane and extracellular membrane (ECM) must occur for migration to ensue and this is accomplished via secretion of proteases and acquisition of migratory/invasive properties (Acloque et al., 2009; Haraguchi et al., 2008). Cabazitaxel-resistant tumor cells exhibited marked alterations in the EMT profile toward induction of EMT and invasive properties (Duran et al., 2014). Recent studies from this laboratory using a dominant-negative TGFβRII × TRAMP mouse model of aggressive advanced prostate cancer progression that prostate tumors undergo EMT to MET with restoration of the normal glandular architecture in response to treatment with pharmacologically relevant doses of Cabazitaxel (Martin, Pu, et al., 2015).

8. OVERCOMING TAXANE RESISTANCE: THE TRANSLATIONAL CHALLENGE

8.1 Intermittent Chemotherapy

Microtubule-targeting taxane chemotherapy is clinically associated with significant toxicity in patients. An approach to minimize the side effects has been the implementation of intermittent chemotherapy. Patients enjoy "drug holidays" or breaks in therapy during which they may be able to recover from cumulative toxicity of prolonged treatment (Madan et al., 2011). Indeed, the clinical experience with prostate cancer patients has indicated that allowing time for patients to resolve drug side effects may allow taxane therapy to extended longer than otherwise would be tolerated by patients (Beer et al., 2008; Bellmunt, Albiol, & Albanell, 2007). Intermittent chemotherapy may prevent selection for "taxane-resistant" cell population and potentially circumventing development of therapeutic resistance (Madan et al., 2011). This intermittent approach was directly evaluated in the ASCENT trial in which 250 patients were administered 36 mg/m^2 Docetaxel weekly with high-dose calcitriol or placebo. The results from this trial provided initial evidence that intermittent taxane chemotherapy could be a valuable approach for increasing survival in select patient populations (Beer et al., 2008). Further investigative efforts provided strong support for the positive clinical outcomes of intermittent chemotherapy (Kelly et al., 2012; Lin, Ryan, & Small, 2007; Ning et al., 2010). Significantly, a Phase II study from the National Cancer Institute investigating the combination of Docetaxel with antiangiogenesis therapeutics reported a median overall survival longer than 28 months with intermittent Docetaxel administration, reflecting a considerable survival advantage (Ning et al., 2010).

9. CHEMOTHERAPY IN COMBINATION: "THE GOLDEN FLEECE" OF CANCER TREATMENT

9.1 Platinum-Based Therapy

An attractive combination therapy is combining the microtubule-stabilizing action of taxanes with the cytotoxic, DNA-alkylating properties of platinum-based chemotherapies such as carboplatin. Two small format studies have investigated the therapeutic combination of Carboplatin + Docetaxel versus Docetaxel alone with modest, but seemingly optimistic results. Twenty percent of Carboplatin + Docetaxel-treated patients experienced delayed disease progression with associated PSA decline (Dayyani, Gallick, Logothetis, & Corn, 2011; Nakabayashi et al., 2008; Regan et al., 2010). The results from the Phase III trials, however, revealed no overall survival advantage of satraplatin (oral platinum therapy) in patients progressing on taxane chemotherapy (Sternberg et al., 2009).

9.2 Abiraterone Acetate

The COU-AA-301 randomized, double-blinded, placebo-controlled Phase III clinical trial investigated the effect of AA in mCRPC patients who had previously received Docetaxel treatment, and found an improvement in overall survival of 4.6 months compared to control patients (De Bono et al., 2011; Fizazi et al., 2012). Interestingly, in retrospectively investigated mCRPC patients who had been treated with enzalutamide alone or in addition to Docetaxel chemotherapy, AA yielded only modest antitumor effects (Loriot et al., 2013; Noonan et al., 2013).

9.3 Enzalutamide

The AFFIRM Phase III, double-blinded, placebo-controlled clinical trial demonstrated that mCRPC patients previously treated with taxanes were definitively responsive to enzalutamide (Hoffman-Censits & Kelly, 2013; Scher et al., 2012). Patients treated with enzalutamide had an overall survival of 18.4 versus 13.6 months for the placebo control-treated patients. Furthermore, enzalutamide-treated patients had reduced PSA levels (by 50% or greater), improved quality of life, longer radiographic progression-free survival, and increased time to first skeletal-related event (Scher et al., 2012). In a direct comparison trial, the efficacy of enzalutamide was proven greater when administered to Docetaxel-naïve patients, as opposed to Docetaxel pretreated patients (Nadal et al., 2014). The therapeutic impact of enzalutamide treatment in patients with mCRPC pretreated with Docetaxel and Abiraterone was not overtly effective. Specifically, 39 patients with mCRPC were selected for a retrospective analysis; although 41% of patients had a PSA decline of 30% or greater, the overall activity of enzalutamide was considerably limited (Bianchini et al., 2014). This study implicates new mechanisms of cross-resistance imposed by the sequential administration of the trio-treatment. Resonating in the clinical setting are findings indicating the limited activity manifested by enzalutamide in mCRPC patients (post-Docetaxel and Abiraterone treatment), although these patients exhibited PSA decline of greater than 30% and overall survival was significantly improved (Brasso et al., 2014).

Radiopharmaceuticals such as ^{153}Sm-ethylenediaminetetramethylenephosphonate (^{153}Sm-EDTMP) may offer a unique combination with Docetaxel. This therapy is intravenously administered and delivers β-emitting radiation to newly remodeled bone such as osteoblastic bone metastases (Goeckeler et al., 1987; Madan et al., 2011). This therapy results in palliation of bone pain in mCRPC patients and has been approved by the FDA (Sartor et al., 2004; Serafini et al., 1998). Early Phase I trial results indicated that among 28 patients, the combination therapy of ^{153}Sm-EDTMP and Docetaxel was well tolerated and induced decline in PSA level, delivering much promise (Serafini et al., 1998).

9.4 Clusterin Targeting (OGX-011)

Clusterin is a protein also known as testosterone-repressed prostate message 2 and sulfated glycoprotein-2 (Gleave & Miyake, 2005; Zellweger et al., 2002) that has been functionally linked to diverse cellular processes including tissue remodeling, reproduction, lipid transportation, and apoptosis (Dayyani et al., 2011). It is overexpressed in prostate, lung, breast tumors, lymphoma, and renal cell carcinoma with increased apoptosis induction (Gleave & Miyake, 2005). Induced overexpression of clusterin renders human prostate cancer xenografts resistant to hormone deprivation and taxane chemotherapy, providing the foundation for its therapeutic targeting. To target clusterin, an antisense oligonucleotide to the clusterin gene was developed and optimized to become Custirsen/OGX-011 through the seminal studies by Gleave et al. (2001). A randomized Phase II trial investigated the effect of Docetaxel + prednisone with or without OGX-011 in 82 mCRPC patients (Chi et al., 2010). Although there was no difference in PSA decline between the two arms of the trial, the results were nevertheless encouraging, since patients treated with OGX-011 tolerated it well and exhibited improved overall survival compared to those treated with Docetaxel + prednisone alone (23.8 vs. 19.6 months; Chi et al., 2010). A randomized Phase III trial concluded that tested Docetaxel + prednisone versus Docetaxel + prednisone with OGX-011 in nearly 1000 patients with overall survival as the primary endpoint (Dayyani et al., 2011).

9.5 Combinations of Promise and Therapeutic Value

Extensive efforts have been invested in identifying potentially efficacious for combination therapy with taxane chemotherapy, but disappointment has been rampant (Antonarakis & Eisenberger, 2013). Antiangiogenic agents represent a realm of therapeutics that hold tremendous promise, but have failed to yield improvement in patient survival. Bevacizumab and Aflibercept in combination with Docetaxel did not confer any improvement in overall survival (Fizazi et al., 2013; Tannock et al., 2013). Agents targeting the bone microenvironment have been tested in Phase II trials; Atrasentan, Zibotentan, and Dasatinib among them did not garner overall survival benefit when combined with Docetaxel compared to Docetaxel alone (Araujo et al., 2013; Kelly et al., 2012; Quinn et al., 2012). Furthermore, immune modulators, GVAX and Lenalidomide, produced overall survival benefit inferior to Docetaxel treatment alone (Petrylak et al., 2012; Small et al.,

2009), thus failing to inspire confidence in their clinically relevant therapeutic sensitization effect.

9.6 The Taxane "Sisterhood" Finds Solace in Microtubules

The TROPIC Phase III clinical trial led to FDA approval of Cabazitaxel (Jevtana, Sanofi-Aventis) for use in mCRPC patients who progressed on Docetaxel chemotherapy (De Bono et al., 2010). Cabazitaxel-treated patients had a 30% relative reduction in the risk of death, doubled rate of progression-free survival, and stronger PSA and tumor response compared to mitoxantrone-treated patients (De Bono et al., 2010). One must critically consider the recently demonstrated therapeutic impact of Cabazitaxel in patients progressing on Docetaxel followed by AA as well as Docetaxel followed by enzalutamide. This clinical evidence supports an action by Cabazitaxel not directly targeting the AR signaling axis, as established for Docetaxel (Nakouzi et al., 2014; Pezaro et al., 2014; Zhu et al., 2010), thereby limiting its place in overcoming cross-resistance with antiandrogen treatment regimes in CRPC.

10. CONCLUSIONS

Docetaxel exerts many effects on prostate tumor cells including stabilization of microtubules toward G2M arrest and apoptosis. Additionally, Docetaxel increases expression of FOXO1 and its sequestration of AR in the nucleus. The first-generation taxanes, Docetaxel and Paclitaxel, can block translocation of AR from the cytoplasm to the nucleus potentially by inhibiting the association of AR with the ATP-dependent motor protein dynein, resulting in reduced AR transcriptional activity and target gene expression. Moreover, Docetaxel treatment can overcome Bcl-2 overexpression toward apoptosis induction (Haldar, Basu, & Croce, 1997). The second-line taxane chemotherapy, Cabazitaxel, enhances apoptosis induction and G2M arrest; in contrast to Docetaxel, it can promote AR nuclear accumulation although it reduces the overall AR levels and FOXO1 expression. Of high mechanistic and translational significance is the ability of Cabazitaxel to downregulate the expression of microtubule-depolymerizing mitotic kinesins, MCAK and HSET (shown to be involved in taxane therapeutic resistance), preventing their ability to depolymerize microtubules and conferring additional sensitivity to taxane treatment. As schematically illustrated in Fig. 4, Docetaxel confers additional therapeutic benefit to prostate cancer patients by targeting AR transport by dynein, while Cabazitaxel targets mitotic kinesins leaving AR stranded within the nucleus and hence, prevents kinesin-mediated microtubule depolymerization, leading to multinucleation (Fig. 4). Thus, one may argue that the clinical efficacy of Cabazitaxel treatment conferring additional survival benefit to patients (Vrignaud et al., 2013) may engage molecular and cellular mechanisms different from Docetaxel action against advanced prostate tumors. With the recent approval of Eg5 inhibitors in clinical trials, it is tempting to consider a combination therapy of Cabazitaxel and Eg5 inhibitors to promote G2M arrest and apoptosis induction, as well as to inhibit mitotic kinesin-dependent microtubule depolymerization (Ding et al., 2014; Rath & Kozielski, 2012).

Intermittent taxane chemotherapy and combination therapy with antiandrogens: enzalutamide and AA have only provided additional survival benefit to patients without a curative impact. Utilization of antiandrogen therapy (enzalutamide and AA) post-taxane treatment has been found incrementally effective for increasing patient survival. Moreover, combination strategies of taxanes with the novel antiandrogen EPI-001 are effective in preclinical models of CRPC (Martin, Banuelos, et al., 2015). Recent clinical evidence revealed that resistance of CRPC patients to enzalutamide and Abiraterone antiandrogen therapies is driven by the ARv7 variant overexpression in circulating tumor cells (Antonarakis et al., 2014), empowering investigators to focus on exploitation of the AR variants for therapeutic targeting, as well as significant biomarkers of CRPC progression and therapeutic resistance in advanced disease.

REFERENCES

Abdulla, A., & Kapoor, A. (2011). Emerging novel therapies in the treatment of castrate-resistant prostate cancer. *Canadian Urological Association Journal, 5*(2), 120–133.

Acloque, H., Adams, M. S., Fishwick, K., Bronner-Fraser, M., & Nieto, M. A. (2009). Epithelial-mesenchymal transitions: The importance of changing cell state in development and disease. *The Journal of Clinical Investigation, 119*(6), 1438–1449.

Andersen, R. J., Mawji, N. R., Wang, J., Wang, G., Haile, S., Myung, J.-K., et al. (2010). Regression of castrate-recurrent prostate cancer by a small-molecule inhibitor of the amino-terminus domain of the androgen receptor. *Cancer Cell, 17*, 535–546.

Antonarakis, E. S., & Eisenberger, M. A. (2013). Phase III trials with docetaxel-based combinations for metastatic castration-resistant prostate cancer: Time to learn from past experiences. *Journal of Clinical Oncology, 31*(14), 1709–1712.

Antonarakis, E. S., Lu, C., Wang, H., Luber, B., Nakazawa, M., Roeser, J. C., et al. (2014). AR-V7 and resistance to enzalutamide and abiraterone in prostate cancer. *The New England Journal of Medicine, 371*(11), 1028–1038.

Aoki, S., Ohta, K., Yamazaki, T., Sugawara, F., & Sakaguchi, K. (2005). Mammalian mitotic centromere associated kinesin (MCAK): A new molecular target of

sulfoquinovosylacylglycerols novel antitumor and immunosuppressive agents. *The FEBS Journal, 272,* 2132–2140.
Araujo, J. C., Trudel, G. C., Saad, F., Armstrong, A. J., Yu, E. Y., Bellmunt, J., et al. (2013). Overall survival (OS) and safety of dasatinib/docetaxel versus docetaxel in patients with metastatic castration-resistant prostate cancer (mCRPC). In *Paper presented at the 2013 genitourinary cancers symposium.*
Attard, G., Cooper, C. S., & de Bono, J. S. (2009). Steroid hormone receptors in prostate cancer: A hard habit to break? *Cancer Cell, 16,* 458–462.
Attard, G., Greystroke, A., Kaye, S., & De Bono, J. (2006). Update on tubulin targeting agents. *Pathologie Biologie, 54,* 72–84.
Azarenko, O., Smiyun, G., Mah, J., Wilson, L., & Jordan, M. A. (2014). Antiproliferative mechanism of action of the novel taxane cabazitaxel as compared with the parent compound docetaxel in MCF7 breast cancer cells. *Molecular Cancer Therapeutics, 13*(8), 2092–2103.
Barrallo-Gimeno, A., & Nieto, M. A. (2005). The Snail genes act as inducers of cell movement and survival: Implications in development and cancer. *Development, 132,* 3151–3161.
Beer, T. M., Ryan, C. W., Venner, P. M., Petrylak, D. P., Chatta, G. S., Ruether, J. D., et al. (2008). Intermittent chemotherapy in patients with metastatic androgen-independent prostate cancer: Results from ASCENT, a double-blinded, randomized comparison of high dose calcitriol plus docetaxel with placebo plus docetaxel. *Cancer, 112,* 326–330.
Bellmunt, J., Albiol, S., & Albanell, J. (2007). Intermittent chemotherapy in metastatic androgen-independent prostate cancer. *BJU International, 100,* 490–492.
Belmont, L. D., & Mitchison, T. J. (1996). Identification of a protein that interacts with tubulin dimers and increases the catastrophe rate of microtubules. *Cell, 84,* 623–631.
Berthold, D. R., Pond, G. R., Soban, F., de Wit, R., Eisenberger, M., & Tannock, I. F. (2008). Docetaxel plus prednisone or mitoxantrone plus prednisone for advanced prostate cancer: Updated survival in the TAX 327 study. *Journal of Clinical Oncology, 26,* 242–245.
Bianchini, D., Lorente, D., Rodriguez-Vida, A., Omlin, A. G., Pezaro, C. J., Ferraldeschi, R., et al. (2014). Antitumor activity of enzalutamide (MDV3100) in patients with metastatic castration resistant prostate cancer (CRPC) pre-treated with docetaxel and abiraterone. *European Journal of Cancer, 50,* 78–84.
Black, B. E., & Paschal, B. M. (2004). Intranuclear organization and function of the androgen receptor. *Trends in Endocrinology and Metabolism, 15,* 411–417.
Brasso, K., Thomsen, F. B., Schrader, A. J., Schmid, S. C., Lorente, D., Retz, M., et al. (2014). Enzalutamide antitumour activity against metastatic castration resistant prostate cancer previously treated with docetaxel and abiraterone: A multicentre analysis. *European Urology.* http://dx.doi.org/10.1016/j.euro.2014.07.028. Epub ahead of print.
Brinkmann, A. O., Blok, L. J., de Ruiter, P. E., Doesburg, P., Steketee, K., Berrevoets, C. A., et al. (1999). Mechanisms of androgen receptor activation and function. *The Journal of Steroid Biochemistry and Molecular Biology, 69*(1), 307–313.
Brodsky, A. S., & Silver, P. A. (1999). Nuclear transport HEATs up. *Nature Cell Biology, 1,* E66–E67.
Bruckheimer, E. M., & Kyprianou, N. (2001). Dihydrotestosterone enhances transforming growth factor beta induced apoptosis in hormone sensitive prostate cancer cells. *Endocrinology, 142,* 2419–2426.
Bruckheimer, E. M., & Kyprianou, N. (2002). BCL-2 antagonizes the combined apoptotic effect of transforming growth factor-beta and dihydrotestosterone in prostate cancer cells. *Prostate, 53,* 133–142.
Bruno, R., & Sanderink, G. J. (1993). Pharmacokinetics and metabolism of Taxotere (docetaxel). *Cancer Surveys, 17,* 305–313.

Chan, S. C., Li, Y., & Dehm, S. M. (2012). Androgen receptor splice variants activate AR target genes and support aberrant prostate cancer cell growth independent of the canonical AR nuclear localization signal. *The Journal of Biological Chemistry, 287*(23), 19736–19749.

Chandhok, N. S., & Pellman, D. (2009). A little CIN may cost a lot: Revisiting aneuploidy and cancer. *Current Opinion in Genetics & Development, 19*, 74–81.

Chang, W., Gruber, D., Chari, S., Kitazawa, H., Hamazumi, Y., Hisanaga, S.-I., et al. (2001). Phosphorylation of MAP4 affects microtubule properties and cell cycle progression. *Journal of Cell Science, 114*, 2879–2887.

Chen, C. D., Welsbie, D. S., Tran, C., Baek, S. H., Chen, R., Vessella, R., et al. (2004). Molecular determinants of resistance to antiandrogen therapy. *Nature Medicine, 10*(1), 33–39.

Chi, K., Hotte, S. J. M., Yu, E. Y., Tu, D., Eigl, B. J., Tannock, I., et al. (2010). Randomized phase II study of docetaxel and prednisone with or without OGX-011 in patients with metastatic castration-resistant prostate cancer. *Journal of Clinical Oncology, 28*(27), 4247–4254.

Chien, A. J., & Moasser, M. M. (2008). Cellular mechanisms of resistance to anthracyclines and taxanes in cancer: Intrinsic and acquired. *Seminars in Oncology, 35*(S2), S1–S14.

Cisternino, S., Bourasset, F., & Archimbaud, Y. (2003). Nonlinear accumulation in the brain of the new taxoid TXD258 following saturation of P-glycoprotein at the blood-brain barrier in mice and rats. *British Journal of Pharmacology, 138*, 1367–1375.

Clark, J., Merson, S., Jhavar, S., Flohr, P., Edwards, S., Foster, C. S., et al. (2007). Diversity of TMPRSS2-ERG fusion transcripts in the human prostate. *Oncogene, 26*, 2667–2673.

Corbett, A. H., & Silver, P. A. (1997). Nucleocytoplasmic transport of macromolecules. *Microbiology and Molecular Biology Reviews, 61*, 193–211.

Cutress, M. L., Whitaker, H. C., Mills, I. G., Stewart, M., & Neal, D. E. (2008). Structural basis for the nuclear import of the human androgen receptor. *Journal of Cell Science, 121*, 957–968.

Darshan, M. S., Loftus, M. S., Thadani-Mulero, M., Levy, B. P., Escuin, D., Zhou, X. K., et al. (2011). Taxane-induced blockade to nuclear accumulation of the androgen receptor predicts clinical responses in metastatic prostate cancer. *Cancer Research, 71*(18), 6019–6029.

Dayyani, F., Gallick, G. E., Logothetis, C. J., & Corn, P. G. (2011). Novel therapies for metastatic castrate-resistant prostate cancer. *Journal of the National Cancer Institute, 103*, 1665–1675.

De Bono, J. S., Logothetis, C. J., Molina, A., Fizazi, K., North, S., Chu, L., et al. (2011). Abiraterone and increased survival in metastatic prostate cancer. *The New England Journal of Medicine, 364*, 1995–2005.

De Bono, J. S., Oudard, S., Ozguroglu, M., Hansen, S., Machiels, J. P., Kocak, I., et al. (2010). Prednisone plus cabazitaxel or mitoxantrone for metastatic castration-resistant prostate cancer progressing after docetaxel treatment: A randomised open-label trial. *Lancet, 376*, 1147–1154.

De, S., Cipriano, R., Jackson, M. W., & Stark, G. R. (2009). Overexpression of kinesins mediates docetaxel resistance in breast cancer cells. *Cancer Research, 69*, 8035–8042.

Debes, J. D., & Tindall, D. J. (2004). Mechanisms of androgen refractory prostate cancer. *The New England Journal of Medicine, 351*, 1488–1490.

Dehm, S. M., Schmidt, L. J., Heemers, H. V., Vessella, R. L., & Tindall, D. J. (2008). Splicing of a novel androgen receptor exon generates a constitutively active androgen receptor that mediates prostate cancer therapy resistance. *Cancer Research, 68*, 5469–5477.

Desai, A., Verma, S., Mitchison, T. J., & Walczak, C. E. (1999). Kin I kinesins are microtubule-destabilizing enzymes. *Cell, 96*, 69–78.

Di Lorenzo, G., Buonerba, C., De Placido, S., & Sternberg, C. N. (2010). Castration-resistant prostate cancer: Current and emerging treatment strategies. *Drugs, 70*(8), 983–1000.

Ding, S., Zhao, Z., Sun, D., Wu, F., Bi, D., Lu, J., et al. (2014). Eg5 inhibitor, a novel potent targeted therapy, induces cell apoptosis in renal cell carcinoma. *Tumor Biology, 35*(8), 7659–7668.

Duran, G. E., Wang, Y. C., Francisco, E. B., Rose, J. C., Martinez, F. J., Coller, J., et al. (2014). Mechanisms of resistance to cabazitaxel. *Molecular Cancer Therapeutics, 14*(1), 193–201. http://dx.doi.org/10.1158/1535-7163.MCT-14-0155.

Feldman, B. J., & Feldman, D. (2001). The development of androgen-independent prostate cancer. *Nature Reviews. Cancer, 1*(1), 34–45. http://dx.doi.org/10.1038/35094009.

Feng, J., Zheng, S. L., Wennuan, L., Isaacs, W. B., & Xu, J. (2011). Androgen receptor signaling in prostate cancer: New twists for an old pathway. *Steroids & Hormonal Science, S2*, 1–7.

Fizazi, K., Higano, C. S., Nelson, J. B., Gleave, M., Miller, K., Morris, T., et al. (2013). Phase III randomized, placebo controlled study of docetaxel in combination with zibotentan in patients with metastatic castration-resistant prostate cancer. *Journal of Clinical Oncology, 31*, 1740–1747.

Fizazi, K., Scher, H. I., Molina, A., Logothetis, C. J., Chi, K., Jones, R. J., et al. (2012). Abiraterone acetate for treatment of metastatic castration resistant prostate cancer: Final overall survival analysis of the COU-AA-301 randomised, double-blind, placebo-controlled phase 3 study. *The Lancet. Oncology, 13*(10), 983–992.

Freedland, S. J., & Moul, J. W. (2007). Prostate specific antigen recurrence after definitive therapy. *The Journal of Urology, 177*(6), 1985–1991.

Friedrich, B., Grongberg, H., Landstrom, M., Gullberg, M., & Bergh, A. (1995). Differentiation stage specific expression of oncoprotein 18 in human and rat prostatic adenocarcinoma. *Prostate, 27*, 102–109.

Galletti, G., Matov, A., Beltran, H., Fontugne, J., Mosquera, J. M., Cheung, C., et al. (2014). ERG induces taxane resistance in castration-resistant prostate cancer. *Nature Communications, 5*, 1–12.

Galsky, M. D., Dritselis, A., Kirkpatrick, P., & Oh, W. K. (2010). Cabazitaxel. *Nature Reviews. Drug Discovery, 9*, 677–678.

Gan, L., Chen, S., Wang, Y., Watahiki, A., Bohrer, L., Sun, Z., et al. (2009). Inhibition of the androgen receptor as a novel mechanisms of taxol chemotherapy in prostate cancer. *Cancer Research, 69*(21), 8386–8394.

Ganguly, A., Yang, H., & Cabral, F. (2011). Overexpression of mitotic centromere associated kinesin stimulates microtubule detachment and confers resistance to paclitaxel. *Molecular Cancer Therapeutics, 10*, 929–937.

Gelmann, E. P. (2002). Molecular biology of the androgen receptor. *Journal of Clinical Oncology, 13*, 3001–3015.

Giannakakou, P., Gussio, R., Nogales, E., Downing, K. H., Zharevitz, D., Bollbuck, B., et al. (2000). A common pharmacophore for epothilone and taxanes: Molecular basis for drug resistance conferred by tubulin mutations in human cancer cells. *PNAS, 97*, 2904–2909.

Giannakakou, P., Sackett, D., Kang, Y.-K., Zhan, Z., Buters, J. T. M., Fojo, T., et al. (1997). Paclitaxel-resistant human ovarian cancer cells have mutant Beta-tubulins that exhibit impaired paclitaxel-driven polymerization. *The Journal of Biological Chemistry, 272*, 17118–17125.

Gleave, M., & Miyake, H. (2005). Use of antisense oligonucleotides targeting the cytoprotective gene, clusterin, to enhance androgen- and chemo-sensitivity in prostate cancer. *World Journal of Urology, 23*, 38–46.

Gleave, M., Miyake, H., Zellweger, T., Chi, K., July, L., Nelson, C., et al. (2001). Use of antisense oligonucleotides targeting the antiapoptotic gene, clusterin/testosterone-repressed prostate message 2, to enhance androgen sensitivity and chemosensitivity in prostate cancer. *Urology*, *58*(2A), 39–49.

Goeckeler, W. F., Edwards, B., Volkert, W. A., Holmes, R. A., Simon, J., & Wilson, D. (1987). Skeletal localization of samarium-153 chelates: Potential therapeutic bone agents. *Journal of Nuclear Medicine*, *28*, 495–504.

Goncalves, A., Braguer, D., Kamath, K., Martello, L., Briand, C., Horwitz, S. B., et al. (2001). Resistance to Taxol in lung cancer cells associated with increased microtubules dynamics. *PNAS*, *98*(20), 11737–11742.

Gorlich, D. (1997). Nuclear protein import. *Current Opinion in Cell Biology*, *9*, 412–419.

Haldar, S., Basu, A., & Croce, C. M. (1997). Bcl2 is the guardian of the microtubule integrity. *Cancer Research*, *57*, 229–233.

Haraguchi, M., Okubo, T., Miyashita, Y., Miyamoto, Y., Hayashi, M., Crotti, T. N., et al. (2008). Snail regulates cell-matrix adhesion by regulation of the expression of integrins and basement membrane proteins. *The Journal of Biological Chemistry*, *283*, 23514–23523.

Heemers, H. V., & Tidall, D. J. (2007). Androgen receptor (AR) coregulators: A diversity of functions converging on and regulating the AR transcriptional complex. *Endocrine Reviews*, *28*(7), 778–808.

Heinlein, C. A., & Chang, C. (2002). Androgen receptor (AR) coregulators: An overview. *Endocrine Reviews*, *23*, 175–200.

Hermans, K. G., Van Marion, R., Van Dekken, H., Jenster, G., Van Weerden, W. M., & Trapman, J. (2006). TMPRSS2:ERG fusion by translocation or interstitial deletion is highly relevant in androgen-dependent prostate cancer, but is bypassed in late stage androgen receptor negative prostate cancer. *Cancer Research*, *66*, 10658–10663.

Hoffman-Censits, J., & Kelly, W. K. (2013). Enzalutamide: A novel antiandrogen for patients with castrate-resistant prostate cancer. *Clinical Cancer Research*, *19*(6), 1335–1339.

Hotte, S. J. M., & Saad, F. (2010). Current management of castrate resistant prostate cancer. *Current Oncology*, *17*, S72–S79.

Hu, R., Denmeade, S. R., & Luo, J. (2010). Molecular processes leading to aberrant androgen receptor signaling and castration resistance in prostate cancer. *Expert Review of Endocrinology and Metabolism*, *5*(5), 753–764.

Huggins, C., & Hodges, C. V. (1941). Studies on prostatic cancer. I. The effect of castration, of estrogen and of androgen injection on serum phosphatase in metastatic carcinoma of the prostate. *Cancer Research*, *1*, 293–297.

Huizing, M. T., Misser, V. H. S., Pieters, R. C., ten Bokkel Huinink, W. W., Veenhof, C. H. N., Vermorken, J. B., et al. (1995). Taxanes: A new class of antitumor agents. *Cancer Investigation*, *13*(4), 381–404.

Huzil, J. T., Chen, K., Kurgan, L., & Tuszynski, J. A. (2007). The roles of beta-tubulin mutations and isotype expression in acquired drug resistance. *Cancer Informatics*, *3*, 159–181.

Kahn, B., Collazo, J., & Kyprianou, N. (2014). Androgen receptor as a driver of therapeutic resistance in advanced prostate cancer. *International Journal of Biological Sciences*, *10*(6), 588–595.

Kalluri, R. (2009). EMT: When epithelial cells decide to become mesenchymal-like cells. *The Journal of Clinical Investigation*, *119*(6), 1417–1419.

Kalluri, R., & Weinberg, R. A. (2009). The basics of epithelial-mesenchymal transition. *The Journal of Clinical Investigation*, *119*(6), 1420–1428.

Kelly, W. K., Halabi, S., Carducci, M. A., George, D. J., Mahoney, J. F., Stadler, W. M., et al. (2012). Randomized, double-blind, placebo-controlled phase III trial comparing docetaxel and prednisone with or without bevacizumab in men with metastatic castration-resistant prostate cancer: CALGB 90401. *Journal of Clinical Oncology*, *30*(13), 1534–1540.

Kraus, L. A., Samuel, S. K., Schmid, S. M., Dykes, D. J., Waud, W. R., & Bissery, M. C. (2003). The mechanism of action of docetaxel (Taxotere) in xenograft models is not limited to bcl-2 phosphorylation. *Investigational New Drugs, 21*, 259–268.

Kwon, M., Godinho, S. A., Chandhok, N. S., Ganem, N. J., Azioune, A., Thery, M., et al. (2008). Mechanisms to suppress multipolar divisions in cancer cells with extra centrosomes. *Genes & Development, 22*(16), 2189–2203.

Lin, A. M., Ryan, C. J., & Small, E. J. (2007). Intermittent chemotherapy for metastatic hormone refractory prostate cancer. *Critical Reviews in Oncology/Hematology, 61*(3), 243–254.

Liu, X., Gong, H., & Huang, K. (2013). Oncogenic role of kinesin proteins and targeting kinesin therapy. *Cancer Science, 104*(6), 651–656.

Liu, P., Li, S., Gan, I., Kao, T. P., & Huang, H. (2008). A transcription-independent function of FOXO1 in inhibition of androgen-independent activation of the androgen receptor in prostate cancer cells. *Cancer Research, 68*, 10290–10299.

Lonergan, P. E., & Tindall, D. J. (2011). Androgen receptor signaling in prostate cancer development and progression. *Journal of Carcinogenesis, 10*(20), 1–12.

Loriot, Y., Bianchini, D., Ileana, E., Sandhu, S., Patrikidou, A., Pezaro, C. J., et al. (2013). Antitumour activity of abiraterone acetate against metastatic castration-resistant prostate cancer progressing after docetaxel and enzalutamide (MDV3100). *Annals of Oncology, 00*, 1–6.

Luadena, R. F. (1998). Multiple forms of tubulin: Different gene products and covalent modifications. *International Review of Cytology, 178*, 207–275.

Madan, R. A., Pal, S. K., Sartor, O., & Dahut, W. L. (2011). Overcoming chemotherapy resistance in prostate cancer. *Clinical Cancer Research, 17*(12), 3892–3902.

Mahon, K. L., Henshall, S. M., Sutherland, R. L., & Horvath, L. G. (2011). Pathways of chemotherapy resistance in castration resistant prostate cancer. *Endocrine-Related Cancer, 18*, R103–R123.

Makarovskiy, A. N., Siryaporn, E., Hixson, D. C., & Akerley, W. (2002). Survival of docetaxel resistant prostate cancer cells in vitro depends on phenotype alterations and continuity of drug exposure. *Cellular and Molecular Life Sciences, 59*, 1198–1211.

Marklund, U., Larsson, N., Gradin, H., Brattsand, G., & Gullberg, M. (1996). Oncoprotein 18 is a phosphorylation-responsive regulator of microtubule dynamics. *The EMBO Journal, 15*, 5290–5298.

Martin, S. K., Banuelos, C. A., Sadar, M. D., & Kyprianou, N. (2015). N-terminal targeting of androgen receptor variant enhances response of castration resistant prostate cancer to taxane chemotherapy. *Molecular Oncology, 9*(3), 628–639.

Martin, S. K., Pu, H., Horbinski, C., Cao, Z., Strup, S. E., & Kyprianou, N. (2015). Targeting mitotic kinesin with next-generation taxane Cabazitaxel to overcome castration resistant prostate cancer. Submitted.

Matuszak, E. A., & Kyprianou, N. (2011). Androgen regulation of epithelial-mesenchymal transition in prostate tumorigenesis. *Expert Review of Endocrinology and Metabolism, 6*(3), 469–482.

Mistry, S. J., & Atweh, G. F. (2006). Therapeutic interactions between stathmin inhibition and chemotherapeutic agents in prostate cancer. *Molecular Cancer Therapeutics, 5*(12), 3248–3257.

Moreno-Bueno, G., Portillo, F., & Cano, A. (2008). Transcriptional regulation of cell polarity in EMT and cancer. *Oncogene, 27*, 6958–6969.

Murphy, M., Hinman, A., & Levine, A. J. (1996). Wild-type p53 negatively regulates the expression of a microtubule-associated protein. *Genes & Development, 10*(23), 2971–2980.

Nadal, R., Zhang, Z., Rahman, H., Schweizer, M. T., Denmeade, S. R., Paller, C. J., et al. (2014). Clinical activity of enzalutamide in docetaxel-naive and docetaxel-pretreated patients with metastatic castration-resistant prostate cancer. *Prostate, 74*, 1560–1568.

Nakabayashi, M., Sartor, O., Jacobus, S., Regan, M. M., McKearn, D., Ross, R. W., et al. (2008). Response to docetaxel/carboplatin-based chemotherapy as first- and second-line therapy in patients with metastatic hormone-refractory prostate cancer. *BJU International, 101*(3), 308–312.

Nakouzi, N. A., Le Moulec, S., Albiges, L., Wang, C., Beuzeboc, P., Gross-Goupil, M., et al. (2014). Cabazitaxel remains active in patients progressing after docetaxel followed by novel androgen receptor pathway targeted therapies. *European Urology*. Epub ahead of print.

Nazareth, L. V., & Weigel, N. L. (1996). Activation of the human androgen receptor through a protein kinase A signaling pathway. *The Journal of Biological Chemistry, 271*, 19900–19907.

Nigg, E. A. (1997). Nucleocytoplasmic transport: Signals, mechanisms, and regulation. *Nature, 386*, 779–787.

Ning, Y. M., Gulley, J. L., Arlen, P. M., Woo, S., Steinberg, S. M., Wright, J. J., et al. (2010). Phase II trial of bevacizumab, thalidomide, docetaxel, and prednisone in patients with metastatic castration resistant prostate cancer. *Journal of Clinical Oncology, 28*(12), 2070–2076.

Noonan, K. L., North, S., Bitting, R. L., Armstrong, A. J., Ellard, S. L., & Chi, K. N. (2013). Clinical activity of abiraterone acetate in patients with metastatic castration resistant prostate cancer progressing after enzalutamide. *Annals of Oncology, 24*, 1802–1807.

Ogawa, T., Nitta, R., Okada, Y., & Hirokawa, N. (2004). A common mechanism for microtubule destabilizers-M type kinesins stabilize curling of the protofilament using the class-specific neck and loops. *Cell, 116*, 591–602.

Ogden, A., Rida, P. C., & Aneja, R. (2012). Let's huddle to prevent a muddle: Centrosome declustering as an attractive anticancer strategy. *Cell Death and Differentiation, 19*(8), 1255–1267.

Ogden, A., Rida, P. C., & Aneja, R. (2013). Heading off with the herd: How cancer cells might maneuver supernumerary centrosomes for directional migration. *Cancer Metastasis Reviews, 32*(1–2), 269–287.

Oliver, C. L., Miranda, M. B., Shangary, S., Land, S., Wang, S., & Johnson, D. E. (2005). (−)-Gossypol acts directly on the mitochondria to overcome Bcl-2 and Bcl-X(L) mediated apoptosis resistance. *Molecular Cancer Therapeutics, 4*, 23–31.

Orr, G. A., Verdier-Pinard, P., McDaid, H., & Band Horwitz, S. (2003). Mechanisms of taxol resistance related to microtubules. *Oncogene, 22*, 7280–7295.

Peindao, H., Olmeda, D., & Cano, A. (2007). Snail, Zeb, bHLH factors in tumor progression: An alliance against the epithelial phenotype? *Nature Reviews. Cancer, 7*, 415–428.

Petrylak, D. P., Fizazi, K., Sternberg, C. N., Budnik, N., De Wit, R., Wiechno, P. J., et al. (2012). A phase III study to evaluate the efficacy and safety of docetaxel and prednisone with or without lenalidomide in patients with castrate resistant prostate cancer: The MAINSAIL trial. In *Paper presented at the meeting of European society of medical oncology, Vienna, Austria*.

Petrylak, D. P., Tangen, C. M., Hussain, M. H., Lara, P. N., Jones, J. A., Taplin, M. E., et al. (2004). Docetaxel and estramustine compared with mitoxantrone and prednisone for advanced refractory prostate cancer. *The New England Journal of Medicine, 351*, 1513–1520.

Pezaro, C. J., Omlin, A. G., Altavilla, A., Lorente, D., Ferraldeschi, R., Bianchini, D., et al. (2014). Activity of cabazitaxel in castration-resistant prostate cancer progressing after docetaxel and next-generation endocrine agents. *European Urology, 66*(3), 459–465.

Picard, D., & Yamamoto, K. R. (1987). Two signals mediate hormone-dependent nuclear localization of the glucocorticoid receptor. *The EMBO Journal, 6*, 3333–3340.

Ploussard, G., Terry, S., Maille, P., Allory, Y., Sirab, N., Kheuang, L., et al. (2010). Class III beta-tubulin expression predicts prostate tumor aggressiveness and patient response to docetaxel based chemotherapy. *Cancer Research, 70*(22), 9253–9264.

Poruchynsky, M. S., Giannakakou, P., Ward, Y., Bulinski, J. C., Telford, W. G., Robey, R. W., et al. (2001). Accompanying protein alterations in malignant cells with a microtubule-polymerizing drug-resistance phenotype and a primary resistance mechanisms. *Biochemical Pharmacology, 62*, 1469–1480.

Poukka, H., Karvonen, U., Yoshikawa, N., Tanaka, H., Palvimo, J. J., & Janne, O. A. (2000). The RING finger protein SNURF modulates nuclear trafficking of the androgen receptor. *Journal of Cell Science, 113*, 2991–3001.

Quinn, D. I., Tangen, C. M., Hussain, M., Lara, P., Goldkorn, A., Garzotto, M., et al. (2012). SWOG SO421: Phase III study of docetaxel and atrasentan versus docetaxel and placebo for men with advanced castrate-resistant prostate cancer. In *Paper presented at the 2012 ASCO annual meeting*.

Rath, O., & Kozielski, F. (2012). Kinesins and cancer. *Nature Reviews. Cancer, 12*, 527–539.

Regan, M. M., O'Donnell, E. K., Kelly, W. K., Halabi, S., Berry, W., Urakami, S., et al. (2010). Efficacy of carboplatin-taxane combinations in the management of castration-resistant prostate cancer: A pooled analysis of seven prospective clinical trials. *Annals of Oncology, 21*(2), 312.

Rickert, K. W. (2008). Discovery and biochemical characterization of selective ATP competitive inhibitors of the human mitotic kinesin KSP. *Archives of Biochemistry and Biophysics, 469*, 220–231.

Roy, A. K., Lavrosky, Y., & Song, C. S. (1999). Regulations of androgen action. *Vitamins and Hormones, 55*, 309–352.

Sadar, M. D. (2011). Small molecule inhibitors targeting the "Achilles' Heel" of androgen receptor activity. *Cancer Research, 71*(4), 1208–1213.

Sanhaji, M., Friel, C. T., Kreis, N.-N., Kramer, A., Martin, C., Howard, J., et al. (2010). Functional and spatial regulation of mitotic centromere associated kinesin by cyclin-dependent kinase 1. *Molecular and Cellular Biology, 30*(11), 2594–2607.

Sanhaji, M., Friel, C. T., Worderman, L., Louwen, F., & Yuan, J. (2011). Mitotic centromere-associated kinesin (MCAK): A potential cancer drug target. *Oncotarget, 2*(12), 935–947.

Sanofi-Aventis, U. S. (2014). *Jevtana® (cabazitaxel): Full prescribing information*. www.jevtana.com.

Sartor, A. O. (2011). Progression of metastatic castrate-resistant prostate cancer: Impact of therapeutic intervention in the post-docetaxel space. *Journal of Hematology & Oncology, 4*, 18.

Sartor, O., Reid, R. H., Hoskin, P., Quick, D. P., Ell, P. J., Coleman, R. E., et al. (2004). Samarium-153-Lexidronam complex for treatment of painful bone metastases in hormone-refractory prostate cancer. *Urology, 63*, 940–945.

Savory, J. G. A., Hsu, B., Laquian, I. R., Giffin, W., Reich, T., Hache, R. J. G., et al. (1999). Discrimination between NL1- and NL2-mediated nuclear localization of the glucocorticoid receptor. *Molecular and Cellular Biology, 19*, 1025–1037.

Schaufele, F., Carbonell, X., Guerbadot, M., Borngraeber, S., Chapman, M. S., Ma, A. A. K., et al. (2005). The structural basis of androgen receptor activation: Intramolecular and intermolecular amino-carboxy interactions. *Proceedings of the National Academy of Sciences of the United States of America, 102*, 9802–9807.

Scher, H. I., Beer, T. M., Higano, C. S., Anand, A., Taplin, M. E., Efstanthiou, E., et al. (2010). Antitumor activity of MDV3100 in castration resistant prostate cancer: A phase 1–2 study. *Lancet, 375*, 1437–1446.

Scher, H. I., Fizazi, K., Saad, F., Taplin, M. E., Sternberg, C. N., Miller, K., et al. (2012). Increased survival with enzalutamide in prostate cancer after chemotherapy. *The New England Journal of Medicine, 367*, 1187–1197.

Schmidt, L. J., & Tindall, D. J. (2011). Steroid 5 alpha reductase inhibitors targeting BPH and prostate cancer. *The Journal of Steroid Biochemistry and Molecular Biology, 125,* 32–38.

Schmizzi, G. V., Currie, J. D., & Rogers, S. L. (2010). Expression levels of a kinesin-13 microtubule depolymerase modulates the effectiveness of anti-microtubule agents. *PloS One, 5,* e11381.

Schutz, F. A., Buzaid, A. C., & Sartor, O. (2014). Taxanes in the management of metastatic castration-resistant prostate cancer: Efficacy and management of toxicity. *Critical Reviews in Oncology/Hematology, 91*(3), 248–256.

Serafini, A. N., Houston, S. J., Resche, I., Quick, D. P., Grund, F. M., Ell, P. J., et al. (1998). Palliation of pain associated with metastatic bone cancer using samarium-153 lexidronam: A double-blind placebo-controlled clinical trial. *Journal of Clinical Oncology, 16,* 1574–1581.

Shen, H. C., & Balk, S. P. (2009). Development of androgen receptor antagonists with promising activity in castration-resistant prostate cancer. *Cancer Cell, 15*(6), 461–463.

Siegsmund, M. J., Kreukler, C., & Steidler, A. (1997). Multidrug resistance in androgen independent growing rat prostate carcinoma cells is mediated by P-glycoprotein. *Urological Research, 25,* 35–41.

Sircar, K., Huang, H., Limei, H., Liu, Y., Dhillon, J., Cogdell, D., et al. (2012). Mitosis phase enrichment with identification of mitotic centromere-associated kinesin as a therapeutic target in castration resistant prostate cancer. *PloS One, 7*(2), 1–8.

Small, E. J., Demkow, T., Gerritsen, W. R., Rolland, F., Hoskin, P., Smith, D. C., et al. (2009). A phase III trial of GVAX immunotherapy for prostate cancer in combination with docetaxel versus docetaxel plus prednisone in symptomatic, castration resistant prostate cancer (CRPC). In *Paper presented at the 2009 genitourinary cancers symposium, Orlando, FL.*

Sprenger, C. C. T., & Plymate, S. R. (2014). The link between androgen receptor splice variants and castration-resistant prostate cancer. *Hormones and Cancer, 5,* 207–217.

Sternberg, C. N., Petrylak, D. P., Sartor, O., Witjes, J. A., Demkow, T., Ferrero, J.-M., et al. (2009). Multinational, double-blind, phase III study of prednisone and either satraplatin or placebo in patients with castrate-refractory prostate cancer progressing after prior chemotherapy: The SPARC trial. *Journal of Clinical Oncology, 27*(32), 5431–5438.

Sullivan, G. F., Yang, J. M., Vassil, A., Yang, J., Bash-Babula, J., & Hait, W. N. (2000). Regulation of expression of the multidrug resistance protein MRP1 by p53 in human prostate cancer cells. *The Journal of Clinical Investigation, 105,* 1261–1267.

Sun, S., Sprenger, C. C. T., Vessella, R. L., Haugk, K., Soriano, K., Mostaghel, E. A., et al. (2010). Castration resistance in human prostate cancer is conferred by a frequently occurring androgen receptor splice variant. *The Journal of Clinical Investigation, 120*(8), 2715–2730.

Takeda, M., Mizokami, A., Mamiya, K., Li, Y. Q., Zhang, J., Keller, E. T., et al. (2007). The establishment of two paclitaxel-resistant prostate cancer cell lines and the mechanisms of paclitaxel resistance with two cell lines. *Prostate, 67,* 955–967.

Tan, M. H., De, S., Bebek, G., Orloff, M. S., Wesolowski, R., Downs-Kelly, E., et al. (2012). Specific kinesin expression profiles associated with taxane resistance in basal-like breast cancer. *Breast Cancer Research and Treatment, 131*(3), 849–858.

Tannock, I., de Wit, R., Berry, W. R., Horti, J., Pluzanska, A., Chi, K. N., et al. (2004). Docetaxel plus prednisone or mitoxantrone and prednisone for advanced prostate cancer. *The New England Journal of Medicine, 351,* 1502–1512.

Tannock, I., Fizazi, K., Ivanov, S., Thellenberg Karlsson, C., Flechon, A., Skoneczna, I. A., et al. (2013). Aflibercept versus placebo in combination with docetaxel/prednisone for first line treatment of men with metastatic castration-resistant prostate cancer (mCRPC):

Results from the multinational phase III trial (VENICE). In *Paper presented at the 2013 genitourinary cancers symposium.*

Terry, S., Ploussard, G., Allory, Y., Nicolaiew, N., Boissiere-Michot, F., Maille, P., et al. (2009). Increased expression of class III beta-tubulin in castration resistant human prostate cancer. *British Journal of Cancer, 101*(6), 951–956.

Thadani-Mulero, M., Portella, L., Sun, S., Sung, M., Matov, A., Vessella, R. L., et al. (2014). Androgen receptor splice variants determine taxane sensitivity in prostate cancer. *Cancer Research, 74*(8), 2270–2282.

Theyer, G., Schirmbock, M., & Thalhammer, T. (1993). Role of the MDR-1 encoded multiple drug resistance phenotype in prostate cancer cell lines. *The Journal of Urology, 150,* 1544–1547.

Thiery, J. P., Acloque, H., Huang, R. Y., & Nieto, M. A. (2009). Epithelial-mesenchymal transitions in development and disease. *Cell, 139*(5), 871–890.

Tomlins, S. A., Laxman, B., Varambally, S., Cao, X., Yu, J., Helgeson, B. E., et al. (2008). Role of the TMPRSS2-ERG gene fusion in prostate cancer. *Neoplasia, 10*(2), 177–188.

Tomlins, S. A., Rhodes, D. R., Perner, S., Dhanasekaran, S. M., Mehra, R., Sun, X.-W., et al. (2005). Recurrent fusion of TMPRSS2 and ETS transcription factor genes in prostate cancer. *Science, 310,* 644–648.

Tran, C., Ouk, S., Clegg, N. J., Chen, Y., Watson, P. A., Arora, V., et al. (2009). Development of a second-generation antiandrogen for treatment of advanced prostate cancer. *Science, 324,* 787–790.

Van Royen, M. E., Cunha, S. M., Brink, M. C., Mattern, K. A., Nigg, A. L., Dubbink, H. J., et al. (2007). Compartmentalization of androgen receptor protein-protein interactions in living cells. *The Journal of Cell Biology, 177,* 63–72.

van Royen, M. E., van Cappellen, W. A., de Vos, C., Houtsmuller, A. B., & Trapman, J. (2012). Stepwise androgen receptor dimerization. *Journal of Cell Science, 125,* 1970–1979.

Vishnu, P., & Tan, W. W. (2010). Update on options for treatment of metastatic castration-resistant prostate cancer. *OncoTargets and Therapy, 3,* 39–51.

Vrignaud, P., Semiond, D., Lejeune, P., Bouchard, H., Calvet, L., Combeau, C., et al. (2013). Preclinical antitumor activity of cabazitaxel, a semisynthetic taxane active in taxane-resistant tumors. *Clinical Cancer Research, 19*(11), 2973–2983.

Walcak, J. R., & Carducci, M. A. (2007). Prostate cancer: A practical approach to current management of recurrent disease. *Mayo Clinic Proceedings, 82,* 243–249.

Watson, P. A., Chen, Y. F., Balbas, M. D., Wongvipat, J., Socci, N. D., Viale, A., et al. (2010). Constitutively active androgen receptor splice variants expressed in castration-resistant prostate cancer require full length androgen receptor. *Proceedings of the National Academy of Sciences of the United States of America, 107,* 16759–16765.

Williams, K., Ghosh, R., Gridhar, P. V., Gu, G., Case, T., Belcher, S. M., et al. (2012). Inhibition of stathmin1 accelerates the metastatic process. *Cancer Research, 72*(20), 5407–5417.

Wilson, J. D. (2001). The role of 5 alpha-reduction in steroid hormone physiology. *Reproduction, Fertility, and Development, 13,* 673–678.

Wiltshire, C., Singh, B. L., Stockley, J., Fleming, J., Doyle, B., Barnetson, R., et al. (2010). Docetaxel-resistant prostate cancer cells remain sensitive to S-trityl-L-cysteine-mediated Eg5 inhibition. *Molecular Cancer Therapeutics, 9*(6), 1730–1739.

Wissing, M. D., De Morree, E. S., Dezentje, V. O., Buijs, J. T., De Krijger, R. R., Smit, V. T. H. B. M., et al. (2014). Nuclear Eg5 (kinesin spindle protein) expression predicts docetaxel response and prostate cancer aggressiveness. *Oncotarget, 5*(17), 7357–7367.

Xie, Y., Xu, K., Linn, D. E., Yang, X., Guo, Z., Shimelis, H., et al. (2008). The 44-kDa Pim-1 kinase phosphorylates BCRP/ABCG2 and thereby promotes its multimerization and drug resistant activity in human prostate cancer cells. *The Journal of Biological Chemistry, 283*(6), 3349–3356.

Yang, C. P., Liu, L., Ikui, A. E., & Horwitz, S. B. (2010). The interaction between mitotic checkpoint proteins, CENP-E and BubR1, is diminished in epothilone-B resistant A549 cells. *Cell Cycle, 9,* 1207–1213.

Yang, J., & Weinberg, R. (2008). Epithelial-mesenchymal transition: At the crossroads of development and tumor metastasis. *Developmental Cell, 14*(6), 818–829.

Yilmaz, M., & Christofori, G. (2009). EMT, the cytoskeleton, and cancer cell invasion. *Cancer Metastasis Reviews, 28,* 15–33.

Zalcberg, J., Hu, X. F., Slater, A., Parisot, J., El-Osta, S., Kantharidis, P., et al. (2000). MRP1 not MDR1 gene expression is the predominant mechanism of acquired multidrug resistance in two prostate carcinoma cell lines. *Prostate Cancer and Prostatic Diseases, 3,* 66–75.

Zellweger, T., Chi, K., Miyake, H., Adomat, H., Kiyama, S., Skov, K., et al. (2002). Enhanced radiation sensitivity in prostate cancer by inhibition of the cell survival protein clusterin. *Clinical Cancer Research, 8,* 3276–3284.

Zhang, X., Lan, W., Ems-McClung, S. C., Stukenberg, P. T., & Walczak, C. E. (2007). Aurora B phosphorylates multiple sites on mitotic centromere-associated kinesin to spatially and temporally regulate its function. *Molecular Biology of the Cell, 18,* 3264–3276.

Zhang, C. C., Yang, J. M., White, E., Murphy, M., Levine, A. J., & Hait, W. N. (1998). The role of MAP4 expression in the sensitivity to paclitaxel and resistance to vinca alkaloids in p53 mutant cells. *Oncogene, 16,* 1617–1624.

Zhou, Z. X., Sar, M., Simental, J. A., Lane, M. V., & Wilson, E. M. (1994). A ligand dependent bipartite nuclear targeting signal in the human androgen receptor. Requirement for the DNA-binding domain and modulation by the NH2 terminal and carboxyl-terminal sequences. *The Journal of Biological Chemistry, 269*(18), 13115–13123.

Zhu, M.-L., Horbinski, C., Garzotto, M., Qian, D. Z., Beer, T. M., & Kyprianou, N. (2010). Tubulin-targeting chemotherapy impairs androgen receptor activity in prostate cancer. *Cancer Research, 70*(20), 7992–8002.

CHAPTER FOUR

Stem Cell-Based Therapies for Cancer

Deepak Bhere, Khalid Shah[1]
Massachusetts General Hospital, Harvard Medical School, Boston, Massachusetts, USA
[1]Corresponding author: e-mail address: kshah@mgh.harvard.edu

Contents

1. Introduction — 159
2. Stem Cell Sources — 160
3. Stem Cell Homing and Migration — 161
4. Therapeutic Stem Cells for Cancer — 164
   - 4.1 Stem Cell - Cytokine Therapy — 165
   - 4.2 Stem Cell - Prodrug Therapy — 168
   - 4.3 Stem Cell - Oncolytic Virus Therapy — 169
   - 4.4 Stem Cell - Antiangiogenic Therapy — 171
   - 4.5 Stem Cell - Proapoptotic Protein Therapy — 172
   - 4.6 Synergistic Approaches Utilizing MSCs-Based Therapeutics with Other Antitumor Agents — 173
5. Encapsulated Stem Cells for Therapy — 176
6. Conclusions — 179
References — 179

Abstract

Stem cell-based therapeutic strategies have emerged as very attractive treatment options over the past decade. Stem cells are now being utilized as delivery vehicles especially in cancer therapy to deliver a number of targeted proteins and viruses. This chapter aims to shed light on numerous studies that have successfully employed these strategies to target various cancer types with a special emphasis on numerous aspects that are critical to the success of future stem cell-based therapies for cancer.

1. INTRODUCTION

In 2015, about 589,430 Americans are expected to die of cancer, and about 1,658,370 new cancer cases may be diagnosed. Cancer is the second most common cause of death in the United States, exceeded only by heart disease, and accounts for nearly one of every four deaths (Source: Cancer Facts and Figure 2015; American Cancer Society). The 5-year relative survival rate for all cancers diagnosed in the recent years was 68%, up from 49% in 1975–1977. This betterment in survival can be attributed to the early diagnostic measures and some newer targeted approaches for management. The short half-life and poor bioavailability of cancer-specific drugs are the key players that contribute to failure of current therapies. Recently, numerous studies have demonstrated the unique inherent tumor-tropic properties of stem cells including but not limited to mesenchymal stem cells (MSCs), neural stem cells (NSCs), and induced pluripotent stem cells (iPSCs) (Jones, Lamb, Goldman, & Di Stasi, 2014; Knoop et al., 2015; Singh, Kalsan, Kumar, Saini, & Chandra, 2015; Stuckey and Shah, 2014). In addition, adult stem cells in their unmodified state have antitumor effects owing to factors that are released and physical interactions that are established between the stem cell and the tumor cell (Schichor et al., 2012). These attributes make stem cells an ideal candidate to treat cancer, acting as effective vehicles for delivering therapies to isolated tumors (Stuckey & Shah, 2014). The improvement in survival reflects both the earlier diagnosis of certain cancers and improvements in treatment. The following sections of this chapter will discuss the use of stem cells as effective therapeutic agents and/or carriers of tumor-targeting therapeutics to treat various cancer types.

2. STEM CELL SOURCES

Stem cells are the natural sources of embryogenic tissue generation and continuous regeneration throughout adult life. Various stem cell types have been extensively studied for use as drug delivery vehicles or for gene therapy, and adult stem cells have been by far the cells of choice in numerous clinical studies being carried out globally (de Lazaro, Yilmazer, & Kostarelos, 2014; Nam, Lee, Nam, & Joo, 2015; Savla, Nelson, Perry, & Adler, 2014).

Embryonic stem cells derived from the inner cell mass of blastocyst-stage embryos, as well as the closely related embryonic germ cells derived from the primordial germ cells in fetal tissues, were the first cell types identified as possible sources of stem cell therapy (Reubinoff, Pera, Fong, Trounson, & Bongso, 2000). However owing to numerous concerns, the major one being ethical concerns, researchers have started to explore adult stem cell types for therapeutic use (Winkler, Hescheler, & Sachinidis, 2005).

Adult stem cells have been successfully isolated from various organs in the body. The sources of these types of stem cells include:

MSCs can be derived from bone marrow, adipose tissue, human umbilical cord blood (UCB), and other tissues (Sun, Williams, Waisbourd, Iacovitti, & Katz, 2015). Although there is some controversy, there are indications that these cells have the potential to differentiate into various other types of cells, including neural cells. Numerous published studies indicated that these cells may have a significant neuroprotective effect on the central nervous system (CNS) because they secrete neurotrophic factors and anti-inflammatory cytokines after transplantation (Torrente & Polli, 2008). MSC-based therapies have been successfully demonstrated in numerous pathological conditions including cancer (Johnson et al., 2010; Park, Suryaprakash, Lao, & Leong, 2015; Shah, 2012).

NSCs, also referred to as neural precursor cells, are a heterogeneous population of mitotically active, self-renewing, and multipotent cells of both the developing and the adult CNS, and show complex patterns of gene expression that vary in space and time (Martino & Pluchino, 2006). NSCs have proven to be an invaluable source for developing cell-based therapies especially for neurodegenerative disorders and also for other conditions that are specific for certain cancer types (Kim, 2004; Kim, Lee, & Kim, 2013; Martino & Pluchino, 2006; Nam et al., 2015; Shah, 2009; Yip & Shah, 2008).

iPSCs have recently emerged as popular and effective stem cell types for therapy. The generation of iPSCs from somatic cells by the ectopic expression of defined transcription factors has provided the regenerative medicine field with a new tool for cell replacement strategies (Abbott, 2012). iPSCs can be derived from a wide variety of starting cells, even though fibroblasts due to their accessibility are the most common source for iPSC generation today. The ability to produce iPSCs that can then be differentiated *in vitro* from individuals suffering from a particular disease is thought to contribute toward development of better disease models for a diverse range of conditions (de Lazaro et al., 2014).

3. STEM CELL HOMING AND MIGRATION

Numerous studies have demonstrated the tumor homing and migratory properties of various stem cell types. Although the mechanisms underlying these properties are not totally understood, various reports have demonstrated the importance of stem cell migration and its role specifically to target various types of brain tumors.

NSCs have been used widely in the treatment of human neurological disorders as cell therapies. SDF-1/CXCR-4 has been shown to be a key player in NSC migration and homing. A very recent study validated the important role SDF-1 plays in nestin positive NSC migration in the brain. Their results suggested that the NSC migration was triggered by the chemotactic effect of SDF-1. This study has direct implications on developing targeted stem cell therapies to the brain and also in understanding the mysteries that still remain about stem cell migration and homing.

Endogenous MSCs have been shown to mobilize from their sites of origin to the peripheral blood under different injury conditions, normoxia, hypoxia, and inflammation (He, Wan, & Li, 2007; Hong et al., 2009). However, it is known that a normal function of MSCs is the ability to migrate to and repair wounded tissue. This wound-healing property originates with migration toward inflammatory signals produced by the wounded environment (Spaeth, Kidd, & Marini, 2012). Many of the same inflammatory mediators that are secreted by wounds are found in the tumor microenvironment and are thought to be involved in attracting MSCs to these sites (Spaeth, Klopp, Dembinski, Andreeff, & Marini, 2008). Extensive studies have shown that migration involves cytokines secreted by MSCs that cooperate with G-protein coupled receptors (GPCRs) and growth factor receptors, as well as the different cytokine/receptor pairs SDF-1/CXCR4, SCF-c-Kit, HGF/c-Met, VEGF/VEGFR, PDGF/PDGFR, MCP-1/CCR2, and HMGB1/RAGE (Momin, Vela, Zaidi, & Quinones-Hinojosa, 2010). Among these cytokine/receptor pairs, stromal cell-derived factor SDF-1 and its receptor CXC chemokine receptor-4 (CXCR4) are important mediators of stem cell recruitment to tumors. The importance of the interaction between secreted SDF-1 and cell surface CXCR4 for stem cell migration has been emphasized by experiments in which the activity of either the receptor or the cytokine has been inhibited (Imitola et al., 2004; Nakamizo et al., 2005; Son et al., 2006). The blockade of both CXCR4 and SDF-1 *in vivo* in diseased mice has been found to markedly reduce the migration of transplanted stem cells toward tumor foci and regions of demyelination, indicating that SDF-1/CXCR4 signaling is essential for effective pathological tropism of therapeutic stem cells (Imitola et al., 2004). Recent data with inhibitors to chemokine receptor CXCR4 and TGFβ receptor suggest that endogenous MSCs homing to tumors, differentiation to myofibroblasts, and/or survival require CXCR4 (Quante et al., 2011).

Apart from CXCR4, MSCs are known to express a broad range of chemokine receptors including CXCR1, CXCR3, CXCR4, CXCR5, CXCR6, chemokine (C–C motif) receptor (CCR)-1, CCR2, CCR3, CCR4, CCR5, CCR9, and others. Recent studies have shown that chemokines, such as CXCL12, CXCL13, CXCL16, and their receptors can enhance the bidirectional migration of MSCs to and from the bone marrow niche (Smith, Whittall, Weksler, & Middleton, 2012). It is known that specific chemokines/receptor pairs are involved in the unidirectional migration of MSCs. CXCL16 (ligand for CXCR6) is effective in the homing of MSCs into the bone marrow, while CCL22 (ligand for CCR4) has the strongest chemotactic effect in mobilizing MSCs from the bone marrow into the circulation (Smith et al., 2012). Both CXCL16 and CCL2 are known to be expressed in tumor tissues such as lung carcinoma, hepatocellular carcinoma (HCC), colorectal cancer (Darash-Yahana et al., 2009), and malignant brain and ovarian cancers (Curiel et al., 2004; Jacobs et al., 2010), and it is plausible that they play a role in the migration of MSCs into tumors.

Cell receptor transactivation is a thoroughly researched mechanism thought to be involved in migration, which involves the activation of one receptor by another. It not only plays an important physiological role in processes such as cellular migration and apoptosis, but also its deregulation can cause pathological states such as cancer. Transactivation of various growth factor receptors, including epidermal growth factor receptor (EGFR), by GPCRs has been documented in multiple cellular model systems, hence demonstrating the potential role of GPCRs in tumor tropism via receptor transactivation (Jagadeesha, Takapoo, Banfi, Bhalla, & Miller, 2012; Maretzky et al., 2011; Yahata et al., 2006). A commonly reported mechanism of receptor transactivation involves the activation of membrane-tethered growth factors, such as EGFR, through direct interaction with GPCRs, such as CXCR4 (Porcile et al., 2005). This process is assisted by MMPs such as MMP-2 and MMP-9 (Roelle et al., 2003), which are proteinases required to proteolytically process precursor proteins such as adhesion molecules, growth factors, cytokines, and their receptors. De Becker and colleagues reported that the migration of MSCs through bone marrow endothelium was mediated by MMP-1 and tissue inhibitor of metalloproteinase-3 (De Becker et al., 2007). In another study, elevated levels of MMP-2 were observed to be responsible for C1q complement protein-mediated migration of cord blood-derived MSCs toward injured tissue and organ (Qiu, Marquez-Curtis, & Janowska-Wieczorek, 2012). A recent report on MSCs behavior indicates that MSCs are attracted to sites of irradiation, and that local irradiation might promote specificity of MSCs migration and engraftment (Francois et al., 2006). Although these findings are not surprising in the light of general stem cell tropism for injured tissues, they do stress the potential synergism between radiotherapy and tumor-specific MSCs targeting in the clinical arena.

The migratory potential for MSCs and NSC has been validated extensively; however, similar behavior of iPSCs has not been adequately evaluated. In a recent study by Yamazoe et al., it was demonstrated that both iPSCs and iPS-NSC present a similarly potent tumor tropism under both *in vivo* and *in vitro* conditions. Additionally, the iPSCs did not differentiate during migration. These findings suggest that iPSCs and their derivatives can be feasible candidates as vehicles in stem cell-based gene therapy for brain tumors (Yamazoe et al., 2015).

4. THERAPEUTIC STEM CELLS FOR CANCER

Naive stem cells have been shown to have antitumor effects both *in vitro* and in different mouse models of cancer. These effects are attributed to the factors released by stem cells that have antitumor properties reducing the proliferation of brain tumor cells (Ding, Chang, Shyu, & Lin, 2015; Hendijani, Haghjooy Javanmard, & Sadeghi-Aliabadi, 2015; Kawabata et al., 2013; Wu, Ju, Du, Zhu, & Liu, 2013; Yang et al., 2014). Human bone marrow-derived MSCs injected intravenously (i.v.) in a mouse model of Kaposi's sarcoma were shown to home to sites of tumorigenesis and potently inhibit tumor growth (Khakoo et al., 2006). MSCs have also been shown to have antiangiogenic effects both *in vitro* and in mouse models of melanoma (Otsu et al., 2009). Direct injection of MSCs into subcutaneous melanoma bearing mice-induced apoptosis and abrogated tumor growth (Otsu et al., 2009). Human UCB-derived MSCs have been used as naïve cells for the treatment of GBM. UCBSC enriched in CD44 and CD133 cells cocultured with GBM cells underwent apoptosis (Gondi et al., 2010). Treatment of glioma cells with hUCBs also inhibited FAK-mediated angiogenesis (Dasari, Kaur, Velpula, Dinh, et al., 2010), upregulated PTEN in glioma cells and in the nude mice tumors and downregulated Akt and PI3K signaling pathway molecules thereby resulting in the inhibition of migration as well as wound-healing property of glioma cells (Dasari, Kaur, Velpula, Gujrati, et al., 2010). This simultaneously resulted in downregulation of XIAP-activating caspase-3 and caspase-9 to trigger apoptosis in glioma cells (Dasari, Velpula, et al., 2010). In another study, UCB-MSCs were shown to inhibit glioma growth, reduce neovascularization, and decrease cyclin-D1

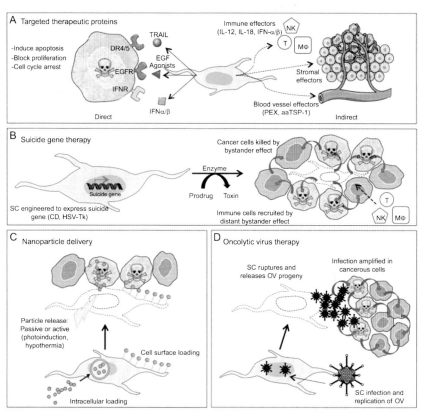

Figure 1 Utilizing stem cells to promote tumor cell death. *Adapted from Stuckey and Shah (2014), Nature Reviews Cancer with permission.* (See the color plate.)

protein expression *in vivo* (Jiao et al., 2011). In a recent comparative study, UCB-MSCs and not AT-MSCs were shown to inhibit GBM growth via the tumor necrosis factor-related apoptosis-inducing ligand (TRAIL) (Akimoto et al., 2013). Very recently, Kawabata and others have discovered that naive rat umbilical cord stem cells (UCMSCs) attenuated mammary tumor growth at least in part by enhancing host antitumor immune responses (Kawabata et al., 2013). Stem cells have been genetically modified mainly to introduce and overexpress target exogenous genes for expression/secretion of a desired therapeutic factor for targeted treatment of different cancer types (Fig. 1).

4.1 Stem Cell - Cytokine Therapy

Interleukins (ILs) are cytokines that regulate inflammatory and immune responses and are known to have antitumor properties via direct tumoricidal effects or positive modulation of the endogenous immune system (Okada &

Pollack, 2004). However, a lack of tumor-targeted delivery has hindered their application for cancer therapy (Zhang et al., 2013). Accordingly, the delivery of ILs by MSCs has been explored. MSCs engineered to express ILs have been utilized to improve the anticancer immune surveillance by activating cytotoxic lymphocytes and natural killer cells (Okada & Pollack, 2004). MSCs expressing IL-12 have been shown to prevent metastasis into the lymph nodes and other internal organs as well as increased tumor cell apoptosis in mice bearing preestablished metastases of melanoma, breast, and hepatoma tumors (Chen et al., 2008). In two previous studies, MSCs expressing IL-12 were shown to have antitumor effects in mice bearing renal cell carcinomas (Gao, Ding, Wu, Jiang, & Fang, 2010) and cervical tumors (Seo et al., 2011). Both studies revealed sustained expression of IL-12 and interferon (IFN)-γ in sera and tumor sites. Furthermore, in a 2011 study, human UCB-derived mesenchymal stem cells (UCB-MSCs) were successfully employed as vehicles to deliver IL-12 to aid in the treatment of malignant glioma (Ryu et al., 2011). Similarly, the transplantation of IL-18 secreting MSCs was previously shown to enhance T cell infiltration and long-term antitumor immunity in mice bearing noninvasive and invasive gliomas (Xu, Yang, Zhang, & Prestwich, 2009). In another study, human UCB-MSCs were engineered to express IL-21 and were shown to have therapeutic efficacy in mice bearing ovarian cancer xenografts (Hu et al., 2011). In a recent study, Zhang et al. evaluated the effects of IL-24 (melanoma differentiation associated gene-7; *mda*-7) (Jiang, Lin, Su, Goldstein, & Fisher, 1995) delivered by MSCs as a therapeutic approach for lung cancer. Human umbilical cord-derived MSCs (UCMSCs) engineered to deliver secretable IL-24 (*mda*-7) were shown to inhibit the growth of lung cancer cells by induction of apoptosis and cell cycle arrest (Zhang et al., 2013). This study also demonstrated that injection of MSCs secreting Il-24 significantly suppressed xenograft tumor growth and had antiangiogenic effects both *in vitro* and *in vivo* (Zhang et al., 2013). Another recent study demonstrated the safety and usability of hUCMSCs-LV-IL-21 in ovarian cancer gene therapy, suggesting the strategy may be a promising new method for clinical treatment of ovarian cancer (Zhang et al., 2014). In a very recent study, it was reported that hUCMSCs secreting IL-18 can suppress the proliferation, migration, and invasion of breast cancer cells *in vitro* and may provide an approach for a novel antitumor therapy in breast cancer (Liu et al., 2015). The results of these experiments indicate that stem cell-delivered ILs have the potential to be used as an alternative strategy for cancer therapy.

IFN-β has been shown to have antiproliferative and proapoptotic effects. However, its *in vivo* therapeutic efficacy has been limited due to toxicity associated with systemic administration. A potential solution to this problem is to engineer human MSCs to express IFN-β, a strategy that has been used for targeted delivery to metastatic breast and melanoma models (Studeny et al., 2002, 2004), gliomas (Nakamizo et al., 2005), and lung metastasis (Nakamizo et al., 2005; Ren, Kumar, Chanda, Chen et al., 2008; Ren, Kumar, Chanda, Kallman, et al., 2008). Recently, Dembinski and colleagues intraperitoneally injected mice bearing ovarian xenografts with IFN-β expressing MSCs which resulted in complete eradication of tumors in 70% of treated mice and an increased survival in others (Dembinski et al., 2013). This study confirmed MSCs potential as a targeted delivery vehicle for the intratumoral production of IFN-β. In 2012, Wang, Zhan, Hu, Wang, and Fu (2012) investigated the use of MSCs engineered to produce IFN-β. Intravenously injected MSC-IFN-β significantly reduced prostate tumor weight and increased animal survival compared with controls (Wang et al., 2012). In another recently published study, amniotic fluid (AF)-derived MSCs were investigated to transport IFN-β to the region of neoplasia in a bladder tumor model. A significant inhibition of tumor growth as well as prolonged survival of mice was observed in the presence of AF-MSC-IFN-β (Bitsika et al., 2012).

Earlier studies include the work of Studeny and colleagues, who engineered human adult MSCs expressing IFN-β and showed their *in vivo* efficacy against solid melanomas in nude mice (Studeny et al., 2002). In 2005, Nakamizo and colleagues investigated the antitumor effects of MSCs expressing IFN-β in CNS tumors, by evaluating whether human MSCs could still track murine brain tumors when administered through the blood stream (Nakamizo et al., 2005). By manipulating MSCs to secrete IFN-β, this tropism could be exploited for antitumor effects, as *in vivo* administration of hMSC-IFN-β resulted in significantly enhanced murine survival. In a related study, Ren, Kumar, Chanda, Kallman, et al. (2008) evaluated the potential of MSCs expressing IFN-β in a model of prostate cancer lung metastasis. Targeted homing of MSCs producing IFN-β was seen at sites of tumors in the lungs with established pulmonary metastases, and this resulted in suppression of tumor growth.

The antitumor effects of a multifunctional regulatory cytokine, IFN-α, has also been investigated. IFN-α is frequently used as a supplementary therapeutic agent to eradicate micrometastatic deposits in patients with a high risk of systemic recurrence (Grander & Einhorn, 1998; Lens, 2006). Sartoris and colleagues found that MSCs expressing IFN-α could be efficiently delivered inside the tumor microenvironment of a mouse plasmacytoma model (Sartoris et al., 2011). Subcutaneous administration of MSC-IFN-α significantly impeded the tumor growth *in vivo* and prolonged the overall survival of mice by inducing apoptosis in tumor cells and by a reduction in tumor vessel density (Sartoris et al., 2011). A similar study in 2006, explored the therapeutic efficacy of MSCs expressing IFN-α for the treatment of lung metastasis in a mouse model of metastatic melanoma. The systemic administration of MSCs expressing IFN-α reduced the growth of melanoma cells and significantly prolonged survival due to an increase in tumor cell apoptosis and a decrease in blood vasculature (Ren, Kumar, Chanda, Chen et al., 2008).

4.2 Stem Cell - Prodrug Therapy

The conversion of nontoxic prodrugs into toxic antimetabolites, known as prodrug activation schemes, is available for selective killing of tumor cells. Herpes simplex virus (HSV)-1 thymidine kinase (TK), cytosine deaminase (CD), and carboxyesterase genes, which confer sensitivity to ganciclovir (GCV), 5-fluorocytosine (5-FC), and camptothecin-11 (CPT-11), respectively, are currently being evaluated in clinical trials (Danks et al., 2007). Human MSCs derived from bone marrow and adipose tissues, as well as NSCs, have been found to be effective vehicles for gene-directed enzyme prodrug therapy (Altaner, 2008).

Activated prodrugs that are not toxic to MSCs and have a bystander tumor-killing effect were initially explored using CD, which can convert the nontoxic "prodrug" 5-FC to the drug, 5-fluorouracil (5-FU), a chemotherapeutic agent that can readily diffuse out of the producer stem cell and into surrounding cells and is selectively toxic to rapidly dividing cells (Aboody et al., 2000). Miletic and colleagues found that MSCs engineered to express herpes simplex virus thymidine kinase (HSV-TK) and injected into the tumor or the vicinity of the tumor, infiltrated solid parts as well as the border of glioma models in rats, and ultimately showed high therapeutic efficacy by significant reduction of tumor volumes through bystander-mediated glioma cell killing (Miletic et al., 2007). In 2012, Choi et al. studied human adipose tissue (hAT)-derived MSCs and prodrug gene therapy against brainstem gliomas in rat models. hAT-MSCs were modified to express rabbit carboxylesterase (rCE) enzyme, which can efficiently convert the prodrug CPT-11 into the active drug SN-38

(7-ethyl-10-hydroxycamptothecin). A significant increase in the survival time of rats treated with hAT-MSC.rCE and CPT-11 was observed than rats treated with CPT-11 alone, demonstrating the therapeutic potential of MSCs as cellular vehicles for prodrug gene therapy in gliomas (Choi et al., 2012). Song and colleagues studied bone marrow-derived MSCs infected with HSV-TK in mouse models of prostate cancer. MSC-HSV-TK significantly inhibited the growth of prostate cancer xenografts in the presence of GCV. Additionally, the MSC-HSV-TK exerted a significant antitumor effect in an animal model of metastatic fibrosarcoma, RIF-1, tumor in the presence of prodrug GCV and had no harmful side effects *in vivo* (Song et al., 2011). In a past study, the ability of AT-MSCs engineered to express the suicide gene CD::uracil phosphoribosyltransferase (CD::UPRT) was explored in mouse models of prostate cancer (Shah, 2012). CD-UPRT converts the relatively nontoxic 5-FC into the highly toxic antitumor 5-FU. Therapeutic AT-MSCs expressing CD::UPRT proved effective in significantly inhibiting prostate cancer tumor growth after intravenous administration in mice bearing tumors and treated with 5-FC (Cavarretta et al., 2010).

In a recent study, we have developed an efficient stem cell-based therapeutic strategy that simultaneously allows killing of GBM tumor cells and assessment and eradication of SCs posttumor treatment. MSCs engineered to coexpress the prodrug-converting enzyme, HSV-TK and S-TRAIL, induced caspase-mediated GBM cell death, and showed selective MSCs sensitization to the prodrug GCV. A significant decrease in tumor growth and a subsequent increase in survival were observed when mice bearing a highly aggressive GBM were treated with MSCs coexpressing S-TRAIL and HSV-TK. Furthermore, the systemic administration of GCV posttumor treatment selectively eliminated therapeutic MSCs expressing HSV-TK *in vitro* and *in vivo*. These findings demonstrate the development and validation of a novel therapeutic strategy that has implications in translating stem cell-based therapies in cancer patients (Martinez-Quintanilla et al., 2013).

4.3 Stem Cell - Oncolytic Virus Therapy

The ability of oncolytic viruses (OVs) to selectively replicate in and destroy tumor cells, while sparing healthy cells makes them strong candidates for antitumor therapy (Aghi & Martuza, 2005; Parato, Senger, Forsyth, & Bell, 2005). Due to clearance of the virus by host defense mechanism and spurious targeting of noncancer tissues through the bloodstream, the systemic administration of OV is often inefficient (Nakashima, Kaur, & Chiocca, 2010). Cell-mediated OV delivery could shield the virus from host defenses and direct them toward tumors. Different stem cell types, including MSCs, have been used as host cells for the replication, transportation, and local release of intact, conditionally replicating oncolytic adenoviruses (CRAd) (Power & Bell, 2007). Human MSCs were shown to support replication of adenovirus bearing TK and to have "bystander effect" against different cancer cell lines (Pereboeva, Komarova, Mikheeva, Krasnykh, & Curiel, 2003). When administered intravenously into murine models of solid ovarian cancer, CRAd-charged MSCs resulted in significantly enhanced antitumor effect and extended survival as compared to direct delivery of CRAd (Komarova, Kawakami, Stoff-Khalili, Curiel, & Pereboeva, 2006).

MSCs have also been employed to deliver adenovirus, which subsequently infected and replicated within malignant cells and eradicated the tumors (Komarova et al., 2006; Stoff-Khalili et al., 2007). MSCs have been utilized to deliver CRAds in a mouse model of intracranial malignant glioma (Sonabend et al., 2008). CRAd-loaded MSCs resulted in efficient adenoviral infection of distant glioma cells confirming the ability of MSCs as carriers for oncolytic adenoviral vectors for the treatment of malignant glioma. In a previous study, delivery and efficacy of oAV, Delta24-RGD by human MSCs has been assessed in mouse models of glioblastomas (Yong et al., 2009). MSC-Delta24 that were injected into the carotid artery of mice-harboring orthotopic glioma xenografts, selectively localized to glioma xenografts and released Delta24-RGD, which subsequently infected glioma cells, inhibited glioma growth, and resulted in eradication of tumors with significant increase in the median survival of treated animals as compared to controls.

The therapeutic approach of engineered adenoviruses has been studied for the treatment of metastatic tumors; however, systemic delivery of these oncolytic adenoviruses lacks metastatic targeting ability (Garcia-Castro et al., 2010). The tumor stroma engrafting property of MSCs may allow them to be used as cellular vehicles for targeted delivery. Garcia-Castro and colleagues explored the safety and efficacy of infusing autologous MSCs infected with ICOVIR-5, a new oncolytic adenovirus, for the treatment of metastatic neuroblastoma. In the study, four children with metastatic neuroblastoma that was resistant to front-line therapies, received several doses of autologous MSCs carrying ICOVIR-5 (Garcia-Castro et al., 2010). The children's tolerance to the treatment was excellent, and a complete clinical response was documented in one case, with the child in remission 3 years after the therapy. These results indicate that MSCs can deliver oncolytic adenoviruses to metastatic tumors with very low systemic toxicity and with beneficial antitumor effects. Recently, MSCs have been shown to serve as carriers to deliver oncolytic measles virus (MV) to ovarian tumors (Mader et al., 2013). MSCs obtained from ovarian cancer patients migrated toward primary ovarian cancer samples in chemotaxis assays and to ovarian tumors in athymic mice. Using noninvasive SPECT-CT imaging, a rapid colocalization of MV-infected MSCs to the ovarian tumors was observed within 5–8 min of their intraperitoneal administration. Furthermore, MV-infected MSCs, but not virus alone, significantly prolonged the survival of measles immune ovarian cancer bearing animals. In a recent study, the potent antitumor activity of systemically delivered MV-infected bone marrow-derived MSCs was explored in human HCC tumors in SCID mice passively immunized with human neutralizing antibodies against MV. MSCs infected with MV homed to the HCC tumors and resulted in a significant inhibition of tumor growth in both measles antibody-naïve and passively immunized SCID mice (Ong et al., 2013).

Our very recent study explored the role of stem cell-delivered oncolytic herpes simplex virus (oHSV) in mouse models of GBM. Our results indicate that human MSCs loaded with different oHSV variants provide a platform to translate OV therapies to clinics in a broad spectrum of GBMs after resection and could also have direct implications in different cancer types (Duebgen et al., 2014). These studies confirm the feasibility of using MSCs as carriers for OV therapy.

4.4 Stem Cell - Antiangiogenic Therapy

Tumor angiogenesis represents a way for cancer cells to function and thrive for self-sustained growth (Jain et al., 2007). Inhibition of tumor-induced angiogenesis may restrict tumor growth and metastasis. However, long-term systemic delivery of angiogenic inhibitors is associated with toxicity, as well as pruning away the vessels, which allows tumors to become more chemoresistant due to inadequate delivery of drugs (Ma, Zhang, Ma, & Yu, 2001). The utility of MSCs as vehicles for antiangiogenic therapeutics has been studied, as they exhibit a tropism to cancer tissue, and may deliver antiangiogenic agents without adverse side effects (Ghaedi et al., 2011). It has also been determined that tumor-mediated angiogenesis results from a dysregulation of both proangiogenic and antiangiogenic factors as well as by various growth factors and molecules of the extracellular matrix, which has led to the study and utilization of targeted antiangiogenic therapy to inhibit tumor proliferation and neovascularization (Folkman, Watson, Ingber, & Hanahan, 1989; Samant & Shevde, 2011).

Recently, Zheng et al. studied endostatin, which is an important endogenous inhibitor of tumor vascularization that has been widely used in antiangiogenic therapy for various cancers (Zheng et al., 2012). Human placenta-derived mesenchymal stem cells (hpMSC) were engineered to deliver endostatin via adenoviral transduction and were injected into nude mice. The hpMSC expressing the human endostatin gene demonstrated preferential homing to the tumor site and significantly decreased the tumor volume without apparent systemic toxic effects. These observations were associated with significantly decreased blood vessel formation, tumor cell proliferation, and increased tumor cell apoptosis (Zheng et al., 2012).

In a previous phase II clinical study, it was found that the delivery of antiangiogenic drugs through vasculature normalized the abnormal structure and function of the blood vessels and resulted in reduction of tumor-associated vasogenic brain edema in most patients (Batchelor et al., 2007). The vessel normalization is associated with a significant decrease in their mean vessel diameter and permeability (Kadambi et al., 2001; Tong et al., 2004) and increased pericyte coating of small vasculature (Hormigo, Gutin, & Rafii, 2007). MSCs are known to localize to tumor vasculature upon intratumoral implantation, thus offering increased abilities for targeting particularly vascularized tumors (Bexell et al., 2009).

Our very recent study demonstrated that the combination of stem cell-delivered antiangiogenic Three type-1 repeat (3TSR) domain of thrombospondin-1 combined with MSC-TRAIL simultaneously act on tumor cells and tumor-associated endothelial cells and offer significant potential to target a broad spectrum of cancers and translate 3TSR/TRAIL therapies into the clinic (Choi et al., 2015).

4.5 Stem Cell - Proapoptotic Protein Therapy

The delivery of the proapoptotic protein, TRAIL via SC offers a promising on-site delivery approach toward tumor cell killing (Shah, 2012). TRAIL is an endogenous member of the TNF ligand family that binds to its death domain containing receptors DR4 and DR5 and induces apoptosis via activation of caspases, preferentially in cancer cells while sparing most other cell types (Walczak & Krammer, 2000). A number of studies have shown the therapeutic efficacy of different adult stem cell types including MSCs engineered to express TRAIL in either cell lines or mouse models of colorectal carcinoma (Mueller et al., 2010), gliomas (Ehtesham, Kabos, Gutierrez, et al., 2002; Ehtesham, Kabos, Kabosova, et al., 2002; Kim et al., 2005), lung, breast, squamous, and cervical cancer (Loebinger, Eddaoudi, Davies, & Janes, 2009) result in induction of apoptosis and a subsequent reduction of tumor cell viability. Grisendi and colleagues found that adipose-derived MSCs armed with TRAIL targeted a variety of tumor cell lines *in vitro*, including human cervical carcinoma, pancreatic cancer, and colon cancer. When injected into mice, AD-MSC-TRAIL migrated to tumors and induced apoptosis in tumor cells, without significant toxicities to normal tissues (Grisendi et al., 2010). In a similar study, AD-MSCs engineered to express TRAIL were found to exhibit antimyeloma activities and significantly induce myeloma cell death *in vitro* (Ciavarella et al., 2012).

Since TRAIL is a type II membrane protein and its release into the microenvironment requires additional cleavage from its cell membrane anchoring site, our lab previously worked on redesigning of the TRAIL protein in order to engineer truly paracrine TRAIL-secreting cells. We have designed a secretable version of TRAIL that consists of fusion between the extracellular domain of TRAIL and the extracellular domain of the hFlt3 ligand, which binds to the Flt3 tyrosine kinase receptor (Shah, 2012). The re-engineered recombinant protein named "secretable TRAIL" (S-TRAIL) is efficiently secreted into the producer cell's immediate microenvironment and exhibits higher cytotoxicity on tumor cells than the native TRAIL protein (Sasportas et al., 2009; Shah et al., 2005; Shah, Tung, Yang, Weissleder, & Breakefield, 2004). In a previous study, we demonstrated that human MSCs are resistant to TRAIL-mediated apoptosis and when engineered to express S-TRAIL, induce caspase-mediated apoptosis in established glioma cell lines as well as glioblastoma stem cells (GBSCs) *in vitro*. Using highly malignant and invasive human glioma models generated from human GBSCs and employing real time imaging, we have shown that MSC-S-TRAIL migrate extensively to tumors in the brain and have profound antitumor effects *in vivo*. This study demonstrates the efficacy of therapeutic S-TRAIL and the potential of human MSCs to be used as delivery vehicles targeting GBSCs *in vivo* (Sasportas et al., 2009).

4.6 Synergistic Approaches Utilizing MSCs-Based Therapeutics with Other Antitumor Agents

Given the heterogeneity of tumors in general, it is unlikely that any one effective strategy will provide a satisfactory treatment regimen for all tumors.

The advent of molecular theranostics and personalized medicine might largely remedy the differences in characteristics and therapeutic resistance between different tumors (Corsten & Shah, 2008; Ozdemir et al., 2006), but cannot provide adequate answers to the existence of profound intratumoral heterogeneity, as is observed, for instance, in gliomas (Noble, 2000). A realist approach would be to combine distinct therapeutic targets, such as those involved in tumor cell growth and apoptosis and the proliferation of tumor-associated vasculature to fully eradicate different tumor types. Given that 50% of tumor lines are resistant to TRAIL, overcoming TRAIL resistance in aggressive tumors, such as GBM, and understanding the molecular dynamics of TRAIL-based combination therapies are critical to using TRAIL as a therapeutic agent. In a recent study, we engineered human MSCs to express S-TRAIL and assessed the ability of MSC-S-TRAIL-mediated medulloblastoma (MB) killing when combined with a small molecule inhibitor of histone-deacetylase, MS-275, in TRAIL-sensitive and -resistant MB *in vitro* and *in vivo* (Nesterenko, Wanningen, Bagci-Onder, Anderegg, & Shah, 2012). In TRAIL-resistant MB, we showed upregulation of DR4/5 levels when pretreated with MS-275 and a subsequent sensitization to MSC-S-TRAIL mediated apoptosis. Using intracranially implanted MB and MSCs lines engineered with different combinations of fluorescent and bioluminescent proteins, we found that MSC-S-TRAIL has significant antitumor effects in mice bearing TRAIL-sensitive and MS-275 pretreated TRAIL-resistant MBs (Nesterenko et al., 2012). In another study, a therapeutic combination of the lipoxygenase inhibitor MK886 and TRAIL-secreting human MSCs was explored. MK886 effectively increased the sensitivity to TRAIL-induced apoptosis via upregulation of the death receptor 5 and downregulation of the antiapoptotic protein survivin in human glioma cell lines and in primary glioma cells. *In vivo* survival experiments and imaging analysis in orthotopic xenografted mice showed that MSC-based TRAIL gene delivery combined with MK886 into the tumors had greater therapeutic efficacy than single-agent treatment (Kim et al., 2012). We have also designed additional supplementary treatments, like microRNA-21 inhibitors (Corsten et al., 2007) and novel PI3-kinase/mTOR inhibitor, PI-103 (Bagci-Onder, Wakimoto, Anderegg, Cameron, & Shah, 2010) to augment the antitumor effect of different stem cell-mediated S-TRAIL therapy *in vivo*. A similar study has demonstrated that the combined approach using systemic MSC-mediated delivery of TRAIL together with XIAP inhibition suppresses metastatic growth of pancreatic carcinomas

(Mohr et al., 2010). These findings offer a preclinical rationale for application of mechanism-based systemically delivered antiproliferative agents and novel stem cell-based proapoptotic therapies to improve treatment of malignant tumors.

In addition to molecular approaches, current clinical treatment regimens such as local radiotherapy might be suited for enhancing stem cell therapy, as their effects on irradiated tissue seem to additionally promote the homing of transplanted stem cells (Francois et al., 2006). Past studies have revealed that tumor irradiation enhances the tumor tropism of human UCB-MSC by increased IL-8 expression on glioma cells (Kim et al., 2010). The sequential treatment with irradiation followed by TRAIL-secreting UCB-MSCs synergistically enhanced apoptosis in glioma cells through upregulation of the expression of DR5 and subsequent caspase activation. *In vivo* survival experiments in orthotopic xenografted mice showed that MSC-based TRAIL gene delivery to irradiated tumors had greater therapeutic efficacy than a single treatment. These results suggest that clinically relevant tumor irradiation increases the therapeutic efficacy of MSC-TRAIL by increasing tropism of MSCs and TRAIL-induced apoptosis, which might be a more useful therapeutic strategy for treating tumors in general and gliomas in particular.

In a recent study, murine MSCs expressing TRAIL were tested in combination with conventional chemotherapeutic drug treatment in colon cancer models (Yu, Deedigan, Albarenque, Mohr, & Zwacka, 2013). A significant decrease in tumor volumes was seen when mice bearing HCT116 colorectal cancer xenografts were cotreated with 5-FU and systemically delivered MSC-S-TRAIL. This antitumor effect was protein 53 (p53) independent and was mediated by TRAIL-receptor 2 (TRAIL-R2) upregulation, demonstrating the applicability of this approach in p53-defective tumors. In another study, Yulyana and colleagues combined the tumor selectivity of TRAIL and tumor-homing properties of MSCs with gap junction (GJ) inhibitory effect of carbenoxolone (CBX) to target orthotopic gliomas (Yulyana et al., 2013). *In vitro* studies revealed that CBX enhanced TRAIL-induced apoptosis through upregulation of death receptor 5, blockade of GJ intercellular communication, and via downregulation of connexin 43. Dual arm therapy using MSC-TRAIL and CBX prolonged the survival of treated mice by $\sim 27\%$ when compared with the controls in an intracranial glioma model. The enhanced efficacy of MSC-TRAIL in combination with either 5-FU or CBX coupled with the minimal cytotoxic nature of these agents suggests their potential clinical implementation.

The ideal stem cell-based combination therapies would entail SC-expressing multitargeted molecules, such as when TRAIL and/or other biomolecules are produced by stem cells. In a recent study, we demonstrated the simultaneous expression of a multifunctional biomolecule by stem cells, specifically, a fusion of EGFR antagonist (EGFR-Nb) and TRAIL. We found that the expression of EGFR-Nb and TRAIL from a single stem cell source caused enhanced killing of tumor cells as opposed to expression of each molecule from stem cells separately (Van de Water et al., 2012). MSCs were engineered to coexpress dodecameric TRAIL and HSV-TK (MSC/dTRAIL-TK) and antitumor effects were assessed in an experimental lung metastasis model (Kim, Kim, et al., 2013). MSC/dTRAIL-TK treatment followed by GCV administrations significantly decreased the number of tumor nodules in the lung as compared to MSC/dTRAIL or MSC/TK treatment alone and resulted in 100% survival of tumor-bearing mice after three injections. Recently, another promising double-target stem cell-based therapeutic system was explored for non-Hodgkin's lymphoma. Human umbilical cord-derived mesenchymal stem cells (HUMSCs) engineered to express scFvCD20-S-TRAIL, which contains a CD20-specific single chain Fv antibody fragment (scFv) and a soluble TRAIL, were shown to significantly inhibit tumor growth in beating lymphoma xenografts (Yan et al., 2013).

5. ENCAPSULATED STEM CELLS FOR THERAPY

Cell encapsulation technology refers to immobilization of cells within biocompatible, semipermeable membranes. The encapsulation of cells instead of therapeutic products allows the delivery of molecules of interest for a longer period of time as cells release these molecules continuously. In addition, cells can be engineered to express any desired protein *in vivo* without the modification of the host's genome (Murua et al., 2008). Cell encapsulation presents an important advantage as compared to encapsulation of proteins, as the former allows a sustained and controlled delivery of therapeutic molecules at a constant rate giving rise to greater physiological concentrations (Murua et al., 2008). Due to their ability to provide a physiologic environment that promotes cell survival and prevent immune response while permitting easy *in vivo* transplantation and cell retention, biodegradable hydrogels, and synthetic extracellular matrix (sECM) to encapsulate stem cells have been utilized in a variety of rodent models (Morris, 1996; Rihova, 2000). A number of different biomaterials such as hyaluronic acid, alginate, agarose, and other polymers have been used for encapsulation. In past studies, sECM acted as the necessary biomechanical substrate for endogenous neuroregeneration in models of intracerebral hypoxia–ischemia and traumatic spinal cord injury, by increasing their stem cell viability and promoting differentiation into neurons (Pan, Ren, Cui, & Xu, 2009; Park, Teng, & Snyder, 2002; Teng et al., 2002). Recently, it was discovered that MSCs encapsulated in fibrinogen-alginate microcapsules possessed a significantly increased survival as compared to unencapsulated cells (Sayyar et al., 2012). In another recent study, Reagan et al. demonstrated the utility of a scaffold-based delivery system for sustained therapeutic MSCs release and their ability to express genetically introduced therapeutic TRAIL. MSCs expressing full-length TRAIL under a doxycycline inducible promoter were encapsulated in porous, biocompatible silk scaffolds and administered to mice using different administration routes. Encapsulated MSC-TRAIL successfully decreased bone and lung metastasis, whereas liver metastasis decreased only with tail vein administration routes upon doxycycline administration (Reagan et al., 2012).

Stem cell encapsulation is an important prospect for the treatment of GBM. The recurrence rates of GBM and associated patient mortality are nearly 100%, which is largely attributed to inefficient delivery of many therapeutic molecules to brain tumor cells, due to the blood–brain barrier (Muldoon et al., 2007) and vascular dysfunction in the tumor (Jain et al., 2007). One of the approaches to overcoming drug delivery problems to intracranial tumors is to develop on-site means to deliver novel tumor-specific agents. However, in order to effectively deliver such therapeutic agents, methods must be developed to introduce stem cells into the resection cavity while preventing the rapid "wash-out" of a significant number of cells by cerebrospinal fluid. Additionally, it is critical to allow efficient secretion of anti-GBM therapies and retains the ability of stem cells to migrate from the resection cavity into the parenchyma toward invasive tumor deposits. In the study from Kauer et al., we investigated a new approach to GBM treatment that would overcome the brain's immune response, using mouse NSCs and human MSCs encapsulated in hyaluronic acid-based sECM. In mouse models of human GBM resection, we found that sECM encapsulation of hMSC and mNSC increased their retention in the tumor resection cavity, permitted tumor-selective migration and release of diagnostic and therapeutic proteins *in vivo*. This study demonstrates the efficacy of encapsulated therapeutic stem cells in mouse models of GBM resection and may have implications for developing effective therapies for GBM (Kauer,

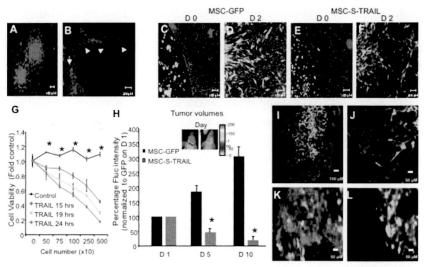

Figure 2 sECM-encapsulated therapeutic human MSCs have antitumor effects in primary invasive human GBMs *in vitro* and *in vivo*: (A and B) Photomicrographs of primary invasive GBM8-mCherry-Fluc grown as neurospheres in a collagen matrix (A) and serial brain section of mice bearing GBM-mCherry-Fluc tumors showing highly invasive nature of GBM8 (B). Arrows indicate site of implantation and arrowheads indicate path of invasion (B). (C–G) MSCs expressing GFP or S-TRAIL were encapsulated in sECMs and placed in the culture dish containing human GBM8-Fluc-mCherry cells. MSCs were followed for migration out of sECMs and GBM8 cells were followed for their response to S-TRAIL secreted by MSCs. Photomicrographs showing sECM-encapsulated hMSC on the day of plating (C and E) and 48 h post-hMSC encapsulation and plating (D and F). (G) Plot showing the GBM8 cell viability at different time points after coculturing with either sECM-encapsulated MSC-GFP or MSC-S-TRAIL ($p < 0.05$ vs. TRAIL). (H–J) Encapsulated MSC-S-TRAIL or MSC-GFP in sECM were implanted intracranially in the tumor resection cavity of mice bearing GBM8-mCherry-Fluc and mice were followed for changes in tumor volume by serial Fluc bioluminescence imaging and correlative immunohistochemistry. Plot and representative figures show the relative mean Fluc signal intensity of sECMs encapsulated MSC-GFP-Fluc or MSC-S-TRAIL-bearing mice (tumor volumes: *$p < 0.05$ vs. controls) (H). (I and J) Low (I) and high (J) magnification photomicrographs from the serial brain sections of mice showing MSCs (green) on day 5 mice post-MSCs implantation in the GBM8 (red) resection cavity. (K and L) Representative images showing cleaved caspase-3 staining (purple) on brain sections from MSC-S-TRAIL (K) and control (L) mice 5 days posttreatment. *Adapted from Kauer et al. (2012) with permission.* (See the color plate.)

Figueiredo, Hingtgen, & Shah, 2012). In the same study, we investigated the therapeutic potential of sECM-encapsulated human bone marrow-derived MSCs, expressing S-TRAIL in mouse resection models of GBM. In sECM hMSC-GBM8 coculture settings, we observed that encapsulated MSCs expressing S-TRAIL migrated out of sECMs and tracked GBM8 cells and also induced tumor cell apoptosis. Furthermore, to assess the therapeutic potential of sECM-encapsulated hMSC-S-TRAIL in mouse resection models of primary GBMs, we tested sECM-encapsulated hMSC-S-TRAIL in a mouse GBM8 model of tumor resection cavity. Our results showed a significant decrease in residual GBM8 cells, as well as a sustained presence of encapsulated hMSC in the tumor resection cavity, and MSCs migration to invading glioma cells (Fig. 2; Kauer et al., 2012).

6. CONCLUSIONS

Stem cells have emerged as attractive candidates for delivery of various proteins/therapeutic agents locally to tumor sites. This is mainly due to their inherent tumor migratory properties that are combined with the relative ease of manipulation of various stem cell types to secrete factors or targeted therapeutics agents. The iPSCs also additionally provide an added clinical advantage of using autologous cell therapy. Although various stem cell types have been shown to have therapeutic benefits in various cancer types, a thorough understanding of their fate and biology is critical to developing successful translatable therapies and thereby improving the quality of lives of many of people affected with cancer.

REFERENCES

Abbott, A. (2012). Cell rewind wins medicine Nobel. *Nature*, *490*(7419), 151–152. http://dx.doi.org/10.1038/490151a.

Aboody, K. S., Brown, A., Rainov, N. G., Bower, K. A., Liu, S., Yang, W., et al. (2000). Neural stem cells display extensive tropism for pathology in adult brain: Evidence from intracranial gliomas. *Proceedings of the National Academy of Sciences of the United States of America*, *97*(23), 12846–12851. http://dx.doi.org/10.1073/pnas.97.23.12846.

Aghi, M., & Martuza, R. L. (2005). Oncolytic viral therapies—The clinical experience. *Oncogene*, *24*(52), 7802–7816.

Akimoto, K., Kimura, K., Nagano, M., Takano, S., To'a Salazar, G., Yamashita, T., et al. (2013). Umbilical cord blood-derived mesenchymal stem cells inhibit, but adipose tissue-derived mesenchymal stem cells promote, glioblastoma multiforme proliferation. *Stem Cells and Development*, *22*(9), 1370–1386. http://dx.doi.org/10.1089/scd.2012.0486.

Altaner, C. (2008). Prodrug cancer gene therapy. *Cancer Letters*, *270*(2), 191–201. http://dx.doi.org/10.1016/j.canlet.2008.04.023.

Bagci-Onder, T., Wakimoto, H., Anderegg, M., Cameron, C., & Shah, K. (2010). A dual PI3K/mTOR inhibitor, PI-103, cooperates with stem cell delivered TRAIL in experimental glioma models. *Cancer Research*, *71*, 154–163.

Batchelor, T. T., Sorensen, A. G., di Tomaso, E., Zhang, W. T., Duda, D. G., Cohen, K. S., et al. (2007). AZD2171, a pan-VEGF receptor tyrosine kinase inhibitor, normalizes tumor vasculature and alleviates edema in glioblastoma patients. *Cancer Cell*, *11*(1), 83–95. http://dx.doi.org/10.1016/j.ccr.2006.11.021, S1535-6108(06)00370-9 [pii].

Bexell, D., Gunnarsson, S., Tormin, A., Darabi, A., Gisselsson, D., Roybon, L., et al. (2009). Bone marrow multipotent mesenchymal stroma cells act as pericyte-like migratory vehicles in experimental gliomas. *Molecular Therapy, 17*(1), 183–190. http://dx.doi.org/10.1038/mt.2008.229, mt2008229 [pii].

Bitsika, V., Roubelakis, M. G., Zagoura, D., Trohatou, O., Makridakis, M., Pappa, K. I., et al. (2012). Human amniotic fluid-derived mesenchymal stem cells as therapeutic vehicles: A novel approach for the treatment of bladder cancer. *Stem Cells and Development, 21*(7), 1097–1111. http://dx.doi.org/10.1089/scd.2011.0151.

Cavarretta, I. T., Altanerova, V., Matuskova, M., Kucerova, L., Culig, Z., & Altaner, C. (2010). Adipose tissue-derived mesenchymal stem cells expressing prodrug-converting enzyme inhibit human prostate tumor growth. *Molecular Therapy, 18*(1), 223–231. http://dx.doi.org/10.1038/mt.2009.237, mt2009237 [pii].

Chen, X., Lin, X., Zhao, J., Shi, W., Zhang, H., Wang, Y., et al. (2008). A tumor-selective biotherapy with prolonged impact on established metastases based on cytokine gene-engineered MSCs. *Molecular Therapy, 16*(4), 749–756.

Choi, S. A., Lee, J. Y., Wang, K. C., Phi, J. H., Song, S. H., Song, J., et al. (2012). Human adipose tissue-derived mesenchymal stem cells: Characteristics and therapeutic potential as cellular vehicles for prodrug gene therapy against brainstem gliomas. *European Journal of Cancer, 48*(1), 129–137. http://dx.doi.org/10.1016/j.ejca.2011.04.033.

Choi, S. H., Tamura, K., Khajuria, R. K., Bhere, D., Nesterenko, I., Lawler, J., et al. (2015). Antiangiogenic variant of TSP-1 targets tumor cells in glioblastomas. *Molecular Therapy, 23*(2), 235–243. http://dx.doi.org/10.1038/mt.2014.214.

Ciavarella, S., Grisendi, G., Dominici, M., Tucci, M., Brunetti, O., Dammacco, F., et al. (2012). *In vitro* anti-myeloma activity of TRAIL-expressing adipose-derived mesenchymal stem cells. *British Journal of Haematology, 157*(5), 586–598. http://dx.doi.org/10.1111/j.1365-2141.2012.09082.x.

Corsten, M. F., Miranda, R., Kasmieh, R., Krichevsky, A. M., Weissleder, R., & Shah, K. (2007). MicroRNA-21 knockdown disrupts glioma growth *in vivo* and displays synergistic cytotoxicity with neural precursor cell delivered S-TRAIL in human gliomas. *Cancer Research, 67*(19), 8994–9000.

Corsten, M. F., & Shah, K. (2008). Therapeutic stem-cells for cancer treatment: Hopes and hurdles in tactical warfare. *The Lancet Oncology, 9*(4), 376–384. http://dx.doi.org/10.1016/S1470-2045(08)70099-8.

Curiel, T. J., Coukos, G., Zou, L., Alvarez, X., Cheng, P., Mottram, P., et al. (2004). Specific recruitment of regulatory T cells in ovarian carcinoma fosters immune privilege and predicts reduced survival. *Nature Medicine, 10*(9), 942–949. http://dx.doi.org/10.1038/nm1093.

Danks, M. K., Yoon, K. J., Bush, R. A., Remack, J. S., Wierdl, M., Tsurkan, L., et al. (2007). Tumor-targeted enzyme/prodrug therapy mediates long-term disease-free survival of mice bearing disseminated neuroblastoma. *Cancer Research, 67*(1), 22–25. http://dx.doi.org/10.1158/0008-5472.CAN-06-3607, 67/1/22 [pii].

Darash-Yahana, M., Gillespie, J. W., Hewitt, S. M., Chen, Y. Y., Maeda, S., Stein, I., et al. (2009). The chemokine CXCL16 and its receptor, CXCR6, as markers and promoters of inflammation-associated cancers. *PLoS One, 4*(8), e6695. http://dx.doi.org/10.1371/journal.pone.0006695.

Dasari, V. R., Kaur, K., Velpula, K. K., Dinh, D. H., Tsung, A. J., Mohanam, S., et al. (2010). Downregulation of Focal Adhesion Kinase (FAK) by cord blood stem cells inhibits angiogenesis in glioblastoma. *Aging (Albany NY), 2*(11), 791–803.

Dasari, V. R., Kaur, K., Velpula, K. K., Gujrati, M., Fassett, D., Klopfenstein, J. D., et al. (2010). Upregulation of PTEN in glioma cells by cord blood mesenchymal stem cells inhibits migration via downregulation of the PI3K/Akt pathway. *PLoS One, 5*(4), e10350. http://dx.doi.org/10.1371/journal.pone.0010350.

Dasari, V. R., Velpula, K. K., Kaur, K., Fassett, D., Klopfenstein, J. D., Dinh, D. H., et al. (2010). Cord blood stem cell-mediated induction of apoptosis in glioma downregulates X-linked inhibitor of apoptosis protein (XIAP). *PLoS One, 5*(7), e11813. http://dx.doi.org/10.1371/journal.pone.0011813.

De Becker, A., Van Hummelen, P., Bakkus, M., Vande Broek, I., De Wever, J., De Waele, M., et al. (2007). Migration of culture-expanded human mesenchymal stem cells through bone marrow endothelium is regulated by matrix metalloproteinase-2 and tissue inhibitor of metalloproteinase-3. *Haematologica, 92*(4), 440–449.

de Lazaro, I., Yilmazer, A., & Kostarelos, K. (2014). Induced pluripotent stem (iPS) cells: A new source for cell-based therapeutics? *Journal of Controlled Release, 185*, 37–44. http://dx.doi.org/10.1016/j.jconrel.2014.04.011.

Dembinski, J. L., Wilson, S. M., Spaeth, E. L., Studeny, M., Zompetta, C., Samudio, I., et al. (2013). Tumor stroma engraftment of gene-modified mesenchymal stem cells as antitumor therapy against ovarian cancer. *Cytotherapy, 15*(1), 20–e32. http://dx.doi.org/10.1016/j.jcyt.2012.10.003, e22.

Ding, D. C., Chang, Y. H., Shyu, W. C., & Lin, S. Z. (2015). Human umbilical cord mesenchymal stem cells: A new era for stem cell therapy. *Cell Transplantation, 24*, 339–347. http://dx.doi.org/10.3727/096368915X686841.

Duebgen, M., Martinez-Quintanilla, J., Tamura, K., Hingtgen, S., Redjal, N., Wakimoto, H., et al. (2014). Stem cells loaded with multimechanistic oncolytic herpes simplex virus variants for brain tumor therapy. *Journal of the National Cancer Institute, 106*(6), dju090. http://dx.doi.org/10.1093/jnci/dju090.

Ehtesham, M., Kabos, P., Gutierrez, M. A., Chung, N. H., Griffith, T. S., Black, K. L., et al. (2002). Induction of glioblastoma apoptosis using neural stem cell-mediated delivery of tumor necrosis factor-related apoptosis-inducing ligand. *Cancer Research, 62*(24), 7170–7174.

Ehtesham, M., Kabos, P., Kabosova, A., Neuman, T., Black, K. L., & Yu, J. S. (2002). The use of interleukin 12-secreting neural stem cells for the treatment of intracranial glioma. *Cancer Research, 62*(20), 5657–5663.

Folkman, J., Watson, K., Ingber, D., & Hanahan, D. (1989). Induction of angiogenesis during the transition from hyperplasia to neoplasia. *Nature, 339*(6219), 58–61. http://dx.doi.org/10.1038/339058a0.

Francois, S., Bensidhoum, M., Mouiseddine, M., Mazurier, C., Allenet, B., Semont, A., et al. (2006). Local irradiation not only induces homing of human mesenchymal stem cells at exposed sites but promotes their widespread engraftment to multiple organs: A study of their quantitative distribution after irradiation damage. *Stem Cells, 24*(4), 1020–1029. http://dx.doi.org/10.1634/stemcells.2005-0260.

Gao, P., Ding, Q., Wu, Z., Jiang, H., & Fang, Z. (2010). Therapeutic potential of human mesenchymal stem cells producing IL-12 in a mouse xenograft model of renal cell carcinoma. *Cancer Letters, 290*(2), 157–166. http://dx.doi.org/10.1016/j.canlet.2009.08.031, S0304-3835(09)00571-0 [pii].

Garcia-Castro, J., Alemany, R., Cascallo, M., Martinez-Quintanilla, J., Arriero Mdel, M., Lassaletta, A., et al. (2010). Treatment of metastatic neuroblastoma with systemic oncolytic virotherapy delivered by autologous mesenchymal stem cells: An exploratory study. *Cancer Gene Therapy, 17*(7), 476–483. http://dx.doi.org/10.1038/cgt.2010.4.

Ghaedi, M., Soleimani, M., Taghvaie, N. M., Sheikhfatollahi, M., Azadmanesh, K., Lotfi, A. S., et al. (2011). Mesenchymal stem cells as vehicles for targeted delivery of anti-angiogenic protein to solid tumors. *The Journal of Gene Medicine, 13*(3), 171–180. http://dx.doi.org/10.1002/jgm.1552.

Gondi, C. S., Veeravalli, K. K., Gorantla, B., Dinh, D. H., Fassett, D., Klopfenstein, J. D., et al. (2010). Human umbilical cord blood stem cells show PDGF-D-dependent glioma cell tropism *in vitro* and *in vivo*. *Neuro-Oncology*, *12*(5), 453–465. http://dx.doi.org/10.1093/neuonc/nop049.

Grander, D., & Einhorn, S. (1998). Interferon and malignant disease—How does it work and why doesn't it always? *Acta Oncologica*, *37*(4), 331–338.

Grisendi, G., Bussolari, R., Cafarelli, L., Petak, I., Rasini, V., Veronesi, E., et al. (2010). Adipose-derived mesenchymal stem cells as stable source of tumor necrosis factor-related apoptosis-inducing ligand delivery for cancer therapy. *Cancer Research*, *70*(9), 3718–3729. http://dx.doi.org/10.1158/0008-5472.CAN-09-1865.

He, Q., Wan, C., & Li, G. (2007). Concise review: Multipotent mesenchymal stromal cells in blood. *Stem Cells*, *25*(1), 69–77. http://dx.doi.org/10.1634/stemcells.2006-0335.

Hendijani, F., Haghjooy Javanmard, S., & Sadeghi-Aliabadi, H. (2015). Human Wharton's jelly mesenchymal stem cell secretome display antiproliferative effect on leukemia cell line and produce additive cytotoxic effect in combination with doxorubicin. *Tissue & Cell*. http://dx.doi.org/10.1016/j.tice.2015.01.005, S0040-8166(15)00012-9 [pii] [Epub ahead of print].

Hong, H. S., Lee, J., Lee, E., Kwon, Y. S., Ahn, W., Jiang, M. H., et al. (2009). A new role of substance P as an injury-inducible messenger for mobilization of CD29(+) stromal-like cells. *Nature Medicine*, *15*(4), 425–435. http://dx.doi.org/10.1038/nm.1909.

Hormigo, A., Gutin, P. H., & Rafii, S. (2007). Tracking normalization of brain tumor vasculature by magnetic imaging and proangiogenic biomarkers. *Cancer Cell*, *11*(1), 6–8. http://dx.doi.org/10.1016/j.ccr.2006.12.008, S1535-6108(06)00375-8 [pii].

Hu, W., Wang, J., Dou, J., He, X., Zhao, F., Jiang, C., et al. (2011). Augmenting therapy of ovarian cancer efficacy by secreting IL-21 human umbilical cord blood stem cells in nude mice. *Cell Transplantation*, *20*(5), 669–680. http://dx.doi.org/10.3727/096368910X536509.

Imitola, J., Raddassi, K., Park, K. I., Mueller, F. J., Nieto, M., Teng, Y. D., et al. (2004). Directed migration of neural stem cells to sites of CNS injury by the stromal cell-derived factor 1alpha/CXC chemokine receptor 4 pathway. *Proceedings of the National Academy of Sciences of the United States of America*, *101*(52), 18117–18122.

Jacobs, J. F., Idema, A. J., Bol, K. F., Grotenhuis, J. A., de Vries, I. J., Wesseling, P., et al. (2010). Prognostic significance and mechanism of Treg infiltration in human brain tumors. *Journal of Neuroimmunology*, *225*(1-2), 195–199. http://dx.doi.org/10.1016/j.jneuroim.2010.05.020.

Jagadeesha, D. K., Takapoo, M., Banfi, B., Bhalla, R. C., & Miller, F. J., Jr. (2012). Nox1 transactivation of epidermal growth factor receptor promotes N-cadherin shedding and smooth muscle cell migration. *Cardiovascular Research*, *93*(3), 406–413. http://dx.doi.org/10.1093/cvr/cvr308.

Jain, R. K., di Tomaso, E., Duda, D. G., Loeffler, J. S., Sorensen, A. G., & Batchelor, T. T. (2007). Angiogenesis in brain tumors. *Nature Reviews. Neuroscience*, *8*(8), 610–622. http://dx.doi.org/10.1038/nrn2175.

Jiang, H., Lin, J. J., Su, Z.-Z., Goldstein, N. I., & Fisher, P. B. (1995). Subtraction hybridization identifies a novel melanoma differentiation associated gene, mda-7, modulated during human melanoma differentiation, growth and progression. *Oncogene*, *11*(12), 2477–2486.

Jiao, H., Guan, F., Yang, B., Li, J., Shan, H., Song, L., et al. (2011). Human umbilical cord blood-derived mesenchymal stem cells inhibit C6 glioma via downregulation of cyclin D1. *Neurology India*, *59*(2), 241–247. http://dx.doi.org/10.4103/0028-3886.79134.

Johnson, T. V., Bull, N. D., Hunt, D. P., Marina, N., Tomarev, S. I., & Martin, K. R. (2010). Neuroprotective effects of intravitreal mesenchymal stem cell transplantation in experimental glaucoma. *Investigative Ophthalmology & Visual Science*, *51*(4), 2051–2059. http://dx.doi.org/10.1167/iovs.09-4509.

Jones, B. S., Lamb, L. S., Goldman, F., & Di Stasi, A. (2014). Improving the safety of cell therapy products by suicide gene transfer. *Frontiers in Pharmacology, 5*, 254. http://dx.doi.org/10.3389/fphar.2014.00254.

Kadambi, A., Mouta Carreira, C., Yun, C. O., Padera, T. P., Dolmans, D. E., Carmeliet, P., et al. (2001). Vascular endothelial growth factor (VEGF)-C differentially affects tumor vascular function and leukocyte recruitment: Role of VEGF-receptor 2 and host VEGF-A. *Cancer Research, 61*(6), 2404–2408.

Kauer, T. M., Figueiredo, J. L., Hingtgen, S., & Shah, K. (2012). Encapsulated therapeutic stem cells implanted in the tumor resection cavity induce cell death in gliomas. *Nature Neuroscience, 15*(2), 197–204. http://dx.doi.org/10.1038/nn.3019.

Kawabata, A., Ohta, N., Seiler, G., Pyle, M. M., Ishiguro, S., Zhang, Y. Q., et al. (2013). Naive rat umbilical cord matrix stem cells significantly attenuate mammary tumor growth through modulation of endogenous immune responses. *Cytotherapy, 15*(5), 586–597. http://dx.doi.org/10.1016/j.jcyt.2013.01.006.

Khakoo, A. Y., Pati, S., Anderson, S. A., Reid, W., Elshal, M. F., Rovira, I. I., et al. (2006). Human mesenchymal stem cells exert potent antitumorigenic effects in a model of Kaposi's sarcoma. *The Journal of Experimental Medicine, 203*(5), 1235–1247. http://dx.doi.org/10.1084/jem.20051921, jem.20051921 [pii].

Kim, S. U. (2004). Human neural stem cells genetically modified for brain repair in neurological disorders. *Neuropathology, 24*(3), 159–171.

Kim, S. K., Cargioli, T. G., Machluf, M., Yang, W., Sun, Y., Al-Hashem, R., et al. (2005). PEX-producing human neural stem cells inhibit tumor growth in a mouse glioma model. *Clinical Cancer Research, 11*(16), 5965–5970.

Kim, S. W., Kim, S. J., Park, S. H., Yang, H. G., Kang, M. C., Choi, Y. W., et al. (2013). Complete regression of metastatic renal cell carcinoma by multiple injections of engineered mesenchymal stem cells expressing dodecameric TRAIL and HSV-TK. *Clinical Cancer Research, 19*(2), 415–427. http://dx.doi.org/10.1158/1078-0432.CCR-12-1568.

Kim, S. U., Lee, H. J., & Kim, Y. B. (2013). Neural stem cell-based treatment for neurodegenerative diseases. *Neuropathology, 33*(5), 491–504. http://dx.doi.org/10.1111/neup.12020.

Kim, S. M., Oh, J. H., Park, S. A., Ryu, C. H., Lim, J. Y., Kim, D. S., et al. (2010). Irradiation enhances the tumor tropism and therapeutic potential of tumor necrosis factor-related apoptosis-inducing ligand-secreting human umbilical cord blood-derived mesenchymal stem cells in glioma therapy. *Stem Cells, 28*(12), 2217–2228.

Kim, S. M., Woo, J. S., Jeong, C. H., Ryu, C. H., Lim, J. Y., & Jeun, S. S. (2012). Effective combination therapy for malignant glioma with TRAIL-secreting mesenchymal stem cells and lipoxygenase inhibitor MK886. *Cancer Research, 72*(18), 4807–4817. http://dx.doi.org/10.1158/0008-5472.CAN-12-0123.

Knoop, K., Schwenk, N., Schmohl, K., Muller, A., Zach, C., Cyran, C., et al. (2015). Mesenchymal stem cell (MSC)-mediated, tumor stroma-targeted radioiodine therapy of metastatic colon cancer using the sodium iodide symporter as theranostic gene. *Journal of Nuclear Medicine, 56*, 600–606. http://dx.doi.org/10.2967/jnumed.114.146662.

Komarova, S., Kawakami, Y., Stoff-Khalili, M. A., Curiel, D. T., & Pereboeva, L. (2006). Mesenchymal progenitor cells as cellular vehicles for delivery of oncolytic adenoviruses. *Molecular Cancer Therapeutics, 5*(3), 755–766.

Lens, M. (2006). Cutaneous melanoma: Interferon alpha adjuvant therapy for patients at high risk for recurrent disease. *Dermatologic Therapy, 19*(1), 9–18. http://dx.doi.org/10.1111/j.1529-8019.2005.00051.x, DTH051 [pii].

Liu, X., Hu, J., Sun, S., Li, F., Cao, W., Wang, Y. U., et al. (2015). Mesenchymal stem cells expressing interleukin-18 suppress breast cancer cells. *Experimental and Therapeutic Medicine, 9*(4), 1192–1200. http://dx.doi.org/10.3892/etm.2015.2286.

Loebinger, M. R., Eddaoudi, A., Davies, D., & Janes, S. M. (2009). Mesenchymal stem cell delivery of TRAIL can eliminate metastatic cancer. *Cancer Research*, *69*(10), 4134–4142. http://dx.doi.org/10.1158/0008-5472.CAN-08-4698, 0008-5472.CAN-08-4698 [pii].

Ma, X., Zhang, L., Ma, H., & Yu, R. (2001). Association of vascular endothelial growth factor expression with angiogenesis and tumor cell proliferation in human lung cancer. *Zhonghua Nei Ke Za Zhi*, *40*(1), 32–35.

Mader, E. K., Butler, G., Dowdy, S. C., Mariani, A., Knutson, K. L., Federspiel, M. J., et al. (2013). Optimizing patient derived mesenchymal stem cells as virus carriers for a phase I clinical trial in ovarian cancer. *Journal of Translational Medicine*, *11*, 20. http://dx.doi.org/10.1186/1479-5876-11-20.

Maretzky, T., Evers, A., Zhou, W., Swendeman, S. L., Wong, P. M., Rafii, S., et al. (2011). Migration of growth factor-stimulated epithelial and endothelial cells depends on EGFR transactivation by ADAM17. *Nature Communications*, *2*, 229. http://dx.doi.org/10.1038/ncomms1232.

Martinez-Quintanilla, J., Bhere, D., Heidari, P., He, D., Mahmood, U., & Shah, K. (2013). In vivo imaging of the therapeutic efficacy and fate of bimodal engineered stem cells in malignant brain tumors. *Stem Cells*, *31*, 1706–1714. http://dx.doi.org/10.1002/stem.1355.

Martino, G., & Pluchino, S. (2006). The therapeutic potential of neural stem cells. *Nature Reviews. Neuroscience*, *7*(5), 395–406. http://dx.doi.org/10.1038/nrn1908.

Miletic, H., Fischer, Y., Litwak, S., Giroglou, T., Waerzeggers, Y., Winkeler, A., et al. (2007). Bystander killing of malignant glioma by bone marrow-derived tumor-infiltrating progenitor cells expressing a suicide gene. *Molecular Therapy*, *15*(7), 1373–1381. http://dx.doi.org/10.1038/mt.sj.6300155, 6300155 [pii].

Mohr, A., Albarenque, S. M., Deedigan, L., Yu, R., Reidy, M., Fulda, S., et al. (2010). Targeting of XIAP combined with systemic mesenchymal stem cell-mediated delivery of sTRAIL ligand inhibits metastatic growth of pancreatic carcinoma cells. *Stem Cells*, *28*(11), 2109–2120. http://dx.doi.org/10.1002/stem.533.

Momin, E. N., Vela, G., Zaidi, H. A., & Quinones-Hinojosa, A. (2010). The oncogenic potential of mesenchymal stem cells in the treatment of cancer: Directions for future research. *Current Immunology Reviews*, *6*(2), 137–148. http://dx.doi.org/10.2174/157339510791111718.

Morris, P. J. (1996). Immunoprotection of therapeutic cell transplants by encapsulation. *Trends in Biotechnology*, *14*(5), 163–167.

Mueller, L. P., Luetzkendorf, J., Widder, M., Nerger, K., Caysa, H., & Mueller, T. (2010). TRAIL-transduced multipotent mesenchymal stromal cells (TRAIL-MSC) overcome TRAIL resistance in selected CRC cell lines in vitro and in vivo. *Cancer Gene Therapy*, *18*, 229–239. http://dx.doi.org/10.1038/cgt.2010.68, cgt201068 [pii].

Muldoon, L. L., Soussain, C., Jahnke, K., Johanson, C., Siegal, T., Smith, Q. R., et al. (2007). Chemotherapy delivery issues in central nervous system malignancy: A reality check. *Journal of Clinical Oncology*, *25*(16), 2295–2305. http://dx.doi.org/10.1200/JCO.2006.09.9861.

Murua, A., Portero, A., Orive, G., Hernandez, R. M., de Castro, M., & Pedraz, J. L. (2008). Cell microencapsulation technology: Towards clinical application. *Journal of Controlled Release*, *132*(2), 76–83. http://dx.doi.org/10.1016/j.jconrel.2008.08.010.

Nakamizo, A., Marini, F., Amano, T., Khan, A., Studeny, M., Gumin, J., et al. (2005). Human bone marrow-derived mesenchymal stem cells in the treatment of gliomas. *Cancer Research*, *65*(8), 3307–3318.

Nakashima, H., Kaur, B., & Chiocca, E. A. (2010). Directing systemic oncolytic viral delivery to tumors via carrier cells. *Cytokine & Growth Factor Reviews*, *21*(2-3), 119–126.

Nam, H., Lee, K. H., Nam, D. H., & Joo, K. M. (2015). Adult human neural stem cell therapeutics: Current developmental status and prospect. *World Journal of Stem Cells*, 7(1), 126–136. http://dx.doi.org/10.4252/wjsc.v7.i1.126.

Nesterenko, I., Wanningen, S., Bagci-Onder, T., Anderegg, M., & Shah, K. (2012). Evaluating the effect of therapeutic stem cells on TRAIL resistant and sensitive medulloblastomas. *PLoS One*, 7(11), e49219. http://dx.doi.org/10.1371/journal.pone.0049219.

Noble, M. (2000). Can neural stem cells be used as therapeutic vehicles in the treatment of brain tumors? *Nature Medicine*, 6(4), 369–370. http://dx.doi.org/10.1038/74610.

Okada, H., & Pollack, I. F. (2004). Cytokine gene therapy for malignant glioma. *Expert Opinion on Biological Therapy*, 4(10), 1609–1620.

Ong, H. T., Federspiel, M. J., Guo, C. M., Lucien Ooi, L., Russell, S. J., Peng, K. W., et al. (2013). Systemically delivered measles virus-infected mesenchymal stem cells can evade host immunity to inhibit liver cancer growth. *Journal of Hepatology*, 59, 999–1006. http://dx.doi.org/10.1016/j.jhep.2013.07.010.

Otsu, K., Das, S., Houser, S. D., Quadri, S. K., Bhattacharya, S., & Bhattacharya, J. (2009). Concentration-dependent inhibition of angiogenesis by mesenchymal stem cells. *Blood*, 113(18), 4197–4205. http://dx.doi.org/10.1182/blood-2008-09-176198, blood-2008-09-176198 [pii].

Ozdemir, V., Williams-Jones, B., Glatt, S. J., Tsuang, M. T., Lohr, J. B., & Reist, C. (2006). Shifting emphasis from pharmacogenomics to theragnostics. *Nature Biotechnology*, 24(8), 942–946.

Pan, L., Ren, Y., Cui, F., & Xu, Q. (2009). Viability and differentiation of neural precursors on hyaluronic acid hydrogel scaffold. *Journal of Neuroscience Research*, 87(14), 3207–3220.

Parato, K. A., Senger, D., Forsyth, P. A., & Bell, J. C. (2005). Recent progress in the battle between oncolytic viruses and tumors. *Nature Reviews. Cancer*, 5(12), 965–976.

Park, J. S., Suryaprakash, S., Lao, Y. H., & Leong, K. W. (2015). Engineering mesenchymal stem cells for regenerative medicine and drug delivery. *Methods*. http://dx.doi.org/10.1016/j.ymeth.2015.03.002, S1046-2023(15)00096-1 [pii] [Epub ahead of print].

Park, K. I., Teng, Y. D., & Snyder, E. Y. (2002). The injured brain interacts reciprocally with neural stem cells supported by scaffolds to reconstitute lost tissue. *Nature Biotechnology*, 20(11), 1111–1117.

Pereboeva, L., Komarova, S., Mikheeva, G., Krasnykh, V., & Curiel, D. T. (2003). Approaches to utilize mesenchymal progenitor cells as cellular vehicles. *Stem Cells*, 21(4), 389–404.

Porcile, C., Bajetto, A., Barbieri, F., Barbero, S., Bonavia, R., Biglieri, M., et al. (2005). Stromal cell-derived factor-1alpha (SDF-1alpha/CXCL12) stimulates ovarian cancer cell growth through the EGF receptor transactivation. *Experimental Cell Research*, 308(2), 241–253. http://dx.doi.org/10.1016/j.yexcr.2005.04.024.

Power, A. T., & Bell, J. C. (2007). Cell-based delivery of oncolytic viruses: A new strategic alliance for a biological strike against cancer. *Molecular Therapy*, 15(4), 660–665.

Qiu, Y., Marquez-Curtis, L. A., & Janowska-Wieczorek, A. (2012). Mesenchymal stromal cells derived from umbilical cord blood migrate in response to complement C1q. *Cytotherapy*, 14(3), 285–295. http://dx.doi.org/10.3109/14653249.2011.651532.

Quante, M., Tu, S. P., Tomita, H., Gonda, T., Wang, S. S., Takashi, S., et al. (2011). Bone marrow-derived myofibroblasts contribute to the mesenchymal stem cell niche and promote tumor growth. *Cancer Cell*, 19(2), 257–272. http://dx.doi.org/10.1016/j.ccr.2011.01.020.

Reagan, M. R., Seib, F. P., McMillin, D. W., Sage, E. K., Mitsiades, C. S., Janes, S. M., et al. (2012). Stem cell implants for cancer therapy: TRAIL-expressing mesenchymal stem cells target cancer cells *in situ*. *Journal of Breast Cancer*, 15(3), 273–282. http://dx.doi.org/10.4048/jbc.2012.15.3.273.

Ren, C., Kumar, S., Chanda, D., Chen, J., Mountz, J. D., & Ponnazhagan, S. (2008). Therapeutic potential of mesenchymal stem cells producing interferon-alpha in a mouse melanoma lung metastasis model. *Stem Cells, 26*(9), 2332–2338. http://dx.doi.org/10.1634/stemcells.2008-0084, 2008-0084 [pii].

Ren, C., Kumar, S., Chanda, D., Kallman, L., Chen, J., Mountz, J. D., et al. (2008). Cancer gene therapy using mesenchymal stem cells expressing interferon-beta in a mouse prostate cancer lung metastasis model. *Gene Therapy, 15*(21), 1446–1453. http://dx.doi.org/10.1038/gt.2008.101, gt2008101 [pii].

Reubinoff, B. E., Pera, M. F., Fong, C. Y., Trounson, A., & Bongso, A. (2000). Embryonic stem cell lines from human blastocysts: Somatic differentiation *in vitro*. *Nature Biotechnology, 18*(4), 399–404. http://dx.doi.org/10.1038/74447.

Rihova, B. (2000). Immunocompatibility and biocompatibility of cell delivery systems. *Advanced Drug Delivery Reviews, 42*(1-2), 65–80.

Roelle, S., Grosse, R., Aigner, A., Krell, H. W., Czubayko, F., & Gudermann, T. (2003). Matrix metalloproteinases 2 and 9 mediate epidermal growth factor receptor transactivation by gonadotropin-releasing hormone. *The Journal of Biological Chemistry, 278*(47), 47307–47318. http://dx.doi.org/10.1074/jbc.M304377200.

Ryu, C. H., Park, S. H., Park, S. A., Kim, S. M., Lim, J. Y., Jeong, C. H., et al. (2011). Gene therapy of intracranial glioma using interleukin 12-secreting human umbilical cord blood-derived mesenchymal stem cells. *Human Gene Therapy, 22*(6), 733–743. http://dx.doi.org/10.1089/hum.2010.187.

Samant, R. S., & Shevde, L. A. (2011). Recent advances in anti-angiogenic therapy of cancer. *Oncotarget, 2*(3), 122–134.

Sartoris, S., Mazzocco, M., Tinelli, M., Martini, M., Mosna, F., Lisi, V., et al. (2011). Efficacy assessment of interferon-alpha-engineered mesenchymal stromal cells in a mouse plasmacytoma model. *Stem Cells and Development, 20*(4), 709–719. http://dx.doi.org/10.1089/scd.2010.0095.

Sasportas, L. S., Kasmieh, R., Wakimoto, H., Hingtgen, S., van de Water, J. A., Mohapatra, G., et al. (2009). Assessment of therapeutic efficacy and fate of engineered human mesenchymal stem cells for cancer therapy. *Proceedings of the National Academy of Sciences of the United States of America, 106*(12), 4822–4827. http://dx.doi.org/10.1073/pnas.0806647106.

Savla, J. J., Nelson, B. C., Perry, C. N., & Adler, E. D. (2014). Induced pluripotent stem cells for the study of cardiovascular disease. *Journal of the American College of Cardiology, 64*(5), 512–519. http://dx.doi.org/10.1016/j.jacc.2014.05.038.

Sayyar, B., Dodd, M., Wen, J., Ma, S., Marquez-Curtis, L., Janowska-Wieczorek, A., et al. (2012). Encapsulation of factor IX-engineered mesenchymal stem cells in fibrinogen-alginate microcapsules enhances their viability and transgene secretion. *Journal of Tissue Engineering, 3*(1). http://dx.doi.org/10.1177/2041731412462018, 2041731412462018.

Schichor, C., Albrecht, V., Korte, B., Buchner, A., Riesenberg, R., Mysliwietz, J., et al. (2012). Mesenchymal stem cells and glioma cells form a structural as well as a functional syncytium *in vitro*. *Experimental Neurology, 234*(1), 208–219. http://dx.doi.org/10.1016/j.expneurol.2011.12.033.

Seo, S. H., Kim, K. S., Park, S. H., Suh, Y. S., Kim, S. J., Jeun, S. S., et al. (2011). The effects of mesenchymal stem cells injected via different routes on modified IL-12-mediated antitumor activity. *Gene Therapy, 18*, 488–495. http://dx.doi.org/10.1038/gt.2010.170, gt2010170 [pii].

Shah, K. (2009). Imaging neural stem cell fate in mouse model of glioma. *Current Protocols in Stem Cell Biology*. http://dx.doi.org/10.1002/9780470151808.sc05a01s8, Chapter 5, Unit 5A 1. 8:5A.1.1–5A.1.11.

Shah, K. (2012). Mesenchymal stem cells engineered for cancer therapy. *Advanced Drug Delivery Reviews, 64*(8), 739–748. http://dx.doi.org/10.1016/j.addr.2011.06.010.

Shah, K., Bureau, E., Kim, D. E., Yang, K., Tang, Y., Weissleder, R., et al. (2005). Glioma therapy and real-time imaging of neural precursor cell migration and tumor regression. *Annals of Neurology*, 57(1), 34–41.

Shah, K., Tung, C. H., Yang, K., Weissleder, R., & Breakefield, X. O. (2004). Inducible release of TRAIL fusion proteins from a proapoptotic form for tumor therapy. *Cancer Research*, 64(9), 3236–3242.

Singh, V. K., Kalsan, M., Kumar, N., Saini, A., & Chandra, R. (2015). Induced pluripotent stem cells: Applications in regenerative medicine, disease modeling, and drug discovery. *Frontiers in Cell and Developmental Biology*, 3, 2. http://dx.doi.org/10.3389/fcell.2015.00002.

Smith, H., Whittall, C., Weksler, B., & Middleton, J. (2012). Chemokines stimulate bidirectional migration of human mesenchymal stem cells across bone marrow endothelial cells. *Stem Cells and Development*, 21(3), 476–486. http://dx.doi.org/10.1089/scd.2011.0025.

Son, B. R., Marquez-Curtis, L. A., Kucia, M., Wysoczynski, M., Turner, A. R., Ratajczak, J., et al. (2006). Migration of bone marrow and cord blood mesenchymal stem cells *in vitro* is regulated by stromal-derived factor-1-CXCR4 and hepatocyte growth factor-c-met axes and involves matrix metalloproteinases. *Stem Cells*, 24(5), 1254–1264. http://dx.doi.org/10.1634/stemcells.2005-0271.

Sonabend, A. M., Ulasov, I. V., Tyler, M. A., Rivera, A. A., Mathis, J. M., & Lesniak, M. S. (2008). Mesenchymal stem cells effectively deliver an oncolytic adenovirus to intracranial glioma. *Stem Cells*, 26(3), 831–841. http://dx.doi.org/10.1634/stemcells.2007-0758, 2007-0758 [pii].

Song, C., Xiang, J., Tang, J., Hirst, D. G., Zhou, J., Chan, K. M., et al. (2011). Thymidine kinase gene modified bone marrow mesenchymal stem cells as vehicles for antitumor therapy. *Human Gene Therapy*, 22(4), 439–449. http://dx.doi.org/10.1089/hum.2010.116.

Spaeth, E. L., Kidd, S., & Marini, F. C. (2012). Tracking inflammation-induced mobilization of mesenchymal stem cells. *Methods in Molecular Biology*, 904, 173–190. http://dx.doi.org/10.1007/978-1-61779-943-3_15.

Spaeth, E., Klopp, A., Dembinski, J., Andreeff, M., & Marini, F. (2008). Inflammation and tumor microenvironments: Defining the migratory itinerary of mesenchymal stem cells. *Gene Therapy*, 15(10), 730–738. http://dx.doi.org/10.1038/gt.2008.39.

Stoff-Khalili, M. A., Rivera, A. A., Mathis, J. M., Banerjee, N. S., Moon, A. S., Hess, A., et al. (2007). Mesenchymal stem cells as a vehicle for targeted delivery of CRAds to lung metastases of breast carcinoma. *Breast Cancer Research and Treatment*, 105, 157–167.

Stuckey, D. W., & Shah, K. (2014). Stem cell-based therapies for cancer treatment: Separating hope from hype. *Nature Reviews Caner*, 14(10), 683–691.

Studeny, M., Marini, F. C., Champlin, R. E., Zompetta, C., Fidler, I. J., & Andreeff, M. (2002). Bone marrow-derived mesenchymal stem cells as vehicles for interferon-beta delivery into tumors. *Cancer Research*, 62(13), 3603–3608.

Studeny, M., Marini, F. C., Dembinski, J. L., Zompetta, C., Cabreira-Hansen, M., Bekele, B. N., et al. (2004). Mesenchymal stem cells: Potential precursors for tumor stroma and targeted-delivery vehicles for anticancer agents. *Journal of the National Cancer Institute*, 96(21), 1593–1603. http://dx.doi.org/10.1093/jnci/djh299, 96/21/1593 [pii].

Sun, Y., Williams, A., Waisbourd, M., Iacovitti, L., & Katz, L. J. (2015). Stem cell therapy for glaucoma: Science or snake oil? *Survey of Ophthalmology*, 60(2), 93–105. http://dx.doi.org/10.1016/j.survophthal.2014.07.001.

Teng, Y. D., Lavik, E. B., Qu, X., Park, K. I., Ourednik, J., Zurakowski, D., et al. (2002). Functional recovery following traumatic spinal cord injury mediated by a unique polymer scaffold seeded with neural stem cells. *Proceedings of the National Academy of Sciences of the United States of America*, 99(5), 3024–3029.

Tong, R. T., Boucher, Y., Kozin, S. V., Winkler, F., Hicklin, D. J., & Jain, R. K. (2004). Vascular normalization by vascular endothelial growth factor receptor 2 blockade induces a pressure gradient across the vasculature and improves drug penetration in tumors. *Cancer Research, 64*(11), 3731–3736. http://dx.doi.org/10.1158/0008-5472.CAN-04-0074, 64/11/3731 [pii].

Torrente, Y., & Polli, E. (2008). Mesenchymal stem cell transplantation for neurodegenerative diseases. *Cell Transplantation, 17*(10–11), 1103–1113.

Van de Water, J. A., B.-O., T., Agarwal, A. S., Wakimoto, H., Kasmieh, R., Roovers, R. C., et al. (2012). Therapeutic stem cells expressing different variants of EGFR-specific nanobodies have anti-tumor effects. *PNAS, 109*, 16642–16647.

Walczak, H., & Krammer, P. H. (2000). The CD95 (APO-1/Fas) and the TRAIL (APO-2L) apoptosis systems. *Experimental Cell Research, 256*(1), 58–66.

Wang, G. X., Zhan, Y. A., Hu, H. L., Wang, Y., & Fu, B. (2012). Mesenchymal stem cells modified to express interferon-beta inhibit the growth of prostate cancer in a mouse model. *The Journal of International Medical Research, 40*(1), 317–327.

Winkler, J., Hescheler, J., & Sachinidis, A. (2005). Embryonic stem cells for basic research and potential clinical applications in cardiology. *Biochimica et Biophysica Acta, 1740*(2), 240–248. http://dx.doi.org/10.1016/j.bbadis.2004.11.018.

Wu, S., Ju, G. Q., Du, T., Zhu, Y. J., & Liu, G. H. (2013). Microvesicles derived from human umbilical cord Wharton's jelly mesenchymal stem cells attenuate bladder tumor cell growth in vitro and in vivo. *PLoS One, 8*(4), e61366. http://dx.doi.org/10.1371/journal.pone.0061366.

Xu, X., Yang, G., Zhang, H., & Prestwich, G. D. (2009). Evaluating dual activity LPA receptor pan-antagonist/autotaxin inhibitors as anti-cancer agents in vivo using engineered human tumors. *Prostaglandins & Other Lipid Mediators, 89*(3-4), 140–146.

Yahata, Y., Shirakata, Y., Tokumaru, S., Yang, L., Dai, X., Tohyama, M., et al. (2006). A novel function of angiotensin II in skin wound healing. Induction of fibroblast and keratinocyte migration by angiotensin II via heparin-binding epidermal growth factor (EGF)-like growth factor-mediated EGF receptor transactivation. *The Journal of Biological Chemistry, 281*(19), 13209–13216. http://dx.doi.org/10.1074/jbc.M509771200.

Yamazoe, T., Koizumi, S., Yamasaki, T., Amano, S., Tokuyama, T., & Namba, H. (2015). Potent tumor tropism of induced pluripotent stem cells and induced pluripotent stem cell-derived neural stem cells in the mouse intracerebral glioma model. *International Journal of Oncology, 46*(1), 147–152. http://dx.doi.org/10.3892/ijo.2014.2702.

Yan, C., Li, S., Li, Z., Peng, H., Yuan, X., Jiang, L., et al. (2013). Human umbilical cord mesenchymal stem cells as vehicles of CD20-specific TRAIL fusion protein delivery: A double-target therapy against non-Hodgkin's lymphoma. *Molecular Pharmaceutics, 10*(1), 142–151. http://dx.doi.org/10.1021/mp300261e.

Yang, C., Lei, D., Ouyang, W., Ren, J., Li, H., Hu, J., et al. (2014). Conditioned media from human adipose tissue-derived mesenchymal stem cells and umbilical cord-derived mesenchymal stem cells efficiently induced the apoptosis and differentiation in human glioma cell lines in vitro. *BioMed Research International, 2014*, 109389. http://dx.doi.org/10.1155/2014/109389.

Yip, S., & Shah, K. (2008). Stem-cell based therapies for brain tumors. *Current Opinion in Molecular Therapeutics, 10*(4), 334–342.

Yong, R. L., Shinojima, N., Fueyo, J., Gumin, J., Vecil, G. G., Marini, F. C., et al. (2009). Human bone marrow-derived mesenchymal stem cells for intravascular delivery of oncolytic adenovirus Delta24-RGD to human gliomas. *Cancer Research, 69*(23), 8932–8940. http://dx.doi.org/10.1158/0008-5472.CAN-08-3873, 0008-5472.CAN-08-3873 [pii].

Yu, R., Deedigan, L., Albarenque, S. M., Mohr, A., & Zwacka, R. M. (2013). Delivery of sTRAIL variants by MSCs in combination with cytotoxic drug treatment leads to p53-independent enhanced antitumor effects. *Cell Death & Disease, 4,* e503. http://dx.doi.org/10.1038/cddis.2013.19.

Yulyana, Y., Endaya, B. B., Ng, W. H., Guo, C. M., Hui, K. M., Lam, P. Y., et al. (2013). Carbenoxolone enhances TRAIL-induced apoptosis through the upregulation of death receptor 5 and inhibition of gap junction intercellular communication in human glioma. *Stem Cells and Development, 22*(13), 1870–1882. http://dx.doi.org/10.1089/scd.2012.0529.

Zhang, Y., Wang, J., Ren, M., Li, M., Chen, D., Chen, J., et al. (2014). Gene therapy of ovarian cancer using IL-21-secreting human umbilical cord mesenchymal stem cells in nude mice. *Journal of Ovarian Research, 7,* 8. http://dx.doi.org/10.1186/1757-2215-7-8.

Zhang, X., Zhang, L., Xu, W., Qian, H., Ye, S., Zhu, W., et al. (2013). Experimental therapy for lung cancer: Umbilical cord-derived mesenchymal stem cell-mediated interleukin-24 delivery. *Current Cancer Drug Targets, 13*(1), 92–102.

Zheng, L., Zhang, D., Chen, X., Yang, L., Wei, Y., & Zhao, X. (2012). Antitumor activities of human placenta-derived mesenchymal stem cells expressing endostatin on ovarian cancer. *PLoS One, 7*(7), e39119. http://dx.doi.org/10.1371/journal.pone.0039119.

CHAPTER FIVE

Emerging Therapeutic Strategies for Overcoming Proteasome Inhibitor Resistance

Nathan G. Dolloff[1]

Department of Cellular and Molecular Pharmacology & Experimental Therapeutics, Medical University of South Carolina, Charleston, South Carolina, USA
[1]Corresponding author: e-mail address: dolloffn@musc.edu

Contents

1. Introduction — 192
2. Mechanisms of Btz Resistance — 193
   2.1 PSMβ5 Gene Mutations — 193
   2.2 Upregulation of Proteasomal Subunits — 196
   2.3 Apoptotic Resistance and Autophagy — 197
3. Approaches to Overcoming Btz Resistance — 200
   3.1 Next-Generation Proteasome Inhibitors — 200
   3.2 Redox Signaling — 201
   3.3 MDM2 Inhibitors — 205
   3.4 Il-6/STAT3 Signaling Axis — 208
   3.5 Therapeutic Monoclonal Antibodies — 209
   3.6 Bromodomain and Other Epigenetic Targets — 210
4. Concluding Remarks — 214
Acknowledgments — 214
References — 214

Abstract

The debut of the proteasome inhibitor bortezomib (Btz; Velcade®) radically and immediately improved the treatment of multiple myeloma (MM), an incurable malignancy of the plasma cell. Therapeutic resistance is unavoidable, however, and represents a major obstacle to maximizing the clinical potential of the drug. To address this challenge, studies have been conducted to uncover the molecular mechanisms driving Btz resistance and to discover new targeted therapeutic strategies and combinations that restore Btz activity. This review discusses the literature describing molecular adaptations that confer Btz resistance with a primary disease focus on MM. Also discussed are the most recent advances in therapeutic strategies that overcome resistance, approaches that include redox-modulating agents, murine double minute 2 inhibitors, therapeutic monoclonal antibodies, and new epigenetic-targeted drugs like bromodomain and extra terminal domain inhibitors.

1. INTRODUCTION

The remarkable activity of the proteasome inhibitor (PI) bortezomib (Btz; Velcade®) was first recognized in an initial phase I clinical trial where Orlowski and colleagues observed a complete response in a multiple myeloma (MM) patient with advanced disease (Orlowski et al., 2002). Accelerated regulatory approval was then granted by the FDA in 2003 following two landmark phase II clinical studies in patients with advanced staged MM (Jagannath et al., 2004; Richardson et al., 2003). In these studies, 35% of patients, all of whom had progressive disease following at least three therapies, achieved a measurable response with the average response duration lasting 1 year. Today, Btz is a cornerstone in the treatment of MM for which it is approved as a first-line therapy and is a ubiquitous component of the multidrug cocktails that are used in the clinical management of MM. Prior to the introduction of Btz, MM was a highly aggressive and deadly form of cancer, and minimal advances in treatment had been made since the first trial of melphalan in the early 1960s (Bergsagel, 2014; Bergsagel, Sprague, Austin, & Griffith, 1962). While MM remains incurable today, the development of novel agents such as Btz has substantially improved survival times and quality of life. The development story of Btz serves as a blueprint for navigating the time and resource-intensive path of bench-to-bedside translational research and is a shining example of success in the era of targeted cancer therapy. The details of this story have been discussed in depth elsewhere and will not be the focus of this review (Allen, 2007; Sanchez-Serrano, 2005, 2006). Rather, the emphasis will be on a rapidly expanding literature of molecular strategies that effectively combat therapeutic resistance to Btz. This is an important topic given that, despite the initial effectiveness of Btz, nearly all patients progress to a refractory stage, and therapeutic resistance has emerged as a clear obstacle to maximizing the clinical benefit of the drug. There are multiple distinct molecular strategies capable of enhancing the activity of Btz and restoring sensitivity to resistant cells, and many of these approaches are positioned for immediate clinical evaluation as they involve existing FDA approved drugs or new molecular entities already in development with established toxicity profiles. Over the years, numerous studies have been conducted on the mechanisms of Btz resistance and scores of molecular-targeted approaches have been evaluated in combination with Btz. Likewise, multiple reviews have been published on the topic. To avoid being duplicative, this review will focus primarily on more recent advances and the treatment options on the horizon for patients with Btz refractory MM. New topics and targets covered include inhibitors of redox regulation, murine double minute 2 (MDM2) inhibitors, and epigenetic modulators like bromodomain and extra terminal domain (BET) inhibitors.

2. MECHANISMS OF BTZ RESISTANCE

The human 26S proteasome is a large (~2.4 MDa) multisubunit protein complex, consisting of a 19S regulatory cap and base and a 20S catalytic core arranged in a cylinder that resembles a stack of rings. The inner two of four stacked rings contain the seven β subunits (β1, β2, β3, etc.), which are the catalytic sites responsible for carrying out the three proteolytic activities of the proteasome. The three enzymatic activities are characterized by their preference for cleaving peptides with specific amino-acid sequence motifs and are named the chymotrypsin-like, trypsin-like, and caspase-like proteolytic activities. For a more in depth review of proteasome structure and function, Adams (2004) and Bhattacharyya, Yu, Mim, and Matouschek (2014) are recommended. The β5 subunit, encoded by the PSMβ5 gene, is the direct molecular binding target of Btz. Binding of Btz to PSMβ5 inhibits the chymotrypsin-like activity of this specific proteasome subunit and is believed to trigger cell death through a host of downstream effects including the inhibition of NFκB signaling via stabilization of IκB, and the activation of multiple stress pathways including the unfolded protein response, endoplasmic reticulum stress, oxidative stress, and the activation of stress signaling kinases like c-Jun N-terminal kinase (JNK; Fig. 1). Cellular models of Btz resistance have shed light on the molecular mechanisms that confer resistance to Btz. Changes in PSMβ5 structure and expression, microenvironmental factors like physical and paracrine interactions with bone stromal cells, and alterations in apoptosis and autophagy signaling are at the core of those changes (Fig. 2). The majority of studies that have investigated these resistance mechanisms have focused on Btz. However, as second- and third-generation PIs activate a common set of downstream pathways and effectors, and in many cases target the same catalytic subunit of the proteasome as Btz, these resistance mechanisms have implications for the use of next-generation PIs as well.

2.1 PSMβ5 Gene Mutations

Gene mutations that alter amino-acid sequences in drug-binding pockets of target proteins are one established mechanism of therapeutic resistance to

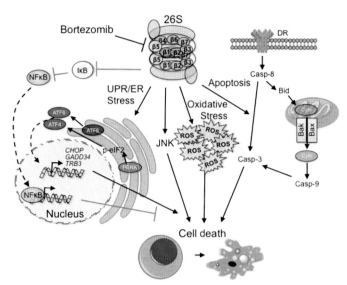

Figure 1 Pleiotropic anti-MM activity of Bortezomib. Inhibition of proteasomal chymotrypsin-like activity by Btz sets in motion a series of events that ultimately lead to the death of MM plasma cells. Btz inhibits NFκB prosurvival signaling by stabilizing the NFκB repressor IκB. Several cellular stress pathways are stimulated, including the unfolded protein response (UPR) and endoplasmic reticulum (ER) stress pathway that culminate in the transcription of proapoptotic genes (i.e., *CHOP, GADD34, TRB3, PUMA, NOXA,* and *BIM*). Btz induces the generation of reactive oxygen species (ROS) resulting in an oxidative shift in cellular redox balance and apoptosis. Stress kinases such as c-Jun N-terminal kinase (JNK) are activated along with the extrinsic and intrinsic apoptosis pathways leading to a cascade of caspase activation, the loss of mitochondrial membrane potential, and release of cytochrome c into the cytosol, further potentiating the activation of the terminal caspase, caspase-3. (See the color plate.)

targeted cancer agents. This was observed in patients with Philadelphia chromosome positive chronic myelogenous leukemia (CML) patients undergoing treatment with c-Abl tyrosine kinase inhibitor imatinib (Gleevec®). Imatinib is highly efficacious in this group of patients due to the expression of a mutant Bcr–Abl fusion protein with constitutive activity. Mutations in the kinase domain of c-Abl near the region of imatinib binding appear in CML patients that have relapsed following chronic imatinib exposure (Branford et al., 2002; Gorre et al., 2001; Roche-Lestienne et al., 2002; Roumiantsev et al., 2002; Shah et al., 2002). It was determined that these point mutations reduce or completely preclude the binding of imatinib to c-Abl. In an analogous situation, mutations in the Btz-binding pocket of PSMβ5 have been identified in MM cell lines following prolonged exposure

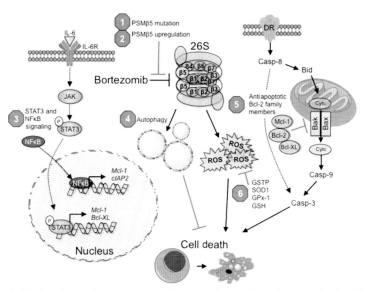

Figure 2 Mechanisms of resistance to Bortezomib. (1) Mutations in the Btz-binding pocket of the PSMβ5 proteasomal subunit appear in cells that have been exposed to Btz for prolonged periods of time. These mutations disrupt Btz binding to PSMβ5 thereby reducing the activity of the drug. (2) Resistant cells upregulate PSMβ5 and other proteasomal subunits as an adaptive response to prolonged Btz treatment. (3) Cellular signaling events such as cytokine (i.e., IL-6) induced or constitutive STAT3 activation and NFκB signaling drive the expression of prosurvival genes, including the antiapoptotic Bcl-2 family member Mcl-1. Mcl-1 is a potent inhibitor of apoptosis known to confer resistance to Btz. (4) Cells may enter into an autophagic state to avoid the cytotoxic effects of Btz. Although autophagy is a catabolic process, it is a protective state under certain conditions and serves as an alternative degradative pathway in the presence of proteasome inhibition. (5) Overexpression of the antiapoptotic Bcl-2 family members, particularly Mcl-1 and Bcl-XL, inhibits the apoptotic pathway and allows cells to escape the cytotoxic effects of Btz. (6) Cellular antioxidants such as GSH neutralize ROS levels to prevent toxicity associated with oxidative stress. The generation of ROS by Btz is critical to the cytotoxic activity of Btz. Redox enzymes such as GSTP, SOD1, and GPx-1 are upregulated in resistant cells, allowing them to neutralize harmful levels of oxidative stress. (See the color plate.)

to Btz (Ri et al., 2010). Similar findings were reported in non-MM cell models of acquired Btz resistance (Lü, Chen, et al., 2008; Lü et al., 2009; Lü, Yang, et al., 2008; Oerlemans et al., 2008; Verbrugge et al., 2013). These mutations also impart cross-resistance to next-generation PIs (Verbrugge et al., 2012). Multiple PSMβ5 mutations have been identified and display varying degrees of resistance (Lü et al., 2009). Mutations, such as the Ala49Thr modification, occur in regions of PSMβ5 that are critical to Btz binding (Groll, Berkers, Ploegh, & Ovaa, 2006). The clinical significance of these mutations has been challenged given that they have not been detected in MM patient samples from patients that have relapsed following Btz treatment. Two published reports failed to detect the same PSMβ5 mutations that were identified in MM cell models in patients, and there was no correlation between patient responsiveness to Btz and PSMβ5 single nucleotide polymorphisms (Lichter et al., 2012; Politou et al., 2006; Ri et al., 2010). The study by Lichter and colleagues reported on sequencing from 16 post-Btz treatment samples, of which three were paired pre- and post-Btz. A potentially confounding variable in this study was that 10 of the 16 post-Btz treatment samples were from patients that were nonresponders to Btz to begin with. However, the fact that none of the mutations identified in cell models were identified in 16 samples casts doubt on the relevance of these mutations in the clinical setting. Additional studies on this topic are needed to boost the statistical power of the clinical data set, and if confirmed, to reconcile the discrepancy between the genetics of Btz resistance in cell models versus patients.

2.2 Upregulation of Proteasomal Subunits

In addition to mutation of the PSMβ5 gene, upregulation of PSMβ5 (wild type and/or mutant) and other proteasomal subunits is associated with Btz resistance. Overexpression of PSMβ5 was detected at the mRNA and protein levels in MM cells that were resistant to Btz and epoxomicin (Balsas, Galán-Malo, Marzo, & Naval, 2012), and upregulation of PSMβ5 along with β1 and β2 subunits and the 11S regulator complex were reported in Btz-resistant MM cell lines (Rückrich et al., 2009). Similar results were reported in cell types of non-MM origin (Lü, Chen, et al., 2008; Lü, Yang, et al., 2008; Oerlemans et al., 2008), although Oerlemans and colleagues did not observe any appreciable upregulation of PSMβ5 at the mRNA level, suggesting a posttranscriptional mechanism. RNAi-mediated repression of PSMβ5 partially restored Btz sensitivity in resistant cells, demonstrating a role as a driver of the resistance phenotype (Oerlemans et al., 2008). Data supporting the existence of this mechanism in Btz refractory patients are limited, but one study confirmed upregulation of PSMβ5 following treatment with a Btz-based regimen (Shuqing, Jianmin, Chongmei, Hui, & Wang, 2011). The precise mechanism by which PSMβ5 upregulation contributes to Btz resistance is not entirely clear, and reports from the literature are somewhat contradictory. Lü, Chen, et al. (2008)

and Lü, Yang, et al. (2008) observed an increase in the cellular chymotrypsin-like proteasome activity in Btz-resistant PSMβ5-overexpressing cells. In this case, it is reasonable to hypothesize that more PSMβ5 and an increase in proteasome activity would necessitate more Btz to inhibit the proteasome to the same degree. However, other studies did not report any changes in chymotrypsin-like proteolytic activity in Btz-resistant cells that overexpress PSMβ5 (Oerlemans et al., 2008). In this study, they did not observe upregulation of other proteasome subunits such as PSMβ1, PSMβ2, or PSMα7, demonstrating that the upregulation of PSMβ5 is not coincident with increased proteasomal density and activity but rather a selective increase in this one particular subunit. They further showed that the increased PSMβ5 proteins did not exist as free subunits in the cytosol but remained in the high molecular weight cellular fraction, suggesting that excess PSMβ5 is associated with the proteasome or forms high molecular weight aggregates. One theoretical role of excess PSMβ5 subunits in mediating resistance is that it acts to scavenge available pools of Btz. However, the existing data from the Oerlemans study do not support this hypothesis as resistant cells showed nearly identical inhibition of chymotrypsin-like proteasome activity when treated with Btz compared to sensitive cells. What also remains controversial is *how* PSMβ5 becomes upregulated. The Oerlemans study concluded that a posttranscriptional mechanism was at play due to no detectable changes in mRNA levels. By comparison, Lu et al. reported an increase in PSMβ5 mRNA, which they concluded was driven by gene amplification as determined by metaphase cytogenetics and fluorescence *in situ* hybridization. Balsas et al. showed that Btz resistance was associated with aneuploidy, a finding that somewhat supports the possibility that changes in PSMβ5 copy number or changes in the chromosomal architecture surrounding the PSMβ5 gene could be the cause of PSMβ5 gene upregulation. These studies present a case that PSMβ5 upregulation at least contributes to Btz resistance, although the mechanistic explanation for that role remains unclear.

2.3 Apoptotic Resistance and Autophagy

The anti-MM activity of Btz is attributed to pleiotropic effects with the induction of programmed cell death/apoptosis being the primary mode of cell death. Early studies into the cytotoxic effects of Btz delineated the sequence of events triggered by Btz leading to MM cell death. Principal among those events was the activation of apoptosis via the intrinsic mitochondrial pathway and/or the extrinsic pathway, which couples death receptors to activation of the apical caspase, caspase-8 (Hideshima et al., 2003; Mitsiades et al., 2002). Treatment of cells with the pan caspase inhibitor ZVAD-FMK partially blocked Btz-induced cell death, demonstrating the contribution of this pathway to the cytotoxic effects of the drug. The importance of apoptotic regulators in modulating Btz sensitivity was further supported in additional studies showing synergy between Btz and apoptotic inducers including tumor necrosis factor-related apoptosis inducing ligand (Mitsiades, Mitsiades, Poulaki, Anderson, & Treon, 2001; Mitsiades, Mitsiades, Poulaki, Chauhan, et al., 2001), and inhibitors of antiapoptotic Bcl-2 family members (Chen et al., 2014; Pei, Dai, & Grant, 2003; Trudel et al., 2007). One of those prosurvival Bcl-2 family members, myeloid cell leukemia-1 (Mcl-1), is a pivotal molecule regulating the sensitivity of MM cells to Btz. Mcl-1 overexpression has been reported as a general trait of MM (Derenne et al., 2002; Wuillème-Toumi et al., 2005), and specifically associated Btz resistance (Balsas et al., 2009; Nencioni et al., 2005). As an important determinant of sensitivity to Btz, Mcl-1 was shown to be cleaved and converted from an antiapoptotic protein of approximately 40 kDa to a proapoptotic 28 kDa form following Btz treatment (Podar et al., 2008). Mcl-1 acts by blocking the activity of the proapoptotic family members, which produce apoptotic signals through the mitochondrial release of cytochrome c and subsequent activation of caspase-9 and caspase-3. To target the prosurvival/antiapoptotic Bcl-2 family members (i.e., Bcl-2, Bcl-XL, and Mcl-1), inhibitors have been designed to block their interaction with proapoptotic members (i.e., Bax, Bak, and Bim). BH3 mimetic Bcl-2 inhibitors include ABT-737, ABT-199, and obatoclax (GX015-070). ABT-737 specifically binds to and inhibits Bcl-2 and Bcl-XL, and ABT-199 targets Bcl-2 only. Both drugs, however, fail to inhibit the activity of Mcl-1, a presumed limitation for the treatment of MM and enhancing the activity of Btz. Despite this limitation, the combination of ABT-737 or ABT-199 with Btz has shown promise in preclinical models of lymphoma (Johnson-Farley, Veliz, Bhagavathi, & Bertino, 2015; Paoluzzi et al., 2008; Touzeau et al., 2011). The fact that similar studies targeting MM are lacking may be an indication that the combination is less active against MM. The characteristically high expression of Mcl-1 in MM cells supports this possibility. Obatoclax, on the other hand, offers activity against all the antiapoptotic family members, including Mcl-1. Obatoclax enhances Btz sensitivity and reverses resistance (Nguyen et al., 2007; Pérez-Galán, Roué, Villamor, Campo, & Colomer, 2007; Pérez-Galán et al., 2008), but these studies too were conducted using non-Hodgkin's lymphoma cell lines. Subsequently, a phase I/II clinical trial of obatoclax and Btz in mantle cell lymphoma (MCL) patients reported disappointing results, calling into question whether the promising preclinical activity of the combination would be translated into clinical benefit. Others have begun to develop specific Mcl-1 inhibitors, including the marinopyrrole drug, maritoclax (Doi et al., 2012), which acts to specifically disrupt the interaction between Mcl-1 and Bim. Future studies evaluating these Mcl-1 targeted drugs in combination with Btz in MM are well justified.

Autophagy is the process by which cells engulf and breakdown organelles and other cytoplasmic components. Although autophagy is a catabolic process, it has been associated with cell survival and resistance to anticancer therapy including Btz. Autophagy as a mechanism of resistance to Btz was first proposed when Btz was found to induce autophagy in MM cells (Hoang, Benavides, Shi, Frost, & Lichtenstein, 2009), and Btz-resistant cells exhibited four times higher levels of autophagy compared to their isogenic sensitive counterparts (Jagannathan, Malek, Vallabhapurapu, Vallabhapurapu, & Driscoll, 2014). Initially, however, it was determined that the use of the autophagy inhibitor 3-methyladenine in combination with Btz was actually antagonistic rather than synergistic (Hoang et al., 2009), suggesting that autophagy contributes to the cytotoxic effects of Btz rather than to resistance. On the other hand, other classes of autophagy inhibitors such as macrolide antibiotics were shown to enhance the activity of Btz (Moriya et al., 2013). Combination of the antimalarial agent and inhibitor of autophagy, hydroxychloroquine, with histone deacetylase (HDAC) inhibitors and the BH3 mimetic, Bcl-2 inhibitor ABT737 together enhanced the activity of Btz dependent on inhibition of protective autophagy (Chen et al., 2014). One clinical trial of hydroxychloroquine combined with Btz in refractory MM patients has been conducted to date (Vogl, Stadtmauer et al., 2014). No conclusions related to efficacy could be made from this single arm study, but responses were observed in 28% of patients. So while autophagy is generally considered a mechanism of resistance to Btz in MM, additional studies are required to determine how this knowledge will be translated into improved therapy for patients. A current limitation may be that hydroxychloroquine and its predecessor, quinacrine, are the only clinically available autophagy inhibitors, and these compounds have autophagy-independent mechanisms of action including the induction of lysosome-mediated apoptosis (Boya et al., 2003; Sui et al., 2013).

3. APPROACHES TO OVERCOMING BTZ RESISTANCE
3.1 Next-Generation Proteasome Inhibitors

The approval and success of Btz has paved the way for the development of second-generation PIs. These new PIs tout improved pharmacology, clinical efficacy, and reduced toxicity as they were designed with increased binding affinity for proteasomal subunits, favorable pharmaceutical properties such as oral bioavailability, and fewer adverse events. New PIs include carfilzomib (Kyprolis®), an epoxomicin derivative and irreversible inhibitor of proteasomal chymotrypsin-like activity; oprozomib (ONX0912), an orally bioavailable derivative of epoxomicin and carfilzomib; ixazomib (MLN9708), an orally bioavailable boronic acid derivative; marizomib (NPI0052; Salinosporamide A), a natural product from the marine bacteria *Salinispora tropica* that binds to PSMβ5 as well as β1 and β2 subunits; and delanzomib (CEP18870), an orally bioavailable and irreversible inhibitor of the proteasomal chymotrypsin-like protease activity. Notably, carfilzomib earned FDA approval for the treatment of refractory MM in 2012 based on phase II data demonstrating a 23.7% response rate, median response duration of 7.8 months, and median overall survival (OS) of 15.6 months (Siegel et al., 2012). The indication for carfilzomib is for patients whose disease has progressed following at least two therapies, including Btz and an immunomodulatory agent (IMiD). The measurable response elicited by carfilzomib in approximately 1/4 of refractory patients suggests that it retains activity in Btz-resistant patients, or at least a portion of them. The major difference in activity between Btz and carfilzomib is that carfilzomib is an irreversible inhibitor of the chymotrypsin-like activity of the 20S proteasome, whereas Btz has reversible binding kinetics. With regard to toxicity, the percentage of patients experiencing peripheral neuropathy, the most common dose-limiting event for Btz, is low with carfilzomib treatment, measured at 13.9% in a cohort of 526 patients from four separate clinical trials, many of whom reported preexisting peripheral neuropathy at baseline (Siegel et al., 2013). Additional studies have shown similar response rates for carfilzomib in multidrug regimens that include lenalidomide and dexamethasone (Stewart et al., 2015). In this trial, roughly 65% of patients were refractory to Btz; however, the specific response rate for the Btz refractory cohort was not separated from the Btz naïve group. Nevertheless, given that the majority of patients enrolled in the study were refractory to Btz, it can be inferred that carfilzomib was effective in at least a portion of patients with Btz-resistant disease. Similar clinical responses were seen with oprozomib, which exhibited measurable but modest activity in Btz refractory patients. In a phase Ib trial, a 14.3% overall response rate (ORR) was observed in Btz refractory patients ($n=7$) and a 25% ORR was observed in phase II ($n=12$). There are other important clinical considerations (i.e., tolerability) of the new drug, but focusing only on response rate, it is clear that while oprozomib provides benefit to a fraction of Btz refractory MM patients, the majority of them remain unresponsive (Vij et al., 2014). These available data for next-generation PIs suggest that resistance mechanism(s) that impact Btz activity may limit the effectiveness of these newer PIs. At the molecular level, this hypothesis is supported by cell models of PI resistance, where cells that have acquired resistance to Btz show cross-resistance to other PIs including carfilzomib and oprozomib (de Wilt et al., 2012; Franke et al., 2012; Stessman et al., 2013, 2014). Therefore, the identification and understanding of Btz resistance mechanisms and strategies to overcome them are not only critical for maximizing the activity of Btz, but will likely benefit next-generation PIs like carfilzomib and oprozomib.

3.2 Redox Signaling

The regulation of reduction and oxidation reactions (i.e., redox) and maintenance of redox homeostasis are critical to the survival and function of all cells. It is particularly important for MM plasma cells, which are naturally specialized for the mass production and secretion of immunoglobulin (Ig) proteins, a process that generates oxidative stress as a by-product. Ig molecules are large multisubunit proteins held together by intra- and interchain disulfide bonds and noncovalent interactions (Liu & May, 2012), and their folding is oxidative by nature. The proper folding of one Ig molecule requires the formation of approximately 100 disulfide bonds. One plasma cell is capable of synthesizing thousands of Ig molecules per second (Shimizu & Hendershot, 2009), meaning that one Ig-producing MM cell may produce roughly 100,000 disulfide bonds per second (Cenci & Sitia, 2007; Hendershot & Sitia, 2005). Molecular oxygen serves as the electron acceptor for each disulfide bond reaction, yielding the production of reactive oxygen species (ROS). To neutralize the increased ROS load associated with Ig synthesis and folding, plasma cell differentiation is accompanied by an intracellular antioxidant response, with the major adaptation being increased synthesis of glutathione (GSH) (Cullinan & Diehl, 2004; Harding et al., 2003), a tripeptide composed of the amino acids cysteine, glutamate, and glycine that is the major endogenous antioxidant of all cells. The thiol functional group of cysteine is critical to the antioxidant properties of GSH, as sulfur is a flexible atom capable of donating electrons to reduce free radicals [i.e., hydrogen peroxide (H_2O_2)] or oxidized proteins, lipids, and nucleic acids. GSH increases the viability and growth of MM cells in culture, and other thiol-containing molecules, such as beta mercaptoethanol, are commonly added to the culture media of plasmacytoma cell lines and antibody-producing hybridomas (de St Groth, 1983; Merten, Keller, Cabanie, Litwin, & Flamand, 1989; Schneider, 1989; Shacter, 1987), further emphasizing the importance of maintaining redox balance in these cell types. Because the natural biology of MM cells as secretory cells predisposes them to high levels of oxidative stress (Cenci & Sitia, 2007), redox signaling is an attractive therapeutic target/pathway for MM.

In addition to being a generally promising therapeutic strategy in the treatment of MM, redox-targeted approaches are also effective Btz-sensitizing agents, capable of restoring sensitivity to resistant cells and enhancing the activity of Btz and other PI therapies. PIs deplete cellular pools of GSH and upregulate the expression of redox enzymes such as glutamate–cysteine ligase, heme oxygenase-1, and GST-pi (Nerini-Molteni, Ferrarini, Cozza, Caligaris-Cappio, & Sitia, 2008; Usami et al., 2005), suggesting that changes in redox-modulating enzymes are an adaptive response to PI therapy. Btz-resistant MM cell lines overexpress important redox-regulating enzymes including copper-zinc superoxide dismutase (CuZnSOD or SOD1), glutathione peroxidase-1 (GPx-1), and GSH (Salem, McCormick, Wendlandt, Zhan, & Goel, 2015). SOD1 and GPx-1 are key antioxidant enzymes involved in scavenging excess levels of ROS by catalyzing reactions that neutralize superoxide anion and H_2O_2, respectively. The ectopic expression of SOD1 leads to Btz resistance in MM cells, confirming that upregulation of this one redox modulator is sufficient to protect cells from Btz-induced cell death (Salem et al., 2015). ROS are generated by Btz treatment in a variety of cancer cell types (Fribley, Zeng, & Wang, 2004; Ling, Liebes, Zou, & Perez-Soler, 2003), and antioxidants such as N-acetyl cysteine (NAC) protect cells from Btz-induced death (Pérez-Galán et al., 2006; Salem et al., 2015; Yu, Rahmani, Dent, & Grant, 2004). Taken together, these studies form a mechanistic link between Btz sensitivity and oxidative stress. They suggest that cells induce a compensatory and protective redox response to block the cytotoxic effects of PIs, and provide rationale for targeting redox signaling as an approach to enhancing the activity of PIs and restoring PI sensitivity to refractory cells. Further evidence of the promise of targeting redox pathways for the treatment of MM comes from studies demonstrating the overexpression of the antioxidant and phase II detoxification enzyme, glutathione S-transferase-pi (GSTP), in >80% of patients with MM and monoclonal gammopathy of undetermined significance (Petrini et al., 1995; Stella et al., 2013). These studies showed that GSTP expression significantly increased following therapy or correlated with therapeutic response to agents that included Btz, suggesting a role in treatment sensitivity/resistance. The GSTP gene is located on the long arm of chromosome 11 (11q13), which is a frequently translocated chromosomal locus in MM due to aberrant and oncogenic Ig heavy-chain gene (IgH, 14q32) translocations. IgH translocations are one of the most common and earliest oncogenic events in MM, and the fact that their breakpoints localize to the locus of an important redox regulatory enzyme with high frequency further implicates the redox pathway in disease pathogenesis. In MCL, a form of non-Hodgkin's lymphoma characterized cytogenetically by the t(11;14) IgH translocation, GSTP expression is highly expressed in histological samples (Bennaceur-Griscelli et al., 2004; Thieblemont et al., 2008), and the inhibition of GSTP enhances the activity of Btz (Rolland, Raharijaona, Barbarat, Houlgatte, & Thieblemont, 2010). Similar examination of GSTP expression in MM clinical samples from patients with and without t(11;14) translocations is needed, but these parallel studies in MCL suggest that GSTP may be a viable molecular drug target for MM and in combination with Btz. GST family members carry out their protective detoxification process via the direct conjugation of GSH to target electrophiles. More recent advances demonstrate that GSTs have broader biological roles unrelated to detoxification. GSTP, for example, associates with and regulates the activity of mitogen-activated protein kinases including the stress signaling kinase c-Jun N-terminal kinase (JNK; Adler et al., 1999; Wang, Arifoglu, Ronai, & Tew, 2001). GSTP catalyzes the conjugation of GSH to protein cysteines (i.e., the process of S-glutathionylation), a posttranslational modification that alters protein function. S-Glutathionylation influences the activity of a variety of proteins involved in diverse cellular processes from the regulation of energy metabolism and calcium homeostasis to signal transduction and redox, indicating the widespread importance of this process (Tew & Townsend, 2011a, 2011b, 2012). With regard to the proteasome, Demasi, Shringarpure, and Davies (2001) were first to show that proteasomal subunits were S-glutathionylated, an effect that specifically affected activity of the chymotrypsin-like protease activity in purified preparations of 20S proteasome extracted from mammalian cells. They concluded from their study that PIs like lactacystin alter global S-glutathionylation levels and specifically enhance S-glutathionylation of the proteasome itself. Additional studies conducted using purified 20S proteasomes from *Saccharomyces cerevisiae* showed that S-glutathionylation of the proteasome was sensitive to redox states and predominantly affected the chymotrypsin-like protease activity relative to trypsin-like and caspase-like activities (Demasi et al., 2013). The specific role of GSH and S-glutathionylation was confirmed as this effect was reversed by enzymes such as glutaredoxin 2 and other oxidoreductases that catalyze deglutathionylation, the reverse reaction of GSTs (Silva et al., 2008). S-Glutathionylation appears to enhance 20S proteasome function by promoting an "open gate" conformation of the structure (Silva et al., 2012), thereby enhancing the proteolytic efficiency of the complex. In contradiction to this finding is that oxidative stress was shown to impair the ATP-dependent activity of the 26S proteasome (Reinheckel, Ullrich, Sitte, & Grune, 2000), and S-glutathionylation of RPN1 and Rpn2, subunits of the 19S regulatory particle of the 26S proteasome, inhibits rather than enhances proteolytic activity of the proteasome (Zmijewski, Banerjee, & Abraham, 2009). These seemingly opposing findings may be reconciled by the fact that oxidative stress is known to disengage the 20S core particle from the 19S regulatory unit, effectively increasing the pool of free 20S proteasome, which are capable of degrading proteins in an ATP-independent manner (Grune et al., 2011; Wang, Yen, Kaiser, & Huang, 2010). It has been proposed that this regulation evolved as an adaptive response to increased oxidative stress, enabling cells to increase their capacity to degrade oxidized proteins (Demasi et al., 2014, 2013). It is clear from this collection of studies that changes in redox and levels of S-glutathionylation have a direct impact on the activity of the proteasome; however, it is not clear if and how this affects the activity of Btz, or if this process contributes to the resistance phenotype in MM. Future studies should address these questions.

Several redox-targeted agents are in preclinical development or are actively used in the clinic for the treatment of cancer (Tew & Townsend, 2011a, 2011b). GST-targeted agents include the GSTP1 inhibitor TLK199 (ezatiostat, Telintra), the GSTP-activated prodrug TLK286 (canfosfamide, Telcyta), nitric oxide (NO) generating prodrugs like JS-K, which has demonstrated promising preclinical activity in MM models and synergized with Btz (Kiziltepe et al., 2007), PABA/NO, mimetics of oxidized GSH (GSSG) like NOV-002, and the metal chelator disulfiram (Antabuse®), which was originally developed in the 1950s to treat alcoholism. Other classes of pro-oxidant chemotherapeutic agents include thiol reactives like arsenic trioxide (As_2O_3; a.k.a. ATO), which has been approved for clinical use in the treatment of acute promyelocytic leukemia (Wang & Chen, 2008). The combination of ATO and Btz was shown to be synergistic in MM and other hematological cancer cell lines (Campbell et al., 2007; Canestraro et al., 2010; Jung, Chen, & McCarty, 2012; Wen et al., 2010; Yan et al., 2007). A phase I/II study combining ATO, Btz, and ascorbic acid in heavily pretreated MM patients showed good tolerability and preliminary signs of efficacy (Berenson et al., 2007), whereas other trials showed no added benefit of combining ATO with Btz (Sharma et al., 2012). Therefore, the benefit of ATO in the treatment of MM and as a enhancer of Btz activity is not strongly supported, although the statistical power of the data sample size has been questioned (He et al., 2014). Alternative strategies for targeting redox to overcome Btz resistance include the inhibition of mucin 1 C-terminal subunit using a novel cell penetrating peptide inhibitor of MUC1-C GO-203 (Yin, Kufe, Avigan, & Kufe, 2014). GO-203 was shown to deplete GSH levels and induce ROS through a mechanism involving downregulation of the p53-inducible regulator of glycolysis and apoptosis (TIGAR).

3.3 MDM2 Inhibitors

MDM2 and its human ortholog, HDM2, are E3 ubiquitin ligases best known for their roles in regulating the stability and activity of p53. Given the critical tumor suppressor function of p53, dubbed the "guardian of the genome" (Lane, 1992), the MDM2–p53 regulatory axis is widely accepted as a promising target for cancer drug development (Chène, 2003). MDM2 directly interacts with p53 and marks it for proteasomal degradation via ligation of ubiquitin tags. Various forms of cellular stress disrupt the MDM2:p53 interaction resulting in the stabilization and derepression of p53. The stabilization of p53 is followed by posttranslational modifications and downstream binding and transactivation of target genes that are involved in a host of cellular processes including cell cycle arrest and apoptosis (El-Deiry, 1998; Meek & Anderson, 2009; Vousden & Prives, 2009). *In lieu* of a physiological stressor that disrupts MDM2 repression of p53 naturally, pharmacological approaches have been devised to interfere with this interaction with the goal of artificially activating p53 for cancer therapy. A number of small-molecule drugs have been designed with high affinity for the p53-binding pocket of MDM2 and act to displace p53 leading to its stabilization and increased transcriptional activity. In addition to being a general approach to cancer therapy, there have been several reports that MDM2 inhibitors are potent Btz sensitizers in MM. Saha and colleagues observed enhanced activity of Btz in MM cell lines and primary patient plasma cells that were co-treated with Nutlin3a (Saha et al., 2010), the first in a class of potent and specific MDM2 inhibitors (Vassilev et al., 2004). With similar results, work by others (Ooi et al., 2009) found that the combination of Nutlin3a and sublethal concentrations of Btz was effective against Btz-sensitive MM cells as well as a variety of epithelial tumor types. Our group showed that in MM cells with acquired resistance to PIs, including Btz, MDM2 inhibition is effective molecular strategy for restoring Btz sensitivity (Stessman et al., 2014). Nutlin3a was also shown to be a augment the activity of Btz in models of MCL, a form of non-Hodgkin's lymphoma that shares cytogenetic anomalies, such as the t(11;14) IgH translocation, with MM (Jin et al., 2010; Tabe et al., 2009). In addition to Nutlins, second-generation MDM2 inhibitors, such as the small-molecule MI63, enhance the activity of Btz in MM cells (Gu et al., 2014). There are now several classes of MDM2 inhibitors in development, and human trials combining them with Btz in refractory MM patients will determine the clinical utility of this approach.

The molecular mechanism(s) that underlie the synergy between MDM2 inhibition and Btz treatment appear to be multifactorial. Given that MDM2 regulates p53 stability through the ubiquitin–proteasome pathway, it is intuitive that the combination of these agents would converge mechanistically to generate a robust anti-MM effect. In support of this theory, early studies investigating the anti-MM activity of Btz showed that p53 upregulation and phosphorylation at the Ser15 residue were initiated by Btz treatment (Hideshima et al., 2003). The Ser15 modification on p53 disrupts the interaction with MDM2 (Shieh, Ikeda, Taya, & Prives, 1997), mimicking the pharmacological activity of MDM2. These results were further supported by Saha et al. (2010) who reported Btz-induced upregulation of the p53 target genes, p21/WAF1, MDM2, PUMA, and Bax, effects that were synergistically enhanced by the cotreatment with Nutlin3a. The activity of Nutlin3a and Btz is most profound in wild-type p53-expressing cells. This is due to the fact that the effects of MDM2 are mitigated in cells that have lost wild-type p53 function, as the stabilization of a p53 protein that lacks functionality would fail to evoke downstream transcriptional events that are essential to the activity of p53 pathway activation. This is potentially a general limitation to the class of MDM2 inhibitors due to the high prevalence of somatic p53 mutations in human cancer (Baker et al., 1989; Hollstein, Sidransky, Vogelstein, & Harris, 1991). Many of these mutations carry loss of function, interfering with the ability of p53 to bind consensus DNA-binding sites and induce transcription of target genes. There are, however, reports that the combination of MDM2 inhibitor and Btz is effective in p53-deficient cells, implicating p53-independent mechanisms. Our group showed synergy between Nutlin3a and Btz and especially with carfilzomib in mutant p53-expressing U266 cells (Stessman et al., 2014). Two- to threefold higher concentrations of Nutlin3a were required to bring out the same effect that was observed in wild-type p53-expressing cells, so it is important to note that p53 mutation, while not an excluding factor, was a limiting factor. Similar results were reported in p53 mutant MCL cells through a mechanism involving posttranscriptional upregulation of the proapoptotic effector NOXA (Jin et al., 2010; Tabe et al., 2009). In MM patients, mutations in p53 are significantly less frequent compared to other tumor types. Genomic studies have shown that p53 mutations are rare in MM, being observed in only 3% newly diagnosed patients (Chng et al., 2007; Preudhomme et al., 1992). In addition to point mutations in the p53 gene that lead to inactivating amino-acid substitutions, complete loss of one or both p53 alleles through chromosomal deletions is also observed in cancer. In MM, deletion of the chromosomal arm where p53 is located (17p) is detected in approximately 10% of newly diagnosed patients and is an indicator of very poor prognosis (Boyd et al., 2011; Chen, Tai, et al., 2012; Chen, Qi, Saha, & Chang, 2012; Fonseca et al., 2003; Lodé et al., 2010). Interestingly, it was shown that in the cohort of patients with 17p deletions, the remaining allele of p53 was prone to mutation with 37% of patients presenting with mutations (Lodé et al., 2010). By comparison, no p53 mutations were detected in patients with an intact chromosome 17p, an observation that has been confirmed by others (Chng et al., 2007). So it seems that loss of p53 function by point mutation or loss of chromosome 17p is a relatively rare event in MM, affecting less than approximately 10% of patients. Thus, p53 deficiency is not likely to be a limiting factor in MM patients, making the use of MDM2 inhibitors as combination therapies with Btz and other PIs a promising approach.

3.4 Il-6/STAT3 Signaling Axis

Early studies investigating the role of MM autocrine and paracrine growth factor and cytokine signaling identified interleukin-6 (IL-6) as a potent inducer of MM plasma cell growth and survival in cell culture (Kawano et al., 1988; Klein et al., 1989). Signal transducer and activator of transcription 3 (STAT3) signaling is a critical effector pathway downstream of interleukin-6 receptor (IL-6R) activation (Lütticken et al., 1994; Wegenka, Buschmann, Lütticken, Heinrich, & Horn, 1993; Zhong, Wen, & Darnell, 1994). In MM cells, STAT3 functions in both an IL-6-dependent and -independent manner, and constitutive STAT3 activation has been associated with oncogenesis and the protection from apoptosis (Bharti et al., 2004; Bromberg et al., 1999; Catlett-Falcone et al., 1999; Dalton & Jove, 1999). STAT3 signaling has been implicated in resistance to several MM therapies (Alas & Bonavida, 2003), and studies have correlated increased STAT3 expression and signaling with Btz resistance. Enforced expression of the CKS1B gene, a gene mapping to the short arm of chromosome 1 (1q21), led to induction of the STAT3 phosphorylation and resistance to Btz (Shi et al., 2010). STAT3 was connected to Btz responsiveness in other studies where inhibition of IL-6 signaling using the IL-6-targeted monoclonal antibody CNTO328 (siltuximab) abrogated STAT3 activity and enhanced Btz sensitivity in MM cells (Voorhees et al., 2007). Despite promising preclinical results and the strong molecular rationale for targeting IL-6 and the IL-6 receptor in combination with Btz, clinical trials conducted to date have not demonstrated overwhelming benefit for combining the two agents. A recent phase I study of single agent siltuximab, an IL-6-targeted monoclonal antibody (MAb), in Japan showed good tolerability and activity in refractory MM patients (Suzuki et al., 2015), although a phase II, double-blind, placebo-controlled trial of siltuximab in combination with Btz showed no significant improvement in progression-free survival (PFS) or OS compared to siltuximab plus placebo (Orlowski et al., 2015). Similar results were observed when siltuximab was added to a multidrug regimen that included Btz. In this study, too, no improvements in clinical outcomes in MM patients were seen (San-Miguel, Bladé, et al., 2014). Other possible approaches include the use of anti-IL-6R targeted mAbs as opposed to blocking the function of the soluble cytokine. Tocilizumab, originally named myeloma receptor antibody due to its promise as an MM therapeutic, is one such molecule. However, there are no published results of clinical trials conducted with tocilizumab in MM, alone or in combination with Btz. The IL-6/IL-6R signaling axis is just one pathway that activates STAT3 in MM cells. An alternative strategy is to disrupt STAT3 signaling through the use of the multikinase inhibitor sorafenib (Nexavar®), a Raf kinase inhibitor that has multiple molecular targets and multifactorial antitumor effects in cells including the inhibition of STAT3 signaling (Ramakrishnan et al., 2010). The inhibition of STAT3 by sorafenib is independent of Raf kinase inhibition, which was determined by Chen et al. (2011) who synthesized sorafenib derivatives that lacked binding affinity for the Raf kinase domain but retained the capacity to inhibit STAT3. Inhibition of STAT3 by sorafenib and derivatives was proposed to occur via activation of the Src homology protein tyrosine phosphatase SHP-1 (Chen, Tai, et al., 2012; Chen, Qi, et al., 2012), which inhibits STAT3 phosphorylation. Numerous preclinical studies have demonstrated a promising anti-MM activity of sorafenib, both as a single agent and in combination with Btz (Kharaziha et al., 2012; Ramakrishnan et al., 2010; Udi et al., 2013). The most consistently observed molecular event triggered by sorafenib in these studies was an inhibition of STAT3 phosphorylation levels and concomitant downregulation of Mcl-1, which is a critical target gene of STAT3 and a known inhibitor of Btz-induced cell death in MM cells (Bhattacharya, Ray, & Johnson, 2005; Carpenter & Lo, 2014; Puthier, Bataille, & Amiot, 1999). Two clinical trials have evaluated sorafenib in MM patients. The first was a phase I study of sorafenib and Btz in patients with advanced malignancies (Kumar et al., 2013). Only 1 of the 14 enrolled patients had MM, and efficacy was not evaluated as an end point in the study, but the regimen was well tolerated. The second trial was a phase II in refractory MM that evaluated sorafenib as a monotherapy (Srkalovic et al., 2014). No responses were detected, results that may discourage additional studies evaluating the activity of sorafenib and Btz. Other strategies to target STAT3 in MM include Janus kinase inhibitors (Li et al., 2010; Monaghan, Khong, Burns, & Spencer, 2011; Ramakrishnan et al., 2010; Scuto et al., 2011), STAT3 peptidomimetics (Turkson et al., 2004), and STAT3-targeted antisense oligonucleotides (Hong et al., 2013).

3.5 Therapeutic Monoclonal Antibodies

The anti-CD38 MAb, daratumumab (JNJ54767414, HuMax® CD38), was recently granted Breakthrough Therapy Designation by the FDA for MM that is refractory to a PI and IMiD. CD38 is a cell surface glycoprotein with cyclic ADP ribose hydrolase activity, but its biological roles are just beginning to be understood. CD38 expressed at high levels in malignant lymphoid tumor cells and especially in MM plasma cells (Lin, Owens, Tricot, &

Wilson, 2004), whereas the majority of normal resting lymphocytes and pluripotent hematopoietic progenitor cells do not express CD38. A predominant path by which daratumumab kills MM cells is through antibody-dependent cell-mediated cytotoxicity (ADCC) and complement-dependent cytotoxicity. Daratumumab also has direct effects on MM cells and was shown to enhance the activity of existing MM therapies including Btz (van der Veer et al., 2011). Ongoing clinical studies combining daratumumab with Btz will determine the utility of this combination in Btz-resistant patients (NCT02136134, NCT02195479, and NCT01998971) although it is clear from the clinical data that daratumumab has significant activity when administered as a single agent in this patient population (Laubach, Tai, Richardson, & Anderson, 2014).

The anti-CS1 MAb elotuzumab is another promising new agent in the treatment of MM, displaying positive data in clinical studies, both alone and in combination with approved drugs like Btz. Cs1 is a cell surface protein belonging to the Ig superfamily. It was first found to be overexpressed in malignant plasma cells compared to normal plasma cells (Hsi et al., 2008), making it a logical target for MM therapy. The anti-MM activity of elotuzumab is attributed to NK cell-dependent ADCC, direct effects on MM cell survival and proliferation, and by blocking the adhesion of MM cells to BMSCs (Collins et al., 2013; Hsi et al., 2008; Tai et al., 2008). A synergistic interaction between Btz and elotuzumab (formerly HuLuc63), as Btz enhanced the ADCC killing of MM cells *in vitro* and enhanced the anti-MM response in preclinical mouse models (van Rhee et al., 2009). In a phase I trial of Elotuzumab and Btz in refractory MM patients, the combination showed a remarkable response rate of 48% and in two of three patients that were refractory to Btz (Jakubowiak et al., 2012). Other promising antibody-based therapeutics for the treatment of MM include antibody–drug conjugates such as the CD138-targeted antibody BT062 (Indatuximab ravtansine), the anti-B cell maturation antigen-targeted antibody GSK2857916 (Tai et al., 2014), and the aforementioned IL-6/IL-6R-targeted siltuximab and tocilizumab.

3.6 Bromodomain and Other Epigenetic Targets

BET family members (BRD2, BRD3, BRD4, and BRDT) are an exciting new class of epigenetic drug targets. These proteins facilitate the initiation and elongation phases of transcription by binding to activated chromatin at acetylated lysine residues. The recognition of activated chromatin by these so-called epigenetic "readers" promotes the recruitment of the RNA polymerase II complex to sites of active transcription. Bromodomain inhibitors, such as JQ1 (Filippakopoulos et al., 2010), were shown to repress the expression and function of c-Myc, which is one of the most dysregulated oncogenes in MM (Affer et al., 2014; Kuehl & Bergsagel, 2012; Shou et al., 2000). The BRD4 bromodomain was found to occupy regions of regulatory DNA termed super enhancers due to the large size and number of bound transcription factors compared to normal gene enhancers. Super enhancers associate with genes that are critical to MM pathogenesis including the aforementioned c-Myc, IRF4, PRDM1, and XBP-1 (Lovén et al., 2013). As a consequence of this, disruption of BRD4 binding to super enhancers by JQ1 is active against MM cell lines and patient plasma cells (Delmore et al., 2011). Evidence supporting the use of bromodomain inhibitors in combination with Btz comes from a clinical study showing that newly diagnosed MM patients with c-Myc gene abnormalities were more likely to develop resistance to Btz plus dexamethasone therapy and exhibited a significantly shorter PFS (Sekiguchi et al., 2014). These findings suggest that c-Myc activity contributes to Btz resistance, providing rationale for the use of bromodomain inhibitors as a strategy to block c-Myc activity. However, a consensus on the extent and precise role of c-Myc in mediating responsiveness to Btz and PI therapy is debatable. The oncogenic role of c-Myc is well accepted, but depending on the context, c-Myc can act as a proapoptotic signal (Fuhrmann et al., 1999). This was shown in the response to Btz, where c-Myc was shown to regulate NOXA-induced apoptosis following Btz treatment (Nikiforov et al., 2007), and to be a key determinant in Btz-induced apoptosis in MM cells (Chen et al., 2010; Nawrocki et al., 2008). Based on these studies, the combination of a bromodomain inhibitor and Btz would be antagonistic rather than synergistic. A caveat to that conclusion is that while c-Myc expression and activity are highly sensitive to treatment with bromodomain inhibitors, c-Myc is not their sole target. BET proteins are global regulators of gene transcription and BET inhibitors affect the recruitment of basal transcriptional machinery to a large set of genes. In fact, MYC-independent molecular signatures in response to the quinolone BET inhibitor I-BET151 have been reported (Chaidos et al., 2014). Two studies support the use of a bromodomain inhibitor in combination with Btz for the treatment of Btz refractory MM. A recent study demonstrated synergistic interaction between Btz and the bromodomain inhibitor CPI203 (Siegel et al., 2014, ASH abstract 4702), and the combination of JQ1 and Btz was more active than either agent alone in serially transplanted Btz-resistant cells from the Vk*MYC transgenic mouse model of MM (Chesi et al., 2012). Additional studies should further evaluate the potential of a bromodomain/Btz combination using MM models of resistance in order to establish rationale for the combination in prospective clinical trials in Btz refractory MM patients.

HDAC enzymes are another class of epigenetic modulator with established potential as anticancer therapeutic targets. HDACs negatively regulate the acetylation of lysine residues on histone tails to alter chromatin structure and ultimately gene transcription. Acetylated histones are generally associated with a less coiled chromatin structure and increased rates of transcription; therefore, HDAC inhibitors, which promote histone acetylation, act by affecting global transcription in tumors cells and impacting on a variety of genes and pathways that are important for cell survival, proliferation, apoptosis, differentiation, and metabolism, among others. HDAC inhibitors are potent anti-MM agents and significantly enhance the effects of Btz. For example, the pan HDAC inhibitors vorinostat and panobinostat synergized with Btz in cell culture and animal models of MM (Chesi et al., 2012; Hideshima, Richardson, & Anderson, 2011; Maiso et al., 2006; Pei, Dai, & Grant, 2004; Stessman et al., 2013). An original phase I clinical trial of vorinostat and bortezomib in relapsed patients showed promise including partial responses achieved in three of nine patients that were refractory to Btz (Badros et al., 2009). Similar signs of efficacy were reported by other groups in phase I trials (Weber et al., 2012), although the results of subsequent double-blind, placebo-controlled studies showed that the addition of vorinostat to Btz only modestly improved PFS in a large randomized cohort of patients (Dimopoulos et al., 2013). Furthermore, there was no significant improvement in PFS in the vorinostat versus control group in patients that had received prior PI therapy. Slightly more positive results were reported for the combination of panobinostat and bortezomib in a phase Ib study (San-Miguel et al., 2013) and a large, multicenter, placebo-controlled study of panobinostat, dexamethasone, and Btz versus placebo, dexamethasone and Btz (San-Miguel, Hungria, et al., 2014). Based on an approximated 4-month improvement in PFS, the FDA recently granted accelerated approval of panobinostat. More recent advances include the development of HDAC6 isoform-specific inhibitors, such as ACY-1215. HDAC6 has been shown to regulate the formation and function of aggresomes, which are cellular structures that degrade and clear polyubiquitinated proteins as an alternative pathway to the proteasome. The combination of HDAC6 gene knockdown or treatment with the HDAC6 inhibitors tubacin and ACY-1215 with Btz was found to be synergistic in preclinical models of MM

(Hideshima et al., 2005; Santo et al., 2012). Clinical trials combining ACY-1215 and Btz are now in progress and preliminary results show good tolerability and evidence of responses in Btz refractory patients (Raje et al., 2012 poster 4061; Vogl, Raje, et al., 2014 poster 4764). Figure 3 provides an overview of the various classes of drugs with potential as Btz sensitizing agents.

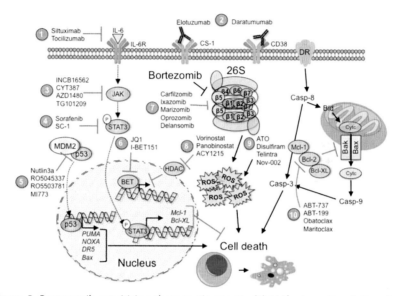

Figure 3 Bortezomib sensitizing therapeutic agents. (1) MAbs targeting IL-6 or IL-6R suppress IL-6 signaling and STAT3 activation, an important signaling network in MM cells. (2) MAbs elotuzumab and daratumumab targeting novel cell surface antigens such as CS-1 and CD38, respectively, have shown promise in patients with Btz refractory MM and have the potential to enhance the activity of Btz. (3) STAT3 is an important prosurvival signal in MM. Strategies to inhibit STAT3 include small-molecule inhibitors of the Janus kinase (JAK), the upstream activator of STAT3, or other means such as the use of (4) sorafenib and its derivatives. (5) MDM2 inhibitors are potent Btz sensitizers that restore Btz sensitivity to resistant cells. (6) BET bromodomain inhibitors are promising new drugs that disrupt the recruitment of transcriptional machinery to super enhancers that regulate the expression of MM oncogenes. (7) Next-generation proteasome inhibitors exhibit clinical activity in a portion of Btz refractory patients. (8) HDAC inhibitors synergize with Btz in preclinical models, although the clinical activity of pan HDAC inhibitors (i.e., vorinostat and panobinostat) has been limited. The HDAC6-selective inhibitor ACY-1215 is currently in clinical trials. (9) Increasing evidence suggests that alterations in redox signaling are key contributors to the Btz resistance phenotype, making redox-modulating agents' prime candidates for trials in Btz refractory patients. (10) Antiapoptotic Bcl-2 family members confer apoptotic resistance and reduce the cytotoxic effects of Btz. BH3 mimetic inhibitors, particularly those that block the activity of Mcl-1, are promising agents for enhancing/restoring the apoptotic effects of Btz treatment. (See the color plate.)

4. CONCLUDING REMARKS

Btz was a revolutionary advancement in the treatment of MM and remains a cornerstone of MM therapy today. A limitation to Btz is that the depth and duration of response vary between patients, and all patients ultimately stop responding due to the emergence of treatment resistance. Studies have shed light on the molecular mechanisms that drive acquired resistance to Btz, establishing the rationale for new targeted therapeutic approaches to be used in combination with Btz. Next-generation PIs retain activity in a portion of Btz refractory patients. However, in those patients that are nonresponders, the strategies that enhance the activity of Btz may be exploited with similar benefit, as Btz and next-generation PIs have the same molecular target and downstream effector pathways. There has been a surge of new agents for the treatment of MM over the past 10 years, providing more treatment options for MM patients than ever before. Currently, there are 1907 clinical studies registered with clinicaltrials.gov for MM, and 183 of those trials incorporate Btz in refractory patients. The expanding preclinical literature on molecular mechanisms of resistance and new targets in Btz-resistant MM will serve as rationale to guide these and future trial designs. Lastly, this area of research has the potential to deliver biomarkers and predictive signatures of response to Btz and PI therapy that will personalize treatment decisions and guide patient selection for new trials.

ACKNOWLEDGMENTS

N.G.D. receives research funding from (1) the American Cancer Society as an ACS Research Scholar (RSG 14-156-01-CDD), (2) as a member of the South Carolina Center of Biomedical Research Excellence (COBRE) in Oxidants, Redox Balance, and Stress Signaling (P20GM103542), and (3) from institutional start-up funds provided by the Medical University of South Carolina and the MUSC Hollings Cancer Center.

REFERENCES

Adams, J. (2004). The proteasome: A suitable antineoplastic target. *Nature Reviews. Cancer, 4*, 349–360.

Adler, V., Yin, Z., Fuchs, S. Y., Benezra, M., Rosario, L., Tew, K. D., et al. (1999). Regulation of JNK signaling by GSTp. *The EMBO Journal, 18*, 1321–1334.

Affer, M., Chesi, M., Chen, W. D., Keats, J. J., Demchenko, Y. N., Tamizhmani, K., et al. (2014). Promiscuous MYC locus rearrangements hijack enhancers but mostly super-enhancers to dysregulate MYC expression in multiple myeloma. *Leukemia, 28*, 1725–1735.

Alas, S., & Bonavida, B. (2003). Inhibition of constitutive STAT3 activity sensitizes resistant non-Hodgkin's lymphoma and multiple myeloma to chemotherapeutic drug-mediated apoptosis. *Clinical Cancer Research*, 9, 316–326.

Allen, S. (2007). *The Velcade story*. Retrieved from: http://www.boston.com/business/healthcare/articles/(2007)/05/06/the_velcade_story/?page=full.

Badros, A., Burger, A. M., Philip, S., Niesvizky, R., Kolla, S. S., Goloubeva, O., et al. (2009). Phase I study of vorinostat in combination with bortezomib for relapsed and refractory multiple myeloma. *Clinical Cancer Research*, 15, 5250–5257.

Baker, S. J., Fearon, E. R., Nigro, J. M., Hamilton, S. R., Preisinger, A. C., Jessup, J. M., et al. (1989). Chromosome 17 deletions and p53 gene mutations in colorectal carcinomas. *Science*, 244, 217–221.

Balsas, P., Galán-Malo, P., Marzo, I., & Naval, J. (2012). Bortezomib resistance in a myeloma cell line is associated to PSMβ5 overexpression and polyploidy. *Leukemia Research*, 36, 212–218.

Balsas, P., López-Royuela, N., Galán-Malo, P., Anel, A., Marzo, I., & Naval, J. (2009). Cooperation between Apo2L/TRAIL and bortezomib in multiple myeloma apoptosis. *Biochemical Pharmacology*, 77, 804–812.

Bennaceur-Griscelli, A., Bosq, J., Koscielny, S., Lefrère, F., Turhan, A., Brousse, N., et al. (2004). High level of glutathione-S-transferase pi expression in mantle cell lymphomas. *Clinical Cancer Research*, 10, 3029–3034.

Berenson, J. R., Matous, J., Swift, R. A., Mapes, R., Morrison, B., & Yeh, H. S. (2007). A phase I/II study of arsenic trioxide/bortezomib/ascorbic acid combination therapy for the treatment of relapsed or refractory multiple myeloma. *Clinical Cancer Research*, 13, 1762–1768.

Bergsagel, P. L. (2014). Where we were, where we are, where we are going: Progress in multiple myeloma. *American Society of Clinical Oncology Educational Book/ASCO*, 199–203.

Bergsagel, D. E., Sprague, C. C., Austin, C., & Griffith, K. M. (1962). Evaluation of new chemotherapeutic agents in the treatment of multiple myeloma. IV. L-Phenylalanine mustard (NSC-8806). *Cancer Chemotherapy Reports*, 21, 87–99.

Bharti, A. C., Shishodia, S., Reuben, J. M., Weber, D., Alexanian, R., Raj-Vadhan, S., et al. (2004). Nuclear factor-kappaB and STAT3 are constitutively active in CD138+ cells derived from multiple myeloma patients, and suppression of these transcription factors leads to apoptosis. *Blood*, 103, 3175–3184.

Bhattacharya, S., Ray, R. M., & Johnson, L. R. (2005). STAT3-mediated transcription of Bcl-2, Mcl-1 and c-IAP2 prevents apoptosis in polyamine-depleted cells. *The Biochemical Journal*, 392, 335–344.

Bhattacharyya, S., Yu, H., Mim, C., & Matouschek, A. (2014). Regulated protein turnover: Snapshots of the proteasome in action. *Nature Reviews. Molecular Cell Biology*, 15, 122–133.

Boya, P., Gonzalez-Polo, R. A., Poncet, D., Andreau, K., Vieira, H. L., Roumier, T., et al. (2003). Mitochondrial membrane permeabilization is a critical step of lysosome-initiated apoptosis induced by hydroxychloroquine. *Oncogene*, 22, 3927–3936.

Boyd, K. D., Ross, F. M., Tapper, W. J., Chiecchio, L., Dagrada, G., Konn, Z. J., et al. (2011). The clinical impact and molecular biology of del(17p) in multiple myeloma treated with conventional or thalidomide-based therapy. *Genes, Chromosomes & Cancer*, 50, 765–774.

Branford, S., Rudzki, Z., Walsh, S., Grigg, A., Arthur, C., Taylor, K., et al. (2002). High frequency of point mutations clustered within the adenosine triphosphate-binding region of BCR/ABL in patients with chronic myeloid leukemia or Ph-positive acute lymphoblastic leukemia who develop imatinib (STI571) resistance. *Blood*, 99, 3472–3475.

Bromberg, J. F., Wrzeszczynska, M. H., Devgan, G., Zhao, Y., Pestell, R. G., Albanese, C., et al. (1999). Stat3 as an oncogene. *Cell, 98*, 295–303.

Campbell, R. A., Sanchez, E., Steinberg, J. A., Baritaki, S., Gordon, M., Wang, C., et al. (2007). Antimyeloma effects of arsenic trioxide are enhanced by melphalan, bortezomib and ascorbic acid. *British Journal of Haematology, 138*, 467–478.

Canestraro, M., Galimberti, S., Savli, H., Palumbo, G. A., Tibullo, D., Nagy, B., et al. (2010). Synergistic antiproliferative effect of arsenic trioxide combined with bortezomib in HL60 cell line and primary blasts from patients affected by myeloproliferative disorders. *Cancer Genetics and Cytogenetics, 199*, 110–120.

Carpenter, R. L., & Lo, H. W. (2014). STAT3 target genes relevant to human cancers. *Cancers, 6*, 897–925.

Catlett-Falcone, R., Landowski, T. H., Oshiro, M. M., Turkson, J., Levitzki, A., Savino, R., et al. (1999). Constitutive activation of Stat3 signaling confers resistance to apoptosis in human U266 myeloma cells. *Immunity, 10*, 105–115.

Cenci, S., & Sitia, R. (2007). Managing and exploiting stress in the antibody factory. *FEBS Letters, 581*, 3652–3657.

Chaidos, A., Caputo, V., Gouvedenou, K., Liu, B., Marigo, I., Chaudhry, M. S., et al. (2014). Potent antimyeloma activity of the novel bromodomain inhibitors I-BET151 and I-BET762. *Blood, 123*, 697–705.

Chen, S., Blank, J. L., Peters, T., Liu, X. J., Rappoli, D. M., Pickard, M. D., et al. (2010). Genome-wide siRNA screen for modulators of cell death induced by proteasome inhibitor bortezomib. *Cancer Research, 70*, 4318–4326.

Chen, M. H., Qi, C. X. Y., Saha, M. N., & Chang, H. (2012). p53 nuclear expression correlates with hemizygous TP53 deletion and predicts an adverse outcome for patients with relapsed/refractory multiple myeloma treated with lenalidomide. *American Journal of Clinical Pathology, 137*, 208–212.

Chen, K. F., Tai, W. T., Hsu, C. Y., Huang, J. W., Liu, C. Y., Chen, P. J., et al. (2012). Blockade of STAT3 activation by sorafenib derivatives through enhancing SHP-1 phosphatase activity. *European Journal of Medicinal Chemistry, 55*, 220–227.

Chen, K. F., Tai, W. T., Huang, J. W., Hsu, C. Y., Chen, W. L., Cheng, A. L., et al. (2011). Sorafenib derivatives induce apoptosis through inhibition of STAT3 independent of Raf. *European Journal of Medicinal Chemistry, 46*, 2845–2851.

Chen, S., Zhang, Y., Zhou, L., Leng, Y., Lin, H., Kmieciak, M., et al. (2014). A Bim-targeting strategy overcomes adaptive bortezomib resistance in myeloma through a novel link between autophagy and apoptosis. *Blood, 124*, 2687–2697.

Chène, P. (2003). Inhibiting the p53-MDM2 interaction: An important target for cancer therapy. *Nature Reviews. Cancer, 3*, 102–109.

Chesi, M., Matthews, G. M., Garbitt, V. M., Palmer, S. E., Shortt, J., Lefebure, M., et al. (2012). Drug response in a genetically engineered mouse model of multiple myeloma is predictive of clinical efficacy. *Blood, 120*, 376–385.

Chng, W. J., Price-Troska, T., Gonzalez-Paz, N., Van Wier, S., Jacobus, S., Blood, E., et al. (2007). Clinical significance of TP53 mutation in myeloma. *Leukemia, 21*, 582–584.

Collins, S. M., Bakan, C. E., Swartzel, G. D., Hofmeister, C. C., Efebera, Y. A., Kwon, H., et al. (2013). Elotuzumab directly enhances NK cell cytotoxicity against myeloma via CS1 ligation: Evidence for augmented NK cell function complementing ADCC. *Cancer Immunology, Immunotherapy, 62*, 1841–1849.

Cullinan, S. B., & Diehl, J. A. (2004). PERK-dependent activation of Nrf2 contributes to redox homeostasis and cell survival following endoplasmic reticulum stress. *The Journal of Biological Chemistry, 279*(2010), 8–17.

Dalton, W. S., & Jove, R. (1999). Drug resistance in multiple myeloma: Approaches to circumvention. *Seminars in Oncology, 26*(5 Suppl. 13), 23–27.

de St Groth, S. F. (1983). Automated production of monoclonal antibodies in a cytostat. *Journal of Immunological Methods, 57*, 121–136.

de Wilt, L. H., Jansen, G., Assaraf, Y. G., van Meerloo, J., Cloos, J., Schimmer, A. D., et al. (2012). Proteasome-based mechanisms of intrinsic and acquired bortezomib resistance in non-small cell lung cancer. *Biochemical Pharmacology, 83*, 207–217.

Delmore, J. E., Issa, G. C., Lemieux, M. E., Rahl, P. B., Shi, J., Jacobs, H. M., et al. (2011). BET bromodomain inhibition as a therapeutic strategy to target c-Myc. *Cell, 146*, 904–917.

Demasi, M., Hand, A., Ohara, E., Oliveira, C. L., Bicev, R. N., Bertoncini, C. A., et al. (2014). 20S proteasome activity is modified via S-glutathionylation based on intracellular redox status of the yeast Saccharomyces cerevisiae: Implications for the degradation of oxidized proteins. *Archives of Biochemistry and Biophysics, 557*, 65–71.

Demasi, M., Netto, L. E., Silva, G. M., Hand, A., de Oliveira, C. L., Bicev, R. N., et al. (2013). Redox regulation of the proteasome via S-glutathionylation. *Redox Biology, 2*, 44–51.

Demasi, M., Shringarpure, R., & Davies, K. J. (2001). Glutathiolation of the proteasome is enhanced by proteolytic inhibitors. *Archives of Biochemistry and Biophysics, 389*, 254–263.

Derenne, S., Monia, B., Dean, N. M., Taylor, J. K., Rapp, M. J., Harousseau, J. L., et al. (2002). Antisense strategy shows that Mcl-1 rather than Bcl-2 or Bcl-x(L) is an essential survival protein of human myeloma cells. *Blood, 100*, 194–199.

Dimopoulos, M., Siegel, D. S., Lonial, S., Qi, J., Hajek, R., Facon, T., et al. (2013). Vorinostat or placebo in combination with bortezomib in patients with multiple myeloma (VANTAGE 088): A multicentre, randomised, double-blind study. *The Lancet Oncology, 14*, 1129–1140.

Doi, K., Li, R., Sung, S. S., Wu, H., Liu, Y., Manieri, W., et al. (2012). Discovery of marinopyrrole A (maritoclax) as a selective Mcl-1 antagonist that overcomes ABT-737 resistance by binding to and targeting Mcl-1 for proteasomal degradation. *The Journal of Biological Chemistry, 287*, 10224–10235.

El-Deiry, W. S. (1998). Regulation of p53 downstream genes. *Seminars in Cancer Biology, 8*, 345–357.

Filippakopoulos, P., Qi, J., Picaud, S., Shen, Y., Smith, W. B., Fedorov, O., et al. (2010). Selective inhibition of BET bromodomains. *Nature, 468*, 1067–1073.

Fonseca, R., Blood, E., Rue, M., Harrington, D., Oken, M. M., Kyle, R. A., et al. (2003). Clinical and biologic implications of recurrent genomic aberrations in myeloma. *Blood, 101*, 4569–4575.

Franke, N. E., Niewerth, D., Assaraf, Y. G., van Meerloo, J., Vojtekova, K., van Zantwijk, C. H., et al. (2012). Impaired bortezomib binding to mutant β5 subunit of the proteasome is the underlying basis for bortezomib resistance in leukemia cells. *Leukemia, 26*, 757–768.

Fribley, A., Zeng, Q., & Wang, C. Y. (2004). Proteasome inhibitor PS-341 induces apoptosis through induction of endoplasmic reticulum stress-reactive oxygen species in head and neck squamous cell carcinoma cells. *Molecular and Cellular Biology, 24*, 9695–9704.

Fuhrmann, G., Rosenberger, G., Grusch, M., Klein, N., Hofmann, J., & Krupitza, G. (1999). The MYC dualism in growth and death. *Mutation Research, 437*, 205–217.

Gorre, M. E., Mohammed, M., Ellwood, K., Hsu, N., Paquette, R., Rao, P. N., et al. (2001). Clinical resistance to STI-571 cancer therapy caused by BCR-ABL gene mutation or amplification. *Science, 293*, 876–880.

Groll, M., Berkers, C. R., Ploegh, H. L., & Ovaa, H. (2006). Crystal structure of the boronic acid-based proteasome inhibitor bortezomib in complex with the yeast 20S proteasome. *Structure, 14*, 451–456.

Grune, T., Catalgol, B., Licht, A., Ermak, G., Pickering, A. M., Ngo, J. K., et al. (2011). HSP70 mediates dissociation and reassociation of the 26S proteasome during adaptation to oxidative stress. *Free Radical Biology & Medicine, 51*, 1355–1364.

Gu, D., Wang, S., Kuiatse, I., Wang, H., He, J., Dai, Y., et al. (2014). Inhibition of the MDM2 E3 ligase induces apoptosis and autophagy in wild-type and mutant p53 models of multiple myeloma, and acts synergistically with ABT-737. *PLoS One, 9*, e103015.

Harding, H. P., Zhang, Y., Zeng, H., Novoa, I., Lu, P. D., Calfon, M., et al. (2003). An integrated stress response regulates amino acid metabolism and resistance to oxidative stress. *Molecular Cell, 11*, 619–633.

He, X., Yang, K., Chen, P., Liu, B., Zhang, Y., Wang, F., et al. (2014). Arsenic trioxide-based therapy in relapsed/refractory multiple myeloma patients: A meta-analysis and systematic review. *OncoTargets and Therapy, 7*, 1593–1599.

Hendershot, L. M., & Sitia, R. (2005). Immunoglobulin assembly and secretion. In T. Honjo, F. W. Alt, & M. Neuberger (Eds.), *Molecular biology of B cells* (pp. 261–273): Amsterdam, The Netherlands: Elsevier Science.

Hideshima, T., Bradner, J. E., Wong, J., Chauhan, D., Richardson, P., Schreiber, S. L., et al. (2005). Small-molecule inhibition of proteasome and aggresome function induces synergistic antitumor activity in multiple myeloma. *Proceedings of the National Academy of Sciences of the United States of America, 102*, 8567–8572.

Hideshima, T., Mitsiades, C., Akiyama, M., Hayashi, T., Chauhan, D., Richardson, P., et al. (2003). Molecular mechanisms mediating antimyeloma activity of proteasome inhibitor PS-341. *Blood, 101*, 1530–1534.

Hideshima, T., Richardson, P. G., & Anderson, K. C. (2011). Mechanism of action of proteasome inhibitors and deacetylase inhibitors and the biological basis of synergy in multiple myeloma. *Molecular Cancer Therapeutics, 10*, 2034–2042.

Hoang, B., Benavides, A., Shi, Y., Frost, P., & Lichtenstein, A. (2009). Effect of autophagy on multiple myeloma cell viability. *Molecular Cancer Therapeutics, 8*, 1974–1984.

Hollstein, M., Sidransky, D., Vogelstein, B., & Harris, C. C. (1991). p53 mutations in human cancers. *Science, 253*, 49–53.

Hong, D. S., Younes, A., Fayad, L., Fowler, N. H., Hagemeister, F. B., Mistry, R., et al. (2013). A phase I study of ISIS 481464 (AZD9150), a first-in-human, first-in-class, antisense oligonucleotide inhibitor of STAT3, in patients with advanced cancers. *Journal of Clinical Oncology, 31*, 8523.

Hsi, E. D., Steinle, R., Balasa, B., Szmania, S., Draksharapu, A., Shum, B. P., et al. (2008). CS1, a potential new therapeutic antibody target for the treatment of multiple myeloma. *Clinical Cancer Research, 14*, 2775–2784.

Jagannath, S., Barlogie, B., Berenson, J., Siegel, D., Irwin, D., Richardson, P. G., et al. (2004). A phase 2 study of two doses of bortezomib in relapsed or refractory myeloma. *British Journal of Haematology, 127*, 165–172.

Jagannathan, S., Malek, E., Vallabhapurapu, S., Vallabhapurapu, S., & Driscoll, J. J. (2014). Bortezomib induces AMPK-dependent autophagosome formation uncoupled from apoptosis in drug resistant cells. *Oncotarget, 5*, 12358–12370.

Jakubowiak, A. J., Benson, D. M., Bensinger, W., Siegel, D. S., Zimmerman, T. M., Mohrbacher, A., et al. (2012). Phase I trial of anti-CS1 monoclonal antibody elotuzumab in combination with bortezomib in the treatment of relapsed/refractory multiple myeloma. *Journal of Clinical Oncology, 30*, 1960–1965.

Jin, L., Tabe, Y., Kojima, K., Zhou, Y., Pittaluga, S., Konopleva, M., et al. (2010). MDM2 antagonist Nutlin-3 enhances bortezomib-mediated mitochondrial apoptosis in TP53-mutated mantle cell lymphoma. *Cancer Letters, 299*, 161–170.

Johnson-Farley, N., 1, Veliz, J., Bhagavathi, S., & Bertino, J. R. (2015). ABT-199, a BH3 mimetic that specifically targets Bcl-2, enhances the antitumor activity of chemotherapy, bortezomib and JQ1 in "double hit" lymphoma cells. *Leukemia & Lymphoma*, 1–7.

Jung, H. J., Chen, Z., & McCarty, N. (2012). Synergistic anticancer effects of arsenic trioxide with bortezomib in mantle cell lymphoma. *American Journal of Hematology, 87*, 1057–1064.

Kawano, M., Hirano, T., Matsuda, T., Taga, T., Horii, Y., Iwato, K., et al. (1988). Autocrine generation and requirement of BSF-2/IL-6 for human multiple myelomas. *Nature, 332*, 83–85.

Kharaziha, P., De Raeve, H., Fristedt, C., Li, Q., Gruber, A., Johnsson, P., et al. (2012). Sorafenib has potent antitumor activity against multiple myeloma in vitro, ex vivo, and in vivo in the 5T33MM mouse model. *Cancer Research, 72*, 5348–5362.

Kiziltepe, T., Hideshima, T., Ishitsuka, K., Ocio, E. M., Raje, N., Catley, L., et al. (2007). JS-K, a GST-activated nitric oxide generator, induces DNA double-strand breaks, activates DNA damage response pathways, and induces apoptosis in vitro and in vivo in human multiple myeloma cells. *Blood, 110*, 709–718.

Klein, B., Zhang, X. G., Jourdan, M., Content, J., Houssiau, F., Aarden, L., et al. (1989). Paracrine rather than autocrine regulation of myeloma-cell growth and differentiation by interleukin-6. *Blood, 73*, 517–526.

Kuehl, W. M., & Bergsagel, P. L. (2012). MYC addiction: A potential therapeutic target in MM. *Blood, 120*, 2351–2352.

Kumar, S. K., Jett, J., Marks, R., Richardson, R., Quevedo, F., Moynihan, T., et al. (2013). Phase 1 study of sorafenib in combination with bortezomib in patients with advanced malignancies. *Investigational New Drugs, 31*, 1201–1206.

Lane, D. (1992). Cancer. p53, guardian of the genome. *Nature, 358*, 15–16.

Laubach, J. P., Tai, Y. T., Richardson, P. G., & Anderson, K. C. (2014). Daratumumab granted breakthrough drug status. *Expert Opinion on Investigational Drugs, 23*, 445–452.

Li, J., Favata, M., Kelley, J. A., Caulder, E., Thomas, B., Wen, X., et al. (2010). INCB16562, a JAK1/2 selective inhibitor, is efficacious against multiple myeloma cells and reverses the protective effects of cytokine and stromal cell support. *Neoplasia, 12*, 28–38.

Lichter, D. I., Danaee, H., Pickard, M. D., Tayber, O., Sintchak, M., Shi, H., et al. (2012). Sequence analysis of β-subunit genes of the 20S proteasome in patients with relapsed multiple myeloma treated with bortezomib or dexamethasone. *Blood, 120*, 4513–4516.

Lin, P., Owens, R., Tricot, G., & Wilson, C. S. (2004). Flow cytometric immunophenotypic analysis of 306 cases of multiple myeloma. *American Journal of Clinical Pathology, 121*, 482–488.

Ling, Y. H., Liebes, L., Zou, Y., & Perez-Soler, R. (2003). Reactive oxygen species generation and mitochondrial dysfunction in the apoptotic response to Bortezomib, a novel proteasome inhibitor, in human H460 non-small cell lung cancer cells. *The Journal of Biological Chemistry, 278*, 33714–33723.

Liu, H., & May, K. (2012). Disulfide bond structures of IgG molecules: Structural variations, chemical modifications and possible impacts to stability and biological function. *MAbs, 4*, 17–23.

Lodé, L., Eveillard, M., Trichet, V., Soussi, T., Wuillème, S., Richebourg, S., et al. (2010). Mutations in TP53 are exclusively associated with del(17p) in multiple myeloma. *Haematologica, 95*, 1973–1976.

Lovén, J., Hoke, H. A., Lin, C. Y., Lau, A., Orlando, D. A., Vakoc, C. R., et al. (2013). Selective inhibition of tumor oncogenes by disruption of super-enhancers. *Cell, 153*, 320–334.

Lü, S., Chen, Z., Yang, J., Chen, L., Gong, S., Zhou, H., et al. (2008). Overexpression of the PSMB5 gene contributes to bortezomib resistance in T-lymphoblastic lymphoma/leukemia cells derived from Jurkat line. *Experimental Hematology, 36*, 1278–1284.

Lü, S., Yang, J., Chen, Z., Gong, S., Zhou, H., Xu, X., et al. (2009). Different mutants of PSMB5 confer varying bortezomib resistance in T lymphoblastic lymphoma/leukemia cells derived from the Jurkat cell line. *Experimental Hematology, 37*, 831–837.

Lü, S., Yang, J., Song, X., Gong, S., Zhou, H., Guo, L., et al. (2008). Point mutation of the proteasome beta5 subunit gene is an important mechanism of bortezomib resistance in bortezomib-selected variants of Jurkat T cell lymphoblastic lymphoma/leukemia line. *The Journal of Pharmacology and Experimental Therapeutics, 326*, 423–431.

Lütticken, C., Wegenka, U. M., Yuan, J., Buschmann, J., Schindler, C., Ziemiecki, A., et al. (1994). Association of transcription factor APRF and protein kinase Jak1 with the interleukin-6 signal transducer gp130. *Science, 263*, 89–92.

Maiso, P., Carvajal-Vergara, X., Ocio, E. M., López-Pérez, R., Mateo, G., Gutiérrez, N., et al. (2006). The histone deacetylase inhibitor LBH589 is a potent antimyeloma agent that overcomes drug resistance. *Cancer Research, 66*, 5781–5789.

Meek, D. W., & Anderson, C. W. (2009). Posttranslational modification of p53: Cooperative integrators of function. *Cold Spring Harbor Perspectives in Biology, 1*, a000950.

Merten, O. W., Keller, H., Cabanie, L., Litwin, J., & Flamand, B. (1989). Development of a serum-free medium for hybridoma fermentor cultures. In R. E. Spier, J. R. Griffiths, J. Stephenné, & P. J. Crooy (Eds.), *Advances in animal cell biology and technology for bioprocesses* (pp. 263–268). United Kingdom: Butterworths.

Mitsiades, N., Mitsiades, C. S., Poulaki, V., Anderson, K. C., & Treon, S. P. (2001). Concepts in the use of TRAIL/Apo2L: An emerging biotherapy for myeloma and other neoplasias. *Expert Opinion on Investigational Drugs, 10*, 1521–1530.

Mitsiades, N., Mitsiades, C. S., Poulaki, V., Chauhan, D., Fanourakis, G., Gu, X., et al. (2002). Molecular sequelae of proteasome inhibition in human multiple myeloma cells. *Proceedings of the National Academy of Sciences of the United States of America, 99*, 14374–14379.

Mitsiades, C. S., Treon, S. P., Mitsiades, N., Shima, Y., Richardson, P., Schlossman, R., et al. (2001). TRAIL/Apo2L ligand selectively induces apoptosis and overcomes drug resistance in multiple myeloma: Therapeutic applications. *Blood, 98*, 795–804.

Monaghan, K. A., Khong, T., Burns, C. J., & Spencer, A. (2011). The novel JAK inhibitor CYT387 suppresses multiple signalling pathways, prevents proliferation and induces apoptosis in phenotypically diverse myeloma cells. *Leukemia, 25*, 1891–1899.

Moriya, S., Che, X. F., Komatsu, S., Abe, A., Kawaguchi, T., Gotoh, A., et al. (2013). Macrolide antibiotics block autophagy flux and sensitize to bortezomib via endoplasmic reticulum stress-mediated CHOP induction in myeloma cells. *International Journal of Oncology, 42*, 1541–1550.

Nawrocki, S. T., Carew, J. S., Maclean, K. H., Courage, J. F., Huang, P., Houghton, J. A., et al. (2008). Myc regulates aggresome formation, the induction of Noxa, and apoptosis in response to the combination of bortezomib and SAHA. *Blood, 112*, 2917–2926.

Nencioni, A., Hua, F., Dillon, C. P., Yokoo, R., Scheiermann, C., Cardone, M. H., et al. (2005). Evidence for a protective role of Mcl-1 in proteasome inhibitor-induced apoptosis. *Blood, 105*, 3255–3262.

Nerini-Molteni, S., Ferrarini, M., Cozza, S., Caligaris-Cappio, F., & Sitia, R. (2008). Redox homeostasis modulates the sensitivity of myeloma cells to bortezomib. *British Journal of Haematology, 141*, 494–503.

Nguyen, M., Marcellus, R. C., Roulston, A., Watson, M., Serfass, L., Murthy Madiraju, S. R., et al. (2007). Small molecule obatoclax (GX15-070) antagonizes MCL-1 and overcomes MCL-1-mediated resistance to apoptosis. *Proceedings of the National Academy of Sciences of the United States of America, 104*, 19512–19517.

Nikiforov, M. A., Riblett, M., Tang, W. H., Gratchouck, V., Zhuang, D., Fernandez, Y., et al. (2007). Tumor cell-selective regulation of NOXA by c-MYC in response to proteasome inhibition. *Proceedings of the National Academy of Sciences of the United States of America, 104*, 19488–19493.

Oerlemans, R., Franke, N. E., Assaraf, Y. G., Cloos, J., van Zantwijk, I., Berkers, C. R., et al. (2008). Molecular basis of bortezomib resistance: Proteasome subunit beta5 (PSMB5) gene mutation and overexpression of PSMB5 protein. *Blood, 112*, 2489–2499.

Ooi, M., Hayden, P. J., Kotoula, V., McMillin, D. W., Charalambous, E., Daskalaki, E., et al. (2009). Interactions of the Hdm2/p53 and proteasome pathways may enhance the antitumor activity of bortezomib. *Clinical Cancer Research, 15*, 7153–7160.

Orlowski, R. Z., Gercheva, L., Williams, C., Sutherland, H., Robak, T., Masszi, T., et al. (2015). A phase 2, randomized, double-blind, placebo-controlled study of siltuximab (anti-IL-6 mAb) and bortezomib versus bortezomib alone in patients with relapsed or refractory multiple myeloma. *American Journal of Hematology, 90*, 42–49.

Orlowski, R. Z., Stinchcombe, T. E., Mitchell, B. S., Shea, T. C., Baldwin, A. S., Stahl, S., et al. (2002). Phase I trial of the proteasome inhibitor PS-341 in patients with refractory hematologic malignancies. *Journal of Clinical Oncology, 20*, 4420–4427.

Paoluzzi, L., Gonen, M., Bhagat, G., Furman, R. R., Gardner, J. R., Scotto, L., et al. (2008). The BH3-only mimetic ABT-737 synergizes the antineoplastic activity of proteasome inhibitors in lymphoid malignancies. *Blood, 112*, 2906–2916.

Pei, X. Y., Dai, Y., & Grant, S. (2003). The proteasome inhibitor bortezomib promotes mitochondrial injury and apoptosis induced by the small molecule Bcl-2 inhibitor HA14-1 in multiple myeloma cells. *Leukemia, 17*, 2036–2045.

Pei, X. Y., Dai, Y., & Grant, S. (2004). Synergistic induction of oxidative injury and apoptosis in human multiple myeloma cells by the proteasome inhibitor bortezomib and histone deacetylase inhibitors. *Clinical Cancer Research, 10*, 3839–3852.

Pérez-Galán, P., Roué, G., López-Guerra, M., Nguyen, M., Villamor, N., Montserrat, E., et al. (2008). BCL-2 phosphorylation modulates sensitivity to the BH3 mimetic GX15-070 (Obatoclax) and reduces its synergistic interaction with bortezomib in chronic lymphocytic leukemia cells. *Leukemia, 22*, 1712–1720.

Pérez-Galán, P., Roué, G., Villamor, N., Campo, E., & Colomer, D. (2007). The BH3-mimetic GX15-070 synergizes with bortezomib in mantle cell lymphoma by enhancing Noxa-mediated activation of Bak. *Blood, 109*, 4441–4449.

Pérez-Galán, P., Roué, G., Villamor, N., Montserrat, E., Campo, E., & Colomer, D. (2006). The proteasome inhibitor bortezomib induces apoptosis in mantle-cell lymphoma through generation of ROS and Noxa activation independent of p53 status. *Blood, 107*, 257–264.

Petrini, M., Di Simone, D., Favati, A., Mattii, L., Valentini, P., & Grassi, B. (1995). GST-pi and P-170 co-expression in multiple myeloma. *British Journal of Haematology, 90*, 393–397.

Podar, K., Gouill, S. L., Zhang, J., Opferman, J. T., Zorn, E., Tai, Y. T., et al. (2008). A pivotal role for Mcl-1 in Bortezomib-induced apoptosis. *Oncogene, 27*, 721–731.

Politou, M., Karadimitris, A., Terpos, E., Kotsianidis, I., Apperley, J. F., & Rahemtulla, A. (2006). No evidence of mutations of the PSMB5 (beta-5 subunit of proteasome) in a case of myeloma with clinical resistance to Bortezomib. *Leukemia Research, 30*, 240–241.

Preudhomme, C., Facon, T., Zandecki, M., Vanrumbeke, M., Laï, J. L., Nataf, E., et al. (1992). Rare occurrence of P53 gene mutations in multiple myeloma. *British Journal of Haematology, 81*, 440–443.

Puthier, D., Bataille, R., & Amiot, M. (1999). IL-6 up-regulates mcl-1 in human myeloma cells through JAK/STAT rather than ras/MAP kinase pathway. *European Journal of Immunology, 29*, 3945–3950.

Raje, N., Hari, P. N., Vogl, D. T., Jagannath, S., Orlowski, R. Z., Supko, J. G., et al. (2012). Rocilinostat (ACY-1215), a selective HDAC6 inhibitor, alone and in combination with bortezomib in multiple myeloma: Preliminary results from the first-in-humans phase I/II study. In *ASH annual meeting*, Abstract 4061.

Ramakrishnan, V., Timm, M., Haug, J. L., Kimlinger, T. K., Wellik, L. E., Witzig, T. E., et al. (2010). Sorafenib, a dual Raf kinase/vascular endothelial growth factor receptor inhibitor has significant anti-myeloma activity and synergizes with common anti-myeloma drugs. *Oncogene, 29*, 1190–1202.

Reinheckel, T., Ullrich, O., Sitte, N., & Grune, T. (2000). Differential impairment of 20S and 26S proteasome activities in human hematopoietic K562 cells during oxidative stress. *Archives of Biochemistry and Biophysics, 377*, 65–68.

Ri, M., Iida, S., Nakashima, T., Miyazaki, H., Mori, F., Ito, A., et al. (2010). Bortezomib-resistant myeloma cell lines: A role for mutated PSMB5 in preventing the accumulation of unfolded proteins and fatal ER stress. *Leukemia, 24*, 1506–1512.

Richardson, P. G., Barlogie, B., Berenson, J., Singhal, S., Jagannath, S., Irwin, D., et al. (2003). A phase 2 study of bortezomib in relapsed, refractory myeloma. *The New England Journal of Medicine, 348*, 2609–2617.

Roche-Lestienne, C., Soenen-Cornu, V., Grardel-Duflos, N., Laï, J. L., Philippe, N., Facon, T., et al. (2002). Several types of mutations of the Abl gene can be found in chronic myeloid leukemia patients resistant to STI571, and they can pre-exist to the onset of treatment. *Blood, 100*, 1014–1018.

Rolland, D., Raharijaona, M., Barbarat, A., Houlgatte, R., & Thieblemont, C. (2010). Inhibition of GST-pi nuclear transfer increases mantle cell lymphoma sensitivity to cisplatin, cytarabine, gemcitabine, bortezomib and doxorubicin. *Anticancer Research, 30*, 3951–3957.

Roumiantsev, S., Shah, N. P., Gorre, M. E., Nicoll, J., Brasher, B. B., Sawyers, C. L., et al. (2002). Clinical resistance to the kinase inhibitor STI-571 in chronic myeloid leukemia by mutation of Tyr-253 in the Abl kinase domain P-loop. *Proceedings of the National Academy of Sciences of the United States of America, 99*, 10700–10705.

Rückrich, T., Kraus, M., Gogel, J., Beck, A., Ovaa, H., Verdoes, M., et al. (2009). Characterization of the ubiquitin-proteasome system in bortezomib-adapted cells. *Leukemia, 23*, 1098–1105.

Saha, M., Jiang, H., Jayakar, J., Reece, D., Branch, D. R., & Chang, H. (2010). MDM2 antagonist nutlin plus proteasome inhibitor velcade combination displays a synergistic anti-myeloma activity. *Cancer Biology & Therapy, 9*, 936–944.

Salem, K., McCormick, M. L., Wendlandt, E., Zhan, F., & Goel, A. (2015). Copper-zinc superoxide dismutase-mediated redox regulation of bortezomib resistance in multiple myeloma. *Redox Biology, 4C*, 23–33.

Sanchez-Serrano, I. (2005). Translational research in the development of bortezomib: A core model. *Discovery Medicine, 5*, 527–533.

Sánchez-Serrano, I. (2006). Success in translational research: Lessons from the development of bortezomib. *Nature Reviews. Drug Discovery, 5*, 107–114.

San-Miguel, J., Bladé, J., Shpilberg, O., Grosicki, S., Maloisel, F., Min, C. K., et al. (2014). Phase 2 randomized study of bortezomib-melphalan-prednisone with or without siltuximab (anti-IL-6) in multiple myeloma. *Blood, 123*, 4136–4142.

San-Miguel, J. F., Hungria, V. T., Yoon, S. S., Beksac, M., Dimopoulos, M. A., Elghandour, A., et al. (2014). Panobinostat plus bortezomib and dexamethasone versus placebo plus bortezomib and dexamethasone in patients with relapsed or relapsed and refractory multiple myeloma: A multicentre, randomised, double-blind phase 3 trial. *The Lancet Oncology, 15*, 1195–1206.

San-Miguel, J. F., Richardson, P. G., Günther, A., Sezer, O., Siegel, D., Bladé, J., et al. (2013). Phase Ib study of panobinostat and bortezomib in relapsed or relapsed and refractory multiple myeloma. *Journal of Clinical Oncology, 31*, 3696–3703.

Santo, L., Hideshima, T., Kung, A. L., Tseng, J. C., Tamang, D., Yang, M., et al. (2012). Preclinical activity, pharmacodynamic, and pharmacokinetic properties of a selective HDAC6 inhibitor, ACY-1215, in combination with bortezomib in multiple myeloma. *Blood, 119*, 2579–2589.

Schneider, Y. J. (1989). Optimization of hybridoma growth and monoclonal antibody secretion in a chemically defined, serum- and protein-free culture medium. *Journal of Immunological Methods, 116*, 65–77.

Scuto, A., Krejci, P., Popplewell, L., Wu, J., Wang, Y., Kujawski, M., et al. (2011). The novel JAK inhibitor AZD1480 blocks STAT3 and FGFR3 signaling, resulting in suppression of human myeloma cell growth and survival. *Leukemia, 25*, 538–550.

Sekiguchi, N., Ootsubo, K., Wagatsuma, M., Midorikawa, K., Nagata, A., Noto, S., et al. (2014). The impact of C-Myc gene-related aberrations in newly diagnosed myeloma with bortezomib/dexamethasone therapy. *International Journal of Hematology, 99*, 288–295.

Shacter, E. (1987). Serum-free medium for growth factor-dependent and -independent plasmacytomas and hybridomas. *Journal of Immunological Methods, 99*, 259–270.

Shah, N. P., Nicoll, J. M., Nagar, B., Gorre, M. E., Paquette, R. L., Kuriyan, J., et al. (2002). Multiple BCR-ABL kinase domain mutations confer polyclonal resistance to the tyrosine kinase inhibitor imatinib (STI571) in chronic phase and blast crisis chronic myeloid leukemia. *Cancer Cell, 2*, 117–125.

Sharma, M., Khan, H., Thall, P. F., Orlowski, R. Z., Bassett, R. L., Jr., Shah, N., et al. (2012). A randomized phase 2 trial of a preparative regimen of bortezomib, high-dose melphalan, arsenic trioxide, and ascorbic acid. *Cancer, 118*, 2507–2515.

Shi, L., Wang, S., Zangari, M., Xu, H., Cao, T. M., Xu, C., et al. (2010). Over-expression of CKS1B activates both MEK/ERK and JAK/STAT3 signaling pathways and promotes myeloma cell drug-resistance. *Oncotarget, 1*, 22–33.

Shieh, S. Y., 1, Ikeda, M., Taya, Y., & Prives, C. (1997). DNA damage-induced phosphorylation of p53 alleviates inhibition by MDM2. *Cell, 91*, 325–334.

Shimizu, Y., & Hendershot, L. M. (2009). Oxidative folding: Cellular strategies for dealing with the resultant equimolar production of reactive oxygen species. *Antioxidants & Redox Signaling, 11*, 2317–2331.

Shou, Y., Martelli, M. L., Gabrea, A., Qi, Y., Brents, L. A., Roschke, A., et al. (2000). Diverse karyotypic abnormalities of the c-myc locus associated with c-myc dysregulation and tumor progression in multiple myeloma. *Proceedings of the National Academy of Sciences of the United States of America, 97*, 228–233.

Shuqing, L., Jianmin, Y., Chongmei, H., Hui, C., & Wang, J. (2011). Upregulated expression of the PSMB5 gene may contribute to drug resistance in patient with multiple myeloma when treated with bortezomib-based regimen. *Experimental Hematology, 39*, 1117–1118.

Siegel, M. B., Davare, M. A., Liu, S. Q., Spurgeon, S. E., Loriaux, M. M., Druker, B. J., et al. (2014). 4702 the bromodomain inhibitor CPI203 demonstrates preclinical synergistic activity with bortezomib in drug resistant myeloma. In *ASH annual meeting*, Abstract 4702.

Siegel, D., Martin, T., Nooka, A., Harvey, R. D., Vij, R., Niesvizky, R., et al. (2013). Integrated safety profile of single-agent carfilzomib: Experience from 526 patients enrolled in 4 phase II clinical studies. *Haematologica, 98*, 1753–1761.

Siegel, D. S., Martin, T., Wang, M., Vij, R., Jakubowiak, A. J., Lonial, S., et al. (2012). A phase 2 study of single-agent carfilzomib (PX-171-003-A1) in patients with relapsed and refractory multiple myeloma. *Blood, 120*, 2817–2825.

Silva, G. M., Netto, L. E., Discola, K. F., Piassa-Filho, G. M., Pimenta, D. C., Bárcena, J. A., et al. (2008). Role of glutaredoxin 2 and cytosolic thioredoxins in cysteinyl-based redox modification of the 20S proteasome. *The FEBS Journal, 275*, 2942–2955.

Silva, G. M., Netto, L. E., Simões, V., Santos, L. F., Gozzo, F. C., Demasi, M. A., et al. (2012). Redox control of 20S proteasome gating. *Antioxidants & Redox Signaling, 16*, 1183–1194.

Srkalovic, G., Hussein, M. A., Hoering, A., Zonder, J. A., Popplewell, L. L., Trivedi, H., et al. (2014). A phase II trial of BAY 43-9006 (sorafenib) (NSC-724772) in patients with relapsing and resistant multiple myeloma: SWOG S0434. *Cancer Medicine, 3*, 1275–1283.

Stella, F., Weich, N., Panero, J., Fantl, D. B., Schutz, N., Fundia, A. F., et al. (2013). Glutathione S-transferase P1 mRNA expression in plasma cell disorders and its correlation with polymorphic variants and clinical outcome. *Cancer Epidemiology, 37*, 671–674.

Stessman, H. A., Baughn, L. B., Sarver, A., Xia, T., Deshpande, R., Mansoor, A., et al. (2013). Profiling bortezomib resistance identifies secondary therapies in a mouse myeloma model. *Molecular Cancer Therapeutics, 12,* 1140–1150.

Stessman, H., Lulla, A., Xia, T., Mitra, A., Harding, T., Mansoor, A., et al. (2014). High-throughput drug screening identifies compounds and molecular strategies for targeting proteasome inhibitor-resistant multiple myeloma. *Leukemia, 28,* 2263–2267.

Stewart, A. K., Rajkumar, S. V., Dimopoulos, M. A., Masszi, T., Špička, I., Oriol, A., et al. (2015). Carfilzomib, lenalidomide, and dexamethasone for relapsed multiple myeloma. *The New England Journal of Medicine, 372,* 142–152.

Sui, X., Chen, R., Wang, Z., Huang, Z., Kong, N., Zhang, M., et al. (2013). Autophagy and chemotherapy resistance: A promising therapeutic target for cancer treatment. *Cell Death & Disease, 4,* e838.

Suzuki, K., Ogura, M., Abe, Y., Suzuki, T., Tobinai, K., Ando, K., et al. (2015). Phase 1 study in Japan of siltuximab, an anti-IL-6 monoclonal antibody, in relapsed/refractory multiple myeloma. *International Journal of Hematology, 101*(3), 286–294.

Tabe, Y., Sebasigari, D., Jin, L., Rudelius, M., Davies-Hill, T., Miyake, K., et al. (2009). MDM2 antagonist nutlin-3 displays antiproliferative and proapoptotic activity in mantle cell lymphoma. *Clinical Cancer Research, 15,* 933–942.

Tai, Y. T., Dillon, M., Song, W., Leiba, M., Li, X. F., Burger, P., et al. (2008). Anti-CS1 humanized monoclonal antibody HuLuc63 inhibits myeloma cell adhesion and induces antibody-dependent cellular cytotoxicity in the bone marrow milieu. *Blood, 112,* 1329–1337.

Tai, Y. T., Mayes, P. A., Acharya, C., Zhong, M. Y., Cea, M., Cagnetta, A., et al. (2014). Novel anti-B-cell maturation antigen antibody-drug conjugate (GSK2857916) selectively induces killing of multiple myeloma. *Blood, 123,* 3128–3138.

Tew, K. D., & Townsend, D. M. (2011a). Regulatory functions of glutathione S-transferase P1-1 unrelated to detoxification. *Drug Metabolism Reviews, 43,* 179–193.

Tew, K. D., & Townsend, D. M. (2011b). Redox platforms in cancer drug discovery and development. *Current Opinion in Chemical Biology, 15,* 156–161.

Tew, K. D., & Townsend, D. M. (2012). Glutathione-s-transferases as determinants of cell survival and death. *Antioxidants & Redox Signaling, 17,* 1728–1737.

Thieblemont, C., Rolland, D., Baseggio, L., Felman, P., Gazzo, S., Callet-Bauchu, E., et al. (2008). Comprehensive analysis of GST-pi expression in B-cell lymphomas: Correlation with histological subtypes and survival. *Leukemia & Lymphoma, 49,* 1403–1406.

Touzeau, C., Dousset, C., Bodet, L., Gomez-Bougie, P., Bonnaud, S., Moreau, A., et al. (2011). ABT-737 induces apoptosis in mantle cell lymphoma cells with a Bcl-2high/Mcl-1low profile and synergizes with other antineoplastic agents. *Clinical Cancer Research, 17,* 5973–5981.

Trudel, S., Li, Z. H., Rauw, J., Tiedemann, R. E., Wen, X. Y., & Stewart, A. K. (2007). Preclinical studies of the pan-Bcl inhibitor obatoclax (GX015-070) in multiple myeloma. *Blood, 109,* 5430–5438.

Turkson, J., Kim, J. S., Zhang, S., Yuan, J., Huang, M., Glenn, M., et al. (2004). Novel peptidomimetic inhibitors of signal transducer and activator of transcription 3 dimerization and biological activity. *Molecular Cancer Therapeutics, 3,* 261–269.

Udi, J., Schüler, J., Wider, D., Ihorst, G., Catusse, J., Waldschmidt, J., et al. (2013). Potent in vitro and in vivo activity of sorafenib in multiple myeloma: Induction of cell death, CD138-downregulation and inhibition of migration through actin depolymerization. *British Journal of Haematology, 161,* 104–116.

Usami, H., Kusano, Y., Kumagai, T., Osada, S., Itoh, K., Ko- bayashi, A., et al. (2005). Selective induction of the tumor marker glutathione S-transferase P1 by proteasome inhibitors. *The Journal of Biological Chemistry, 280,* 25267–25276.

van der Veer, M. S., de Weers, M., van Kessel, B., Bakker, J. M., Wittebol, S., Parren, P. W., et al. (2011). The therapeutic human CD38 antibody daratumumab improves the antimyeloma effect of newly emerging multi-drug therapies. *Blood Cancer Journal, 1,* e41.

van Rhee, F., Szmania, S. M., Dillon, M., van Abbema, A. M., Li, X., Stone, M. K., et al. (2009). Combinatorial efficacy of anti-CS1 monoclonal antibody elotuzumab (HuLuc63) and bortezomib against multiple myeloma. *Molecular Cancer Therapeutics, 8,* 2616–2624.

Vassilev, L. T., Vu, B. T., Graves, B., Carvajal, D., Podlaski, F., Filipovic, Z., et al. (2004). In vivo activation of the p53 pathway by small-molecule antagonists of MDM2. *Science, 303,* 844–848.

Verbrugge, S. E., Al, M., Assaraf, Y. G., Niewerth, D., van Meerloo, J., Cloos, J., et al. (2013). Overcoming bortezomib resistance in human B cells by anti-CD20/rituximab-mediated complement-dependent cytotoxicity and epoxyketone-based irreversible proteasome inhibitors. *Experimental Hematology & Oncology, 2,* 2.

Verbrugge, S. E., Assaraf, Y. G., Dijkmans, B. A., Scheffer, G. L., Al, M., den Uyl, D., et al. (2012). Inactivating PSMB5 mutations and P-glycoprotein (multidrug resistance-associated protein/ATP-binding cassette B1) mediate resistance to proteasome inhibitors: Ex vivo efficacy of (immuno)proteasome inhibitors in mononuclear blood cells from patients with rheumatoid arthritis. *The Journal of Pharmacology and Experimental Therapeutics, 341,* 174–182.

Vij, R., Savona, M., Siegel, D. S., Kaufman, J. L., Badros, A., Ghobrial, I. M., et al. (2014). Clinical profile of single-agent oprozomib in patients (Pts) with multiple myeloma (MM): Updated results from a multicenter, open-label, dose escalation phase 1b/2 study [abstract]. *Blood, 124,* 34.

Vogl, D. T., Raje, N., Hari, P., Jones, S. S., Supko, J. G., Leone, G., et al. (2014). Phase 1B results of ricolinostat (ACY-1215) combination therapy with bortezomib and dexamethasone in patients with relapsed or relapsed and refractory multiple myeloma (MM). In *ASH annual meeting*, Abstract 4764.

Vogl, D. T., Stadtmauer, E. A., Tan, K. S., Heitjan, D. F., Davis, L. E., Pontiggia, L., et al. (2014). Combined autophagy and proteasome inhibition: A phase 1 trial of hydroxychloroquine and bortezomib in patients with relapsed/refractory myeloma. *Autophagy, 10,* 1380–1390.

Voorhees, P. M., Chen, Q., Kuhn, D. J., Small, G. W., Hunsucker, S. A., Strader, J. S., et al. (2007). Inhibition of interleukin-6 signaling with CNTO 328 enhances the activity of bortezomib in preclinical models of multiple myeloma. *Clinical Cancer Research, 13,* 6469–6478.

Vousden, K. H., & Prives, C. (2009). Blinded by the light: The growing complexity of p53. *Cell, 137,* 413–431.

Wang, T., Arifoglu, P., Ronai, Z., & Tew, K. D. (2001). Glutathione S-transferase P1-1 (GSTP1-1) inhibits c-Jun N-terminal kinase (JNK1) signaling through interaction with the C terminus. *The Journal of Biological Chemistry, 276,* 20999–21003.

Wang, Z. Y., & Chen, Z. (2008). Acute promyelocytic leukemia: From highly fatal to highly curable. *Blood, 111,* 2505–2515.

Wang, X., Yen, J., Kaiser, P., & Huang, L. (2010). Regulation of the 26S proteasome complex during oxidative stress. *Science Signaling, 3,* ra88.

Weber, D. M., Graef, T., Hussein, M., Sobecks, R. M., Schiller, G. J., Lupinacci, L., et al. (2012). Phase I trial of vorinostat combined with bortezomib for the treatment of relapsing and/or refractory multiple myeloma. *Clinical Lymphoma, Myeloma & Leukemia, 12,* 319–324.

Wegenka, U. M., Buschmann, J., Lütticken, C., Heinrich, P. C., & Horn, F. (1993). Acute-phase response factor, a nuclear factor binding to acute-phase response elements, is rapidly activated by interleukin-6 at the posttranslational level. *Molecular and Cellular Biology, 13,* 276–288.

Wen, J., Feng, Y., Huang, W., Chen, H., Liao, B., Rice, L., et al. (2010). Enhanced antimyeloma cytotoxicity by the combination of arsenic trioxide and bortezomib is further potentiated by p38 MAPK inhibition. *Leukemia Research, 34*, 85–92.

Wuillème-Toumi, S., Robillard, N., Gomez, P., Moreau, P., Le Gouill, S., Avet-Loiseau, H., et al. (2005). Mcl-1 is overexpressed in multiple myeloma and associated with relapse and shorter survival. *Leukemia, 19*, 1248–1252.

Yan, H., Wang, Y. C., Li, D., Wang, Y., Liu, W., Wu, Y. L., et al. (2007). Arsenic trioxide and proteasome inhibitor bortezomib synergistically induce apoptosis in leukemic cells: The role of protein kinase Cdelta. *Leukemia, 21*, 1488–1495.

Yin, L., Kufe, T., Avigan, D., & Kufe, D. (2014). Targeting MUC1-C is synergistic with bortezomib in downregulating TIGAR and inducing ROS-mediated myeloma cell death. *Blood, 123*, 2997–3006.

Yu, C., Rahmani, M., Dent, P., & Grant, S. (2004). The hierarchical relationship between MAPK signaling and ROS generation in human leukemia cells undergoing apoptosis in response to the proteasome inhibitor Bortezomib. *Experimental Cell Research, 295*, 555–566.

Zhong, Z., Wen, Z., & Darnell, J. E., Jr. (1994). Stat3: A STAT family member activated by tyrosine phosphorylation in response to epidermal growth factor and interleukin-6. *Science, 264*, 95–98.

Zmijewski, J. W., Banerjee, S., & Abraham, E. (2009). S-glutathionylation of the Rpn2 regulatory subunit inhibits 26 S proteasomal function. *The Journal of Biological Chemistry, 284*, 22213–22221.

CHAPTER SIX

Influence of Bone Marrow Microenvironment on Leukemic Stem Cells: Breaking Up an Intimate Relationship

Puneet Agarwal, Ravi Bhatia[1]

Division of Hematology-Oncology, Department of Medicine, University of Alabama Birmingham, Birmingham, Alabama, USA
[1]Corresponding author: e-mail address: rbhatia@uabmc.edu

Contents

1. Introduction — 228
2. Hematopoietic Stem Cells — 229
3. Leukemia Stem Cells — 230
4. Bone Marrow Microenvironment — 231
    4.1 Cellular Components of the BMM — 232
5. Microenvironment Leukemia Crosstalk — 236
    5.1 Critical Signaling Pathways Altered in Leukemia — 237
6. *In Vivo* Models to Study BMM and Leukemia Interactions — 241
7. Potential Therapeutic Avenue: Targeting the Seed and Soil? — 242
8. Conclusions and Future Directions — 243
References — 245

Abstract

The bone marrow microenvironment (BMM) plays a critical role in hematopoietic stem cells (HSCs) maintenance and regulation. There is increasing interest in the role of the BMM in promoting leukemia stem cell (LSC) maintenance, resistance to conventional chemotherapy and targeted therapies, and ultimately disease relapse. Recent studies have enhanced our understanding of how the BMM regulates quiescence, self-renewal, and differentiation of LSC. In this comprehensive review, we discuss recent advances in our understanding of the crosstalk between the BMM and LSC, and the critical signaling pathways underlying these interactions. We also discuss potential approaches to exploit these observations to create novel strategies for targeting therapy-resistant LSC to achieve relapse-free survival in leukemic patients.

1. INTRODUCTION

Hematopoietic stem cells (HSCs) are multipotent stem cells that have the ability to self-renew as well as differentiate into multiple lineages; thus, giving rise to all mature blood cells. Starting from embryonic development, hematopoiesis takes place in the placenta, fetal liver, and then eventually in the bone marrow, which is the site for hematopoiesis in the adult (Tavian & Peault, 2005). The process of hematopoiesis is very dynamic and tightly regulated to maintain steady-state levels of the blood cells in circulation throughout the life of an individual. An exquisite balance between self-renewal, proliferation and differentiation of HSCs allows for long-term maintenance of primitive HSC at the apex of the hematopoietic hierarchy, despite the daily production of billions of differentiated blood cells (Reya, Morrison, Clarke, & Weissman, 2001). It is now clear that the HSC homeostasis is controlled by the combination of cell intrinsic and extrinsic mechanisms from nonhematopoietic cells present in the bone marrow microenvironment (BMM). With the development of variety of new mouse models, and improved imaging technology, the nature of the HSC niche within the BMM has been increasingly well characterized, and several cell types, including sinusoidal and arteriolar vascular endothelial cells, subendothelial cells, and osteoblastic cells, contributing to HSC regulation have been identified. In addition, these and other mouse models have allowed the regulatory mechanisms by which the BMM influences HSC fate to be better understood.

As with normal hematopoiesis, several leukemias, including chronic myelogenous leukemia (CML) and acute myeloid leukemia (AML) are maintained by a pool of leukemia stem cells (LSCs) within the bone marrow. Analogous to HSC, LSC possess the ability to remain quiescent, self-renew, and differentiate (Warner, Wang, Hope, Jin, & Dick, 2004). There is evidence that LSC can resist elimination by anticancer treatments and are responsible for therapy resistance and disease relapse (Essers & Trumpp, 2010). There is increasing evidence that LSC is also regulated by the cellular and molecular components of the microenvironment, and that microenvironmental interactions can protect LSC from conventional and targeted therapy. In addition, leukemic-induced alterations in the BMM may lead to altered hematopoietic regulatory function, and may confer a competitive growth advantage to LSCs over normal HSCs. Therefore, a clear understanding of the role of BMM in the pathogenesis of leukemia is of considerable clinical significance, since understanding mechanistic cues from the BMM responsible for LSC maintenance could provide alternative strategies to improve their elimination and enhance survival of patients. Among the topics discussed in this chapter are (a) advances in understanding the complex nature of the BMM, (b) the role of the BMM in influencing LSC survival and disease maintenance, (c) the key microenvironmental signaling pathways that are deregulated in leukemia, (d) the utility of murine models to study this bidirectional crosstalk between leukemia cells and the BMM, and (e) the potential of niche-targeting strategies in combination with conventional therapies as a new paradigm for therapy of leukemia.

2. HEMATOPOIETIC STEM CELLS

HSC make an excellent model system for studying adult stem cells (Weissman, 2000) since significant progress has been made to homogenously purify adult murine HSC. In fact, a single long-term HSC prospectively isolated from the bone marrow can reconstitute the entire hematopoietic system of a lethally irradiated mouse (Kiel et al., 2005; Osawa, Hanada, Hamada, & Nakauchi, 1996; Wagers, Sherwood, Christensen, & Weissman, 2002). Adult hematopoiesis is preferentially located to the bone marrow compartment, where it follows a strict differentiation program to give rise to all the mature blood cells present in the immune system and peripheral blood.

The physical association of HSC and BM was first noted when researchers observed that HSC localized close to the endosteal bone surface (Lord, Testa, & Hendry, 1975). Subsequent studies suggested that HSC resided in distinct physiological niches, including the endosteal niche and vascular niche (Perry & Li, 2007). The endosteal niche includes osteoblasts (bone-forming cells) forming the inner lining of the bone cavity, and the vascular niche includes sinusoidal endothelial cells lining the blood vessels. It has been suggested that the endosteum is involved in the regulation of quiescence or slow cycling of HSC over long periods of time, whereas HSC in the vascular niche may be more actively dividing and generating progeny (Calvi et al., 2003; Passegue, Wagers, Giuriato, Anderson, & Weissman, 2005), and that the coordinated cycling between the two distinct niches regulated the self-renewal and differentiation of HSC such that quiescent HSC from the endosteum move toward the vasculature where they were activated and left the BM site through the blood vessels (Scadden, 2006).

However, as discussed below recent studies have revealed considerable additional layers of complexity in HSC regulation by the BMM.

3. LEUKEMIA STEM CELLS

CML results from the reciprocal translocation of chromosomes 9 and 22, t(9; 22), leading to formation of the *BCR-ABL* oncogene by fusion of N-terminus sequences of the *BCR* gene with much of the *c-ABL* tyrosine kinase. CML cells express the BCR-ABL protein that has constitutive tyrosine kinase activity (Rowley, 1973). CML develops in three phases, starting from chronic phase, progressing to accelerated phase, and finally a terminal acute leukemia-like blast crisis (BC) phase which is almost uniformly fatal. A great deal of success has been achieved in the treatment of CML with the development of tyrosine kinase inhibitor (TKI) treatment. The first TKI to enter treatment was Imatinib (also known as Gleevec) (Druker et al., 2001). The vast majority of CML patients treated with Imatinib achieve sustained clinical remissions and major reduction in the *BCR-ABL* transcript levels (Hughes et al., 2003). Although Imatinib is quite effective in targeting the leukemic populations which are actively cycling (Holtz et al., 2002), it has limited efficacy against the small population of LSCs with self-renewing and leukemia-initiating capacity, which remain quiescent (Corbin et al., 2011; Warner et al., 2004). As a result, LSC can persist in the BM of CML patients who are in deep remission upon Imatinib treatment (Chu et al., 2011). CML LSC is able to survive despite effective inhibition of BCR-ABL tyrosine kinase inhibition following Imatinib treatment, suggesting that kinase-independent mechanisms likely contribute to drug resistance (Hamilton et al., 2012). The lack of LSC targeting by TKI may be the result of intrinsic LSC properties or could be related to extrinsic survival and maintenance signals provided by the BMM.

AML is the most common form of acute leukemia with an overall incidence of 3.8 cases per 100,000 in the United States, primarily occurring in adults over 60 years of age. (Dores, Devesa, Curtis, Linet, & Morton, 2012). When compared to CML, AML is more aggressive in nature, manifesting with abnormal hematopoiesis, accumulation of immature blasts in the bone marrow and blood, anemia, and thrombocytopenia. The standard first-line treatment for AML consists of induction chemotherapy using cytarabine and an anthracycline, such as daunorubicin (Showel & Levis, 2014). Although disease remission is often achieved, leukemia eventually relapses in most patients despite additional consolidation chemotherapy (Byrd et al.,

2002). Although bone marrow transplant is an effective option for high-risk patients, it is associated with considerable morbidity and mortality, a significant proportion of patients still relapse posttransplant (Cornelissen et al., 2007). AML shows considerable genetic complexity with number of different cytogenetic abnormalities and molecular lesions modulating disease manifestations and response to treatment. In addition to genetic heterogeneity, AML is also characterized by heterogeneity of cell type. Bonnet and Dick, 1997 elegantly showed the presence of a small population of LSC amongst the bulk of AML cells that were capable of generating leukemia after transplantation into nonobese diabetic severe combined immunodeficient mice. AML LSC could be prospectively isolated and shown to be present within the CD34+ CD38− cell compartment. However, recent studies indicate considerable heterogeneity in phenotype of AML LSC. For example, Taussig et al. showed that the leukemia-initiating capacity could also be present in the CD34− population (Taussig et al., 2008, 2010), and Goardon et al. reported the presence of a previously uncharacterized leukemic population, which is similar to lymphoid-primed multipotent progenitors (Goardon et al., 2011). It is now believed that LSC can either be derived from HSC or from progenitor cells which acquire self-renewal property.

4. BONE MARROW MICROENVIRONMENT

The BM contains cells of both hematopoietic and nonhematopoietic lineages. As opposed to the epithelial organs, which are well organized with the basement membrane separating the epithelial and niche compartment, the BM contains hematopoietic cells of the myeloid and lymphoid lineage which are surrounded by mesenchymal cells. Raymond Schofield, who formally proposed the term—stem cell niche articulating in 1978, described the functional attributes of a specialized microenvironment that regulated stem cell function *in vivo*. (Schofield, 1978). He postulated that stem cells were located in physical sites where they were uniquely regulated rather than autonomous, and that the niche could restrict stem cell entry into cycle or differentiation, and could impose the stem cell state on more differentiated cells. HSC were found to be present in the endosteum region of the BM (Gong, 1978). Since then, several studies have analyzed the composition and function of the complex BM compartment, and shown that it consists of several cell types such as endothelial cells, osteoblasts, osteoclasts, and mesenchymal stem cells (MSCs) which further differentiate into adipocytes, chondrocytes, myocytes, and fibroblasts (Yin & Li, 2006). It was traditionally conceived that the BM niche includes an endosteal niche comprising osteoblasts lining the endosteal bone surface (Calvi et al., 2003) and a vascular niche comprising endothelial cells lining sinusoids (Nilsson, Johnston, & Coverdale, 2001). Immunofluorescence visualization shows that the majority of HSC in their native BMM are localized at the endosteal bone surfaces, where they are directly associated with the vasculature (Kiel et al., 2005; Kunisaki et al., 2013). Quiescent HSC are localized to a periarteriolar niche comprised of arteriolar endothelial cells, perivascular mesenchymal stem cells, sympathetic nerves, and nonmyelinating Schwann cells. CXCL12-abundant reticular (CAR) cells are mesenchymal progenitors with osteogenic and adipogenic potential that are localized near both arteriolar and sinusoidal endothelium. These nonhematopoietic cells of the BMM produce a range of different ligands, cytokines, soluble factors, and chemokines, which are responsible for the adhesion, quiescence, self-renewal, proliferation, and differentiation of HSC and LSC (Lane, Scadden, & Gilliland, 2009; Suda, Arai, & Hirao, 2005; Wilson & Trumpp, 2006).

4.1 Cellular Components of the BMM
4.1.1 Endosteal Osteoblasts
Osteoblasts are bone-forming cells present in the endosteal region along the bone lining. The role of osteoblasts in leukemia has been actively studied for several years (Calvi et al., 2003; Cordeiro-Spinetti, Taichman, & Balduino, 2015). OBs are reported to provide signals required for HSC quiescence, long-term maintenance and BM retention. Visnjic et al., 2004 showed that ablation of osteoblasts leads to loss of BM cellularity, decreased numbers of HSCs and progenitors in the BM, and increased extramedullary hematopoiesis. Other studies showed that osteoblasts expressing CD166 were reported to play an important role in supporting quiescent long-term repopulating cells, and that osteoblasts in trabecular bone expressing high levels of jagged-1 supported a high frequency of HSC. Increase in numbers of spindle-shaped N-cadherin+ osteoblasts following conditional inactivation of BMP was associated with an increase in HSC numbers. However, other studies did not support a role for osteoblasts in regulating HSC. For example, HSC numbers were not affected when stem cell factor (SCF), which is essential for HSC maintenance, was conditionally deleted from osteoblasts. Deletion of CXCL12 from endosteal osteoblasts did not significantly change the HSC homeostasis. Deletion of SCF and CXCL12 from OB did not affect HSC maintenance. However, these studies, while excluding a role for OB in contributing these specific factors required for HSC regulation,

do not rule out a role for other OB-mediated mechanisms in HSC regulation, and the interaction of several BMM populations may be required for HSC maintenance. Recently Bowers et al., 2015 showed that osteoblast depletion in this model reduced long-term engraftment and secondary repopulating capacity of HSC, associated with loss of quiescence. They showed that osteoblast ablation in a CML mouse model was associated with significantly accelerated development of leukemia and reduced survival. Their studies supported a potential role for osteoblast-expressed jagged-1 in this osteoblast-mediated HSC regulation. Recently, Krevvata et al. reported that osteoblasts act as tumor suppressors by inhibiting leukemia cell engraftment and disease progression (Krevvata et al., 2014). They showed that the number of osteoblasts significantly decreased in the MDS/AML patients. Further, when the osteoblasts were ablated, the tumor engraftment in the BM was drastically increased due to the increased blast cells. The survival of mice was also decreased. This was coupled with a deregulation of the differentiation program of LSC as myelopoiesis was increased while the lymphopoiesis decreased. However, when the osteoblast numbers were stabilized using an inhibitor of the synthesis of duodenal serotonin, there was significant reduction in leukemic burden and increased survival. Therefore, osteoblasts may play a critical role in LSC regulation by the BMM.

4.1.2 Vascular Endothelial Cells
Recent studies indicate that the BM vasculature is characterized by a network of arterioles and sinusoids that are in close proximity to the endosteum surface (Nombela-Arrieta et al., 2013). Vascular endothelial cells at the interface of blood vessels and the stromal microenvironment are identified via the endothelial-specific markers CD31, MECA-32, VE-cadherin, and VEGFR2. ECs express factors that support HSC including SCF and CXCL12, various Notch ligands (Butler, Kobayashi, & Rafii, 2010) and E-selectin (Winkler et al., 2012). The BM vascular network supports low rates of blood flow which creates relatively hypoxic conditions in the microenvironment (Spencer et al., 2014). Quiescent HSCs preferentially localize near arteriolar endothelial cells. An important role for ECs in HSC maintenance was shown by studies in which SCF or CXCL12 expression in ECs was deleted using Tie2-Cre (Ding, Saunders, Enikolopov, & Morrison, 2012; Ding and Morrison, 2013; Greenbaum et al., 2013). LSC home to the vascular niche and are then retained there through cellular processes mediated by VLA4. AML leukemia cells expressing VLA4 exhibit resistance to apoptosis induced by therapeutic drugs. VLA4-targeting

antibodies have been shown to reverse this resistance and enhance efficacy of targeting of leukemia cells. Therefore, the continued characterization of the vasculature in terms of its regulation of LSC maintenance may be applied to newer therapeutic targets.

4.1.3 Perivascular MSCs

MSCs are cells within the bone marrow that are capable of generating bone, cartilage, fat cells, and fibroblast-like cells in the stromal microenvironment. BM MSCs are located wrapped around arteriole and sinusoid vessels. Nestin is an intermediate filament protein originally identified as a marker of neural progenitors, subsequently found to be expressed in a wide range of other progenitor cells and endothelial cells. Perivascular MSCs have been visualized using transgenic mice in which GFP is expressed under the control of the neural specific regulatory elements of the Nes (Nestin) gene (Mendez-Ferrer et al., 2010). Perivascular MSC is also characterized by expression of other markers including Sca-1, CD51, and CD140a (Pinho et al., 2013; Winkler et al., 2010). They are also reported to express the leptin receptor (Zhou, Yue, Murphy, Peyer, & Morrison, 2014). There is heterogeneity of MSC populations in terms of multipotency, self-renewal, distribution in the BM cavity, and association with specific blood vessels, but also functional overlap of MSC populations defined by different markers (Frenette, Pinho, Lucas, & Scheiermann, 2013; Morrison & Scadden, 2014). For example, Nes dim MSCs are periarteriolar and located close to the endosteal bone surface, and Nes bright MSCs are perisinusoidal and located more centrally (Kunisaki et al., 2013). There is evidence that quiescent HSCs colocalize with periarteriolar MSCs, whereas activated HSCs are located close to perisinusoidal MSCs, suggesting their differential roles in maintaining HSC quiescence or promoting HSC proliferation. These observations are consistent with previous descriptions of an endosteal localization of quiescent HSC and central location of cycling HSC in the BMM. Perivascular MSCs express many HSC-supportive factors including SCF and CXCL12. A subset of perivascular MSCs called CAR cells has been identified which will be described subsequently (Omatsu et al., 2010; Sugiyama, Kohara, Noda, & Nagasawa, 2006). Recent studies using Nes-Cre, LepR-Cre or Prrx-Cre-mediated deletion of KitL or CXCL12 in perivascular MSC support the direct role of this cell population in HSC maintenance by the BMM (Ding et al., 2012; Ding & Morrison, 2013; Greenbaum et al., 2013; Mendez-Ferrer et al., 2010; Omatsu et al., 2010).

4.1.4 CAR Cells

CXCR4 is commonly expressed by leukemic cells of both myeloid and lymphoid lineage and SDF-1 (usually called CXCL12) is its ligand which is released by the BMM. This CXCL12/CXCR4 interaction has been shown to be critical in the retention of LSC in the BMM (Kremer et al., 2014; Tavor et al., 2004). CXCR4 antagonist such as Plerixafor (AMD3100) has shown promising results in disruption of LSC from the BM lodgment, hence making them more accessible to conventional therapeutics (Becker, 2012). Studies of mice with the GFP gene knocked into the CXCL12 locus identified reticular cells with high expression of CXCL12 termed CAR cells, scattered throughout the bone marrow with long processes creating a network. CAR cells have the ability to differentiate into adipocytes as well as osteoblasts and to act as niches for HSC. Ablation studies indicate that they have the ability to support proliferation of B cells, erythroid cells and HSC, and to maintain HSC in the undifferentiated state. CAR cells localize perivascularly next to the BM sinusoidal endothelium in human BM; CD146 expressing subendothelial cells appear to be the counterpart of CAR cells.

4.1.5 Sympathetic Neuronal Cells

The BM nestin+ MSCs are innervated by sympathetic nerve fibers in order to regulate normal HSC. The sympathetic nervous system communicates through release of catecholamines including norepinephrine, which binds to β3-adrenergic receptors expressed by BM stromal components such as MSCs and OBs (Asada et al., 2013). Circadian, rhythmic secretion of norepinephrine leads to rhythmic downregulation of CXCL12 by BM stromal cells, leading to the release of HSCs from the BM niche (Mendez-Ferrer, Lucas, Battista, & Frenette, 2008). Nonmyelinating Schwann cells contribute to keeping HSCs quiescent by activating latent TGF-β1 found in the surrounding microenvironment (Yamazaki et al., 2011). Arranz et al. recently reported the critical role of the sympathetic neurons in the pathogenesis of myeloproliferative neoplasms (MPN) (Arranz et al., 2014). In the BM of MPN patients and the mouse model of MPN, there was a significant decrease in the number of nestin+ MSC, sympathetic neurons, and Schwann cells due to the neuronal damage in the BM and the Schwann cell death from secretion of IL-1β from mutant HSC. Depletion of the nestin+ MSC results in a significant increase in the LSC. When mice were treated with neuroprotective drugs, the expansion of LSC was abrogated. Therefore, LSC drive deregulation of MSC and neuronal cells, which in turn contribute to the disease, suggesting a therapeutic target could be developed in the pathogenesis of MPN.

5. MICROENVIRONMENT LEUKEMIA CROSSTALK

Over the years, our understanding of the BMM composition and its role in the regulation of LSCs has greatly increased. Unlike solid cancers, which metastasize into the BM compartment in the advanced stages of the disease, CML and AML originate in the BM. Genetic lesions in leukemia cells provide cell-autonomous growth signals to leukemic cells but also alter how LSC interact with and respond to the BMM. Gordon et al. described that CML progenitors have altered adhesion to stroma (Gordon, Dowding, Riley, Goldman, & Greaves, 1987). LSCs may demonstrate altered dependency on BMM signals compared to normal HSC. LSCs in murine models of CML and AML become insensitive to TGF-β-mediated inhibition (Krause et al., 2013; Santaguida et al., 2009), less dependent on Wnt signals for localization to the marrow compared to normal HSCs (Lane et al., 2011), and more dependent on CD44 and different selectins for homing compared to normal HSCs (Jin, Hope, Zhai, Smadja-Joffe, & Dick, 2006; Krause, Lazarides, von Andrian, & Van Etten, 2006; Krause, Lazarides, Lewis, von Andrian, & Van Etten, 2014). It is also becoming evident that leukemic cells can induce changes in the BMM to enhance support of leukemic hematopoiesis at the expense of normal hematopoiesis. A series of elegant experiments have shown that LSC disrupt normal HSC homeostasis by significantly remodeling the BMM within the same ecosystem (Colmone et al., 2008; Schepers et al., 2013; Zhang et al., 2012).

A newer hypothesis has already starting to emerge suggesting that an abnormal BMM can by itself lead to the development of hematopoietic malignancies by inducing hematopoietic dysfunction (Kim et al., 2008; Raaijmakers et al., 2010; Shiozawa & Taichman, 2010; Walkley, Olsen, et al., 2007a; Walkley, Shea, Sims, Purton, & Orkin, 2007b). The microenvironment can influence and perhaps even induce the development of myeloid leukemias. Mice with deletions of the retinoblastoma (Rb) develop myeloproliferation, which requires deletion of Rb in both hematopoietic cells and the microenvironment (Walkley et al., 2007b). On the other hand, retinoic acid receptor-gamma (RARγ) deletions in the BMM can induce hematopoietic abnormalities leading to myeloproliferative disease. Transplantation studies revealed that leukemia did not originate from RAR$\gamma-/-$ hematopoietic cells, but developed when wild-type hematopoietic cells were transplanted into a RAR$\gamma-/-$ hosts (Walkley et al.,

2007a). The Scadden group has also shown that deletion of microRNA processing enzymes or the Shwachman–Bodian–Diamond syndrome protein in immature osteolineage mesenchymal cells results in disordered hematopoiesis, leading to the development of a myeloid leukemias with clonal genetic abnormalities (Raaijmakers et al., 2010). Recently activation of β-catenin in mouse osteoblasts was also shown to lead to development of myeloid leukemia related to increased Notch signaling (Kode et al., 2014). Additionally, secondary leukemias originating from donor cells can occur in some patients receiving allogeneic bone marrow transplantation. These data indicate that microenvironmental abnormalities can induce or promote disordered growth of hematopoietic cells that can evolve to myeloproliferation and acute myeloid leukemia with clonal genetic abnormalities.

The above description indicates that a comprehensive understanding of mechanisms and significance of the complex crosstalk between the BMM and HSC or LSC in the regulation of normal and malignant hematopoiesis will be essential for the development of novel therapeutic strategies aimed at targeting the BM niche to improve outcomes of leukemia patients.

5.1 Critical Signaling Pathways Altered in Leukemia
5.1.1 Wnt/β-Catenin Pathway

Wnt ligands are a family of secreted glycoproteins that interact with membrane-associated receptors to activate signaling cascades. The Wnt pathway includes 19 different Wnt ligands, 10 frizzled receptors, and multiple signaling intermediates. Wnt signaling is involved in embryonic development, cell differentiation, proliferation, and polarity. Many Wnt proteins are expressed in hematopoietic cells, and aberrant activation of Wnt signaling is reported in several hematologic malignancies.

Considerable research has been performed to understand the role of Wnt pathway in the regulation of normal and malignant hematopoiesis in the context of the BMM (Heidel et al., 2012; Hu, Chen, Douglas, & Li, 2009; Lane et al., 2011; Luis, Ichii, Brugman, Kincade, & Staal, 2012; Luis & Staal, 2009; Nemeth, Mak, Yang, & Bodine, 2009; Schreck, Bock, Grziwok, Oostendorp, & Istvanffy, 2014; Sugimura et al., 2012; Wang et al., 2010; Zhang et al., 2013). Several factors including developmental stage, Wnt dosage, and microenvironmental factors modulate the effect of Wnt signaling. The canonical Wnt signaling pathway is mediated by β-catenin stabilization and translocation into the nucleus, complex formation with TCF/LEF, and transcription of target genes. Conditional inactivation of *CTNNB1* in adult animals did not alter HSC repopulation and self-renewal (Cobas et al., 2004), whereas Vav-Cre-mediated deletion of *CTNNB1* resulted in decreased long-term capacity of HSCs (Zhao et al., 2007). Since Vav-Cre exerts its effect during fetal hematopoiesis, these observations suggest a greater requirement for β-catenin in developmental compared to adult HSCs. Overexpression of constitutively active β-catenin in adult HSC causes severe disruptions in HSC function and a differentiation block. Studies by Luis et al. combining different *APC* conditional knockouts to create five different levels of Wnt signaling *in vivo* indicated that levels of Wnt signaling affect the balance of HSC self-renewal versus differentiation (Luis et al., 2011). A decrease in Wnt signaling leads to loss of HSC quiescence and self-renewal, mild increases in Wnt activation supports HSC maintenance and increased hematopoietic reconstitution, and high levels of Wnt activation impairs HSC self-renewal and differentiation. Other studies have supported a critical role for Wnt signaling in the BMM, demonstrating a requirement for canonical Wnt signaling in the BM niche to support hematopoiesis, and its function in maintaining HSC quiescence. As compared to canonical Wnt signaling, the role of noncanonical Wnt signaling in HSCs is not well defined. Noncanonical Wnt signaling mediated by Frizzled (Fz8) has been shown to antagonize canonical Wnt signaling, contributing to quiescence of HSCs through suppression of the NFAT–IFNγ pathway (Sugimura et al., 2012).

Enhanced Wnt-β-catenin activity is reported to contribute to LSC transformation in BC CML, although its role in CP CML is less clear (Jamieson et al., 2004; Minami et al., 2008). In a BCR-ABL transduction–transplantation model, expression of BCR-ABL in β-catenin deleted HSC resulted in reduced LSC self-renewal and leukemia development (Hu et al., 2009; Zhao et al., 2007). Deletion of β-catenin in established leukemia synergized with IM to eliminate CML stem cells (Heidel et al., 2012). Potential mechanisms underlying increased β-catenin in CML cells include BCR-ABL-mediated β-catenin phosphorylation, leading to protein stabilization and activation of nuclear signaling (Coluccia et al., 2007), and reduced β-catenin degradation related to GSK3β inactivation downstream of BCR-ABL (Samanta, Lin, Sun, Kantarjian, & Arlinghaus, 2006), or missplicing of GSK3β as reported in BC CML LSC(Coluccia et al., 2007). Zhang et al. showed that N-Cadherin-mediated adhesion of CML LSC to mesenchymal cells resulted in increased cytoplasmic N-Cadherin-β-catenin complex formation, as well as enhanced β-catenin nuclear translocation and transcriptional activity. Increased exogenous Wnt-mediated β-catenin signaling played an important role in protection of CML progenitors from TKI treatment, indicating a close interplay between N-Cadherin and the Wnt/β-catenin pathway in protecting CML LSC during TKI treatment. Importantly, these results reveal novel mechanisms of resistance of CML LSC to TKI treatment, and new targets for treatment to eradicate LSC in CML patients.

Although Wnt signaling is recognized as a key regulator of HSC and LSC fate, the complexity of this regulatory mechanism has not been fully resolved. The role of both canonical and noncanonical pathways in HSC, and the interaction between the two branches of Wnt signaling needs to be better understood.

5.1.2 Hedgehog Pathway

Hedgehog signaling was initially discovered in 1980 by a genetic screen for investigating the factors which influence the embryonic patterning in Drosophila (Nusslein-Volhard & Wieschaus, 1980). Hh pathway is comprised of three different isoforms of the original Hh ligand, Indian Hh (IHH), Sonic Hh (SHH), and Desert Hh (DHH). These ligands upon cleavage, and palmitoylation produce the active secretory protein. All the three isoforms are similar in structure, and hence have redundant functions. The central players of the pathway are the transmembrane proteins patched (PTCH), smoothened (SMO), and glioma zinc finger transcription factors (GLI1, GLI2, and GLI3). Although PTCH is a negative regulator, the SMO receptor is the central protein, which positively controls signaling. IHH is produced in hematopoietic tissue, which is the BM and is dispensable for normal hematopoiesis (Gao et al., 2009; Hofmann et al., 2009). In CML, Hh signaling is substantially increased in leukemic stem and progenitor cells as the disease progresses from chronic phase to accelerated and BC (Dierks et al., 2008; Irvine & Copland, 2012; Zhao et al., 2009). Two groups of researchers investigated the effect of *Smo* deletion in transgenic mouse models of CML. While the studies used different experimental approaches, the results were quite similar in that inhibition of Hh signaling reduced the number of LSC and, hence, decreased the incidence of leukemia in primary and secondary transplantation experiments. Mice treated with pharmacological inhibitors of Hh showed increased survival, and lower LSC functional activity. Therefore, Hh signaling appears essential for the maintenance and expansion of CML LSC and represents a promising strategy to target CML LSC (Katagiri et al., 2013). In AML, HHIP expression in the BMM was correlated with the SMO expression in leukemic cells (Kobune et al., 2012).

Therefore, targeting the BMM-mediated Hh signaling may represent a novel strategy to target AML LSC (Boyd et al., 2014).

5.1.3 Notch Pathway

Notch signaling is a highly conserved pathway that regulates many developmental decisions such as cell fate, proliferation, and differentiation. It was first identified in Drosophila and named for the phenotypic appearance of the notched wings. The central components of the Notch pathway include the four Notch receptors (Notch1–4) and five cognate ligands Delta-like (Dll1, Dll3, and Dll4), Jagged (Jag1, Jag2). Like other secretory proteins of the developmental pathways, the Notch ligands are secretory proteins which are palmitoylated and then secreted out of the cell where they interact with the respective receptors to initiate downstream signaling (Evans & Calvi, 2015).

The role of the Notch pathway has been controversial since the effects of Notch differ depending on cell conditions and are modified by other stemness-related signals. Notch activation can lead to quiescence or the stemness of the target cell in the context of the BMM whereas in other scenarios it may be associated with loss of quiescence and increased proliferative state. As described above, osteoblasts in trabecular bone areas express high levels of the Notch ligand Jagged-1, which interact with Notch receptors on HSCs. HSCs that are not bound to Jagged-1 have reduced repopulating capacity. Bowers et al. showed that Jagged-1 inhibits HSC entry into cell cycle, and that loss of Jagged-1 induced signaling may contribute to loss of HSC quiescence following osteoblast ablation. Jagged-1 was overexpressed in OB from BCR-ABL compared to normal mice. As with normal HSC, Jagged-1 exposure inhibited cell cycling of CML LSC, suggesting that loss of Jagged-1 signaling.

AML cells express Notch and Notch ligands, but activating Notch1 mutations are rare in AML. Notch ligand stimulation generally suppresses in vitro growth of AML cells, whereas knockdown of Notch1 and Notch2 does not affect the growth of AML cells. Lobry et al. showed that Notch signaling is silenced in human AML samples, and in LSC in a murine AML model (Lobry et al., 2013). *In vivo* activation of Notch signaling in AML LSC-induced rapid cell cycle arrest, differentiation, and apoptosis. Kannan et al showed that although human AML samples have robust expression of Notch receptors; however, Notch pathway activation is low (Kannan et al., 2013). Induced activation of Notch receptors resulted in growth arrest and apoptosis, and reduced AML growth *in vivo*. A Notch agonist peptide led to significant apoptosis in AML patient samples. These data demonstrate a tumor suppressor role for Notch signaling in AML and the potential use of Notch receptor agonists in treatment of AML. Further investigations to achieve a detailed understanding of the Notch pathway in the regulation of HSC and LSC by the BMM are warranted.

6. *IN VIVO* MODELS TO STUDY BMM AND LEUKEMIA INTERACTIONS

Genetically modified mouse models (Politi & Pao, 2011) and patient-derived xenografts (Williams, Anderson, Santaguida, & Dylla, 2013) have helped us to unravel several important regulatory mechanisms in the field of cancer biology. These valuable tools coupled with reporters and sophisticated imaging techniques have yielded better understanding of LSC–BMM crosstalk and applied the findings to the development of strategies to target the BMM in leukemia. Seminal studies have shown that LSC hijack the BM niche from the normal HSC for their own survival and drug resistance (Colmone et al., 2008). To better understand the spatial distribution between HSC and LSC within the BMM, Boyd et al. used a xenografted immunodeficient mouse model (Boyd et al., 2014). In this study, it was shown that the normal hematopoietic stem and progenitor cells can outcompete the leukemia-initiating cells (LIC) to occupy the vacant niches within the BMM. When competitive BM transplantation experiments were performed using primary cells from AML patients or healthy individuals, the long-term self-renewal of LSC was compromised. Therefore, such *in vivo* models may be used to test whether dissociation of LIC-niche crosstalk may lead to a competitive disadvantage against normal HSC, suggesting that such strategies could increase the likelihood of LSC elimination in leukemia patients. To investigate the role of endogenous BMM in the determination of lineage fate of LSC, Wei et al. used a retroviral transduction and transplantation model with established oncogenes such as MLL-AF9. This fusion gene is the result of the chromosomal translocation between t(9;11) and is frequently observed in AML patients. It was found that the leukemia generated as a result of this mutation required BMM signals for efficient phenotypic differentiation (Wei et al., 2008). The phenotype of leukemia that develops following transplantation of MLL–AF9 gene expressing human CD34+ cells in immunodeficient mice was dependent on whether or not the mouse strain used for transplantation expressed human GM-CSF, SCF and IL-3, or not. Furthermore, numerous genetic strains are already available and have been extensively used to ablate the hematopoietic cells and the niche components in order to study the interactions and their mechanisms (Joseph et al., 2013). Similarly, it will interesting to employ these various Cre mice strains to understand the role of BMM in the protection of LSC from therapeutic drugs. While we still have some way to go, the *in vivo* models are highly evolved and provide exciting opportunities to study the autocrine and paracrine mechanisms regulating LSC in the context of BMM.

7. POTENTIAL THERAPEUTIC AVENUE: TARGETING THE SEED AND SOIL?

One of the major challenges that experimental hematologists face is the translation of the biological discoveries related to stem cell biology into the clinic, which will benefit leukemic patients. Identification of underlying mutations and availability of drugs to target the resultant deregulated pathways is already experiencing varying success in the treatment of patients with leukemia. Examples are the success of BCR-ABL TKI in CML. JAK2 inhibitors in myeloproliferative neoplasms and FLT3 inhibitors in FLT3-ITD AML. On the other hand, there is emerging evidence that the therapies targeting the BMM may provide an additional source of drug resistance in addition to cell-autonomous signaling taking place in LSC responsible for drug resistance, such as multidrug resistance *MDR1* expression and activity (Mahadevan & List, 2004) and selection of resistant mutant clones for which the therapy would be ineffective (Heidel et al., 2006). Therefore, there has been interest in developing methods to target the BMM as an adjuvant treatment in addition to direct targeting of leukemia cells, to overcome drug resistance, and to shift the competitive advantage to normal HSC over LSC to achieve deeper remissions.

As discussed earlier, preclinical studies have shown that targeting adhesion via CD44 and VLA-4 or the CXCL12–CXCR4 axis can dislodge leukemic cells from BM niches. CXCR4 inhibition mobilized leukemic cells from their niches in xenograft models and enhanced disease eradication *in vivo* (Nervi et al., 2009; Weisberg et al., 2012). Clinical trials combining plerixafor with cytotoxic chemotherapy in patients with relapsed AML have been conducted and have found this approach to be feasible, although the added benefit of this approach is not clear as yet. In addition, the anti-CD44v6 monoclonal antibody Bivatuzumab and the anti-a4 integrin monoclonal antibody Natalizumab are already FDA approved for other indications and are candidates for testing in clinical trials.

There is increasing amount of evidence that leukemia cells can modify the BMM to provide critical signals to the LSC for their self-renewal and that LSC exploit the mechanisms, which are responsible for the normal maintenance of the hematopoietic system. A number of studies have demonstrated that BMM can promote resistance of LSC from therapeutic drugs in both CML and AML (Ben-Batalla et al., 2013; Dillmann et al., 2009; Kojima et al., 2011; MacLean, Lo Celso, & Stumpf, 2013; Nair et al., 2012; Quintarelli et al., 2014; Taussig et al., 2008; Weisberg, Azab, et al., 2012; Weisberg, Liu, et al., 2012; Weisberg et al., 2013; Yamamoto-Sugitani et al., 2011; Yang, Sexauer, & Levis, 2014). In acute lymphoblastic leukemia (ALL), Duan et al. have identified a novel therapy-induced niche, which protects leukemia-propagating cells (LPC) from chemotherapeutic drugs (Duan et al., 2014). When mice engrafted with ALL cells and were treated with standard treatment (cytarabine and daunorubicin), a small population of LPC persisted in the BMM surrounded by sheaths of supporting cells that comprise a protective niche. This dynamic niche was in fact formed as a result of the secretion of CCL3 by the resistant LPC. Disrupting the formation of the protective niche may enhance the therapeutic efficacy of chemotherapeutic agents. Kumano and colleagues recently reported that leukemia-induced protective niche is found in CML as well (Kobayashi et al., 2014). Using a CML-like myeloproliferative mouse model, it was found that the CML LIC secrete Th2 cytokines (such as IL-4, IL-6, and IL-13) which are responsible for the constitution of a LIC niche. IL-2 from the BMM interacted with CD25 expressed on CML LIC resulting in protective signaling. These results suggest that targeting this LIC-induced niche could be exploited for the potentiation of CML therapy. In AML, studies have already reported that neovascularization in the BM compartment is protective for leukemic cells (Sennino & McDonald, 2012). Although the therapeutic efficacy of inhibitors targeting the angiogenic process have not had very much promise so far (Ossenkoppele et al., 2012; Zahiragic et al., 2007), the conceptual framework of niche-directed therapy in combination with conventional chemotherapeutic agents for treating leukemia patients is compelling and deserves further evaluation in the clinical setting.

8. CONCLUSIONS AND FUTURE DIRECTIONS

In this review, we have summarized the discrete architecture of the BMM and discussed their important role in regulating HSC and LSC. We have also discussed the function of critical signaling pathways that can be exploited in the future to offer the novel therapeutic avenues to target LSC. These strategies are an alternative to the conventional therapies, which solely are concentrated on targeting the cell-autonomous signaling of LSC. The standard cell line based assays such as cell viability and cell death assays cannot identify the compounds that are able to target LSC seeded in their soil, which is the BMM. Therefore, to better target the bidirectional crosstalk and to better identify plausible targets or candidate genes, well-designed high-through *in vitro* coculture systems such as those described by Hartwell and colleagues will be required (Hartwell et al., 2013). These screenings will enable modeling of complex microenvironment consisting of leukemic cells and their counterpart stromal cells so that relevant molecules can be identified in a high-throughput manner and with reproducible readout. However, targets identified in this manner require careful validation *in vivo*. LSC are quite similar to HSC in several of their biological characteristics and BMM interactions, and selectivity of therapies toward leukemic compared to normal stem cells requires careful evaluation (Roboz, 2011).

In conclusion, the approach of targeting the seed and soil together still poses some critical questions that require consideration before successful application to the clinic. Further testing of drugs targeting homing and adhesion of leukemic cells are required to determine whether the strategy of releasing LSCs from their BM niches is an effective approach. It is not known whether targeting the BMM will results in LSC moving to a different microenvironmental niche to maintain their survival? Seminal studies have already shown that the spleens of both CML and AML mice are capable of supporting LIC that are able to efficiently induce disease upon transplantation into normal recipient mice (Huntly et al., 2004; Lane et al., 2009; Schemionek et al., 2012). Furthermore, the spleen leukemic cells were relatively less sensitive to Imatinib when compared to the leukemic cells present in the BM. These results suggest either that disease phenotype is influenced by the environment or that leukemic cells from the BM can move to a new location for their survival. Another critical challenge will be the correct timing and the combination of appropriate compounds. Clearly, improved understanding of the bidirectional crosstalk between LSC and BMM, and further testing of strategies to target these interactions will substantially benefit the development of new therapeutic approaches to better target leukemia.

REFERENCES

Arranz, L., Sanchez-Aguilera, A., Martin-Perez, D., Isern, J., Langa, X., Tzankov, A., et al. (2014). Neuropathy of haematopoietic stem cell niche is essential for myeloproliferative neoplasms. *Nature, 512*(7512), 78–81.

Asada, N., Katayama, Y., Sato, M., Minagawa, K., Wakahashi, K., Kawano, H., et al. (2013). Matrix-embedded osteocytes regulate mobilization of hematopoietic stem/progenitor cells. *Cell Stem Cell, 12*(6), 737–747.

Becker, P. S. (2012). Dependence of acute myeloid leukemia on adhesion within the bone marrow microenvironment. *Scientific World Journal, 2012*, 856467.

Ben-Batalla, I., Schultze, A., Wroblewski, M., Erdmann, R., Heuser, M., Waizenegger, J. S., et al. (2013). Axl, a prognostic and therapeutic target in acute myeloid leukemia mediates paracrine crosstalk of leukemia cells with bone marrow stroma. *Blood, 122*(14), 2443–2452.

Bonnet, D., & Dick, J. E. (1997). Human acute myeloid leukemia is organized as a hierarchy that originates from a primitive hematopoietic cell. *Nature Medicine, 3*(7), 730–737.

Bowers, M., Zhang, B., Ho, Y., Agarwal, P., Chen, C. C., & Bhatia, R. (2015). Osteoblast ablation reduces normal long-term hematopoietic stem cell self-renewal but accelerates leukemia development. *Blood, 125*(17), 2678–2688.

Boyd, A. L., Campbell, C. J., Hopkins, C. I., Fiebig-Comyn, A., Russell, J., Ulemek, J., et al. (2014). Niche displacement of human leukemic stem cells uniquely allows their competitive replacement with healthy HSPCs. *Journal of Experimental Medicine, 211*(10), 1925–1935.

Butler, J. M., Kobayashi, H., & Rafii, S. (2010). Instructive role of the vascular niche in promoting tumour growth and tissue repair by angiocrine factors. *Nature Reviews Cancer, 10*(2), 138–146.

Byrd, J. C., Mrozek, K., Dodge, R. K., Carroll, A. J., Edwards, C. G., Arthur, D. C., et al. (2002). Pretreatment cytogenetic abnormalities are predictive of induction success, cumulative incidence of relapse, and overall survival in adult patients with de novo acute myeloid leukemia: Results from Cancer and Leukemia Group B (CALGB 8461). *Blood, 100*(13), 4325–4336.

Calvi, L. M., Adams, G. B., Weibrecht, K. W., Weber, J. M., Olson, D. P., Knight, M. C., et al. (2003). Osteoblastic cells regulate the haematopoietic stem cell niche. *Nature, 425*(6960), 841–846.

Chu, S., McDonald, T., Lin, A., Chakraborty, S., Huang, Q., Snyder, D. S., et al. (2011). Persistence of leukemia stem cells in chronic myelogenous leukemia patients in prolonged remission with imatinib treatment. *Blood, 118*(20), 5565–5572.

Cobas, M., Wilson, A., Ernst, B., Mancini, S. J., MacDonald, H. R., Kemler, R., et al. (2004). Beta-catenin is dispensable for hematopoiesis and lymphopoiesis. *Journal of Experimental Medicine, 199*(2), 221–229.

Colmone, A., Amorim, M., Pontier, A. L., Wang, S., Jablonski, E., & Sipkins, D. A. (2008). Leukemic cells create bone marrow niches that disrupt the behavior of normal hematopoietic progenitor cells. *Science, 322*(5909), 1861–1865.

Coluccia, A. M., Vacca, A., Dunach, M., Mologni, L., Redaelli, S., Bustos, V. H., et al. (2007). Bcr-Abl stabilizes beta-catenin in chronic myeloid leukemia through its tyrosine phosphorylation. *EMBO Journal, 26*(5), 1456–1466.

Corbin, A. S., Agarwal, A., Loriaux, M., Cortes, J., Deininger, M. W., & Druker, B. J. (2011). Human chronic myeloid leukemia stem cells are insensitive to imatinib despite inhibition of BCR-ABL activity. *Journal of Clinical Investigation, 121*(1), 396–409.

Cordeiro-Spinetti, E., Taichman, R. S., & Balduino, A. (2015). The bone marrow endosteal niche: How far from the surface? *Journal of Cellular Biochemistry, 116*(1), 6–11.

Cornelissen, J. J., van Putten, W. L., Verdonck, L. F., Theobald, M., Jacky, E., Daenen, S. M., et al. (2007). Results of a HOVON/SAKK donor versus no-donor analysis of myeloablative HLA-identical sibling stem cell transplantation in first remission acute myeloid leukemia in young and middle-aged adults: Benefits for whom? *Blood*, *109*(9), 3658–3666.

Dierks, C., Beigi, R., Guo, G. R., Zirlik, K., Stegert, M. R., Manley, P., et al. (2008). Expansion of Bcr-Abl-positive leukemic stem cells is dependent on Hedgehog pathway activation. *Cancer Cell*, *14*(3), 238–249.

Dillmann, F., Veldwijk, M. R., Laufs, S., Sperandio, M., Calandra, G., Wenz, F., et al. (2009). Plerixafor inhibits chemotaxis toward SDF-1 and CXCR4-mediated stroma contact in a dose-dependent manner resulting in increased susceptibility of BCR-ABL+ cell to Imatinib and Nilotinib. *Leukemia and Lymphoma*, *50*(10), 1676–1686.

Ding, L., & Morrison, S. J. (2013). Haematopoietic stem cells and early lymphoid progenitors occupy distinct bone marrow niches. *Nature*, *495*(7440), 231–235.

Ding, L., Saunders, T. L., Enikolopov, G., & Morrison, S. J. (2012). Endothelial and perivascular cells maintain haematopoietic stem cells. *Nature*, *481*(7382), 457–462.

Dores, G. M., Devesa, S. S., Curtis, R. E., Linet, M. S., & Morton, L. M. (2012). Acute leukemia incidence and patient survival among children and adults in the United States, 2001-2007. *Blood*, *119*(1), 34–43.

Druker, B. J., Sawyers, C. L., Capdeville, R., Ford, J. M., Baccarani, M., & Goldman, J. M. (2001). Chronic myelogenous leukemia. *Hematology/the Education Program of the American Society of Hematology*, 87–112.

Duan, C. W., Shi, J., Chen, J., Wang, B., Yu, Y. H., Qin, X., et al. (2014). Leukemia propagating cells rebuild an evolving niche in response to therapy. *Cancer Cell*, *25*(6), 778–793.

Essers, M. A., & Trumpp, A. (2010). Targeting leukemic stem cells by breaking their dormancy. *Molecular Oncology*, *4*(5), 443–450.

Evans, A. G., & Calvi, L. M. (2015). Notch signaling in the malignant bone marrow microenvironment: Implications for a niche-based model of oncogenesis. *Annals of the New York Academy of Sciences*, *1335*(1), 63–77.

Frenette, P. S., Pinho, S., Lucas, D., & Scheiermann, C. (2013). Mesenchymal stem cell: keystone of the hematopoietic stem cell niche and a stepping-stone for regenerative medicine. *Annual Review of Immunology*, *31*, 285–316.

Gao, J., Graves, S., Koch, U., Liu, S., Jankovic, V., Buonamici, S., et al. (2009). Hedgehog signaling is dispensable for adult hematopoietic stem cell function. *Cell Stem Cell*, *4*(6), 548–558.

Goardon, N., Marchi, E., Atzberger, A., Quek, L., Schuh, A., Soneji, S., et al. (2011). Coexistence of LMPP-like and GMP-like leukemia stem cells in acute myeloid leukemia. *Cancer Cell*, *19*(1), 138–152.

Gong, J. K. (1978). Endosteal marrow: A rich source of hematopoietic stem cells. *Science*, *199*(4336), 1443–1445.

Gordon, M. Y., Dowding, C. R., Riley, G. P., Goldman, J. M., & Greaves, M. F. (1987). Altered adhesive interactions with marrow stroma of haematopoietic progenitor cells in chronic myeloid leukaemia. *Nature*, *328*(6128), 342–344.

Greenbaum, A., Hsu, Y. M., Day, R. B., Schuettpelz, L. G., Christopher, M. J., Borgerding, J. N., et al. (2013). CXCL12 in early mesenchymal progenitors is required for haematopoietic stem-cell maintenance. *Nature*, *495*(7440), 227–230.

Hamilton, A., Helgason, G. V., Schemionek, M., Zhang, B., Myssina, S., Allan, E. K., et al. (2012). Chronic myeloid leukemia stem cells are not dependent on Bcr-Abl kinase activity for their survival. *Blood*, *119*(6), 1501–1510.

Hartwell, K. A., Miller, P. G., Mukherjee, S., Kahn, A. R., Stewart, A. L., Logan, D. J., et al. (2013). Niche-based screening identifies small-molecule inhibitors of leukemia stem cells. *Nature Chemical Biology*, *9*(12), 840–848.

Heidel, F. H., Bullinger, L., Feng, Z., Wang, Z., Neff, T. A., Stein, L., et al. (2012). Genetic and pharmacologic inhibition of beta-catenin targets imatinib-resistant leukemia stem cells in CML. *Cell Stem Cell, 10*(4), 412–424.

Heidel, F., Solem, F. K., Breitenbuecher, F., Lipka, D. B., Kasper, S., Thiede, M. H., et al. (2006). Clinical resistance to the kinase inhibitor PKC412 in acute myeloid leukemia by mutation of Asn-676 in the FLT3 tyrosine kinase domain. *Blood, 107*(1), 293–300.

Hofmann, I., Stover, E. H., Cullen, D. E., Mao, J., Morgan, K. J., Lee, B. H., et al. (2009). Hedgehog signaling is dispensable for adult murine hematopoietic stem cell function and hematopoiesis. *Cell Stem Cell, 4*(6), 559–567.

Holtz, M. S., Slovak, M. L., Zhang, F., Sawyers, C. L., Forman, S. J., & Bhatia, R. (2002). Imatinib mesylate (STI571) inhibits growth of primitive malignant progenitors in chronic myelogenous leukemia through reversal of abnormally increased proliferation. *Blood, 99*(10), 3792–3800.

Hu, Y., Chen, Y., Douglas, L., & Li, S. (2009). Beta-Catenin is essential for survival of leukemic stem cells insensitive to kinase inhibition in mice with BCR-ABL-induced chronic myeloid leukemia. *Leukemia, 23*(1), 109–116.

Hughes, T. P., Kaeda, J., Branford, S., Rudzki, Z., Hochhaus, A., Hensley, M. L., et al. (2003). Frequency of major molecular responses to imatinib or interferon alfa plus cytarabine in newly diagnosed chronic myeloid leukemia. *New England Journal of Medicine, 349*(15), 1423–1432.

Huntly, B. J., Shigematsu, H., Deguchi, K., Lee, B. H., Mizuno, S., Duclos, N., et al. (2004). MOZ-TIF2, but not BCR-ABL, confers properties of leukemic stem cells to committed murine hematopoietic progenitors. *Cancer Cell, 6*(6), 587–596.

Irvine, D. A., & Copland, M. (2012). Targeting hedgehog in hematologic malignancy. *Blood, 119*(10), 2196–2204.

Jamieson, C. H., Ailles, L. E., Dylla, S. J., Muijtjens, M., Jones, C., Zehnder, J. L., et al. (2004). Granulocyte-macrophage progenitors as candidate leukemic stem cells in blast-crisis CML. *New England Journal of Medicine, 351*(7), 657–667.

Jin, L., Hope, K. J., Zhai, Q., Smadja-Joffe, F., & Dick, J. E. (2006). Targeting of CD44 eradicates human acute myeloid leukemic stem cells. *Nature Medicine, 12*(10), 1167–1174.

Joseph, C., Quach, J. M., Walkley, C. R., Lane, S. W., Lo Celso, C., & Purton, L. E. (2013). Deciphering hematopoietic stem cells in their niches: A critical appraisal of genetic models, lineage tracing, and imaging strategies. *Cell Stem Cell, 13*(5), 520–533.

Kannan, S., Sutphin, R. M., Hall, M. G., Golfman, L. S., Fang, W., Nolo, R. M., et al. (2013). Notch activation inhibits AML growth and survival: A potential therapeutic approach. *Journal of Experimental Medicine, 210*(2), 321–337.

Katagiri, S., Tauchi, T., Okabe, S., Minami, Y., Kimura, S., Maekawa, T., et al. (2013). Combination of ponatinib with Hedgehog antagonist vismodegib for therapy-resistant BCR-ABL1-positive leukemia. *Clinical Cancer Research, 19*(6), 1422–1432.

Kiel, M. J., Yilmaz, O. H., Iwashita, T., Yilmaz, O. H., Terhorst, C., & Morrison, S. J. (2005). SLAM family receptors distinguish hematopoietic stem and progenitor cells and reveal endothelial niches for stem cells. *Cell, 121*(7), 1109–1121.

Kim, Y. W., Koo, B. K., Jeong, H. W., Yoon, M. J., Song, R., Shin, J., et al. (2008). Defective Notch activation in microenvironment leads to myeloproliferative disease. *Blood, 112*(12), 4628–4638.

Kobayashi, C. I., Takubo, K., Kobayashi, H., Nakamura-Ishizu, A., Honda, H., Kataoka, K., et al. (2014). The IL-2/CD25 axis maintains distinct subsets of chronic myeloid leukemia-initiating cells. *Blood, 123*(16), 2540–2549.

Kobune, M., Iyama, S., Kikuchi, S., Horiguchi, H., Sato, T., Murase, K., et al. (2012). Stromal cells expressing hedgehog-interacting protein regulate the proliferation of myeloid neoplasms. *Blood Cancer Journal, 2*, e87.

Kode, A., Manavalan, J. S., Mosialou, I., Bhagat, G., Rathinam, C. V., Luo, N., et al. (2014). Leukaemogenesis induced by an activating beta-catenin mutation in osteoblasts. *Nature*, *506*(7487), 240–244.

Kojima, K., McQueen, T., Chen, Y., Jacamo, R., Konopleva, M., Shinojima, N., et al. (2011). p53 activation of mesenchymal stromal cells partially abrogates microenvironment-mediated resistance to FLT3 inhibition in AML through HIF-1alpha-mediated down-regulation of CXCL12. *Blood*, *118*(16), 4431–4439.

Krause, D. S., Fulzele, K., Catic, A., Sun, C. C., Dombkowski, D., Hurley, M. P., et al. (2013). Differential regulation of myeloid leukemias by the bone marrow microenvironment. *Nature Medicine*, *19*(11), 1513–1517.

Krause, D. S., Lazarides, K., Lewis, J. B., von Andrian, U. H., & Van Etten, R. A. (2014). Selections and their ligands are required for homing and engraftment of BCR-ABL1$^+$ leukemic stem cells in the bone marrow niche. *Blood*, *123*(9), 1361–1371.

Krause, D. S., Lazarides, K., von Andrian, U. H., & Van Etten, R. A. (2006). Requirement for CD44 in homing and engraftment of BCR-ABL-expressing leukemic stem cells. *Nature Medicine*, *12*(10), 1175–1180.

Kremer, K. N., Dudakovic, A., McGee-Lawrence, M. E., Philips, R. L., Hess, A. D., Smith, B. D., et al. (2014). Osteoblasts protect AML cells from SDF-1-induced apoptosis. *Journal of Cellular Biochemistry*, *115*(6), 1128–1137.

Krevvata, M., Silva, B. C., Manavalan, J. S., Galan-Diez, M., Kode, A., Matthews, B. G., et al. (2014). Inhibition of leukemia cell engraftment and disease progression in mice by osteoblasts. *Blood*, *124*(18), 2834–2846.

Kunisaki, Y., Bruns, I., Scheiermann, C., Ahmed, J., Pinho, S., Zhang, D., et al. (2013). Arteriolar niches maintain haematopoietic stem cell quiescence. *Nature*, *502*(7473), 637–643.

Lane, S. W., Scadden, D. T., & Gilliland, D. G. (2009). The leukemic stem cell niche: Current concepts and therapeutic opportunities. *Blood*, *114*(6), 1150–1157.

Lane, S. W., Wang, Y. J., Lo Celso, C., Ragu, C., Bullinger, L., Sykes, S. M., et al. (2011). Differential niche and Wnt requirements during acute myeloid leukemia progression. *Blood*, *118*(10), 2849–2856.

Lobry, C., Ntziachristos, P., Ndiaye-Lobry, D., Oh, P., Cimmino, L., Zhu, N., et al. (2013). Notch pathway activation targets AML-initiating cell homeostasis and differentiation. *Journal of Experimental Medicine*, *210*(2), 301–319.

Lord, B. I., Testa, N. G., & Hendry, J. H. (1975). The relative spatial distributions of CFUs and CFUc in the normal mouse femur. *Blood*, *46*(1), 65–72.

Luis, T. C., Ichii, M., Brugman, M. H., Kincade, P., & Staal, F. J. (2012). Wnt signaling strength regulates normal hematopoiesis and its deregulation is involved in leukemia development. *Leukemia*, *26*(3), 414–421.

Luis, T. C., Naber, B. A., Roozen, P. P., Brugman, M. H., de Haas, E. F., Ghazvini, M., et al. (2011). Canonical wnt signaling regulates hematopoiesis in a dosage-dependent fashion. *Cell Stem Cell*, *9*(4), 345–356.

Luis, T. C., & Staal, F. J. (2009). WNT proteins: Environmental factors regulating HSC fate in the niche. *Annals of the New York Academy of Sciences*, *1176*, 70–76.

MacLean, A. L., Lo Celso, C., & Stumpf, M. P. (2013). Population dynamics of normal and leukaemia stem cells in the haematopoietic stem cell niche show distinct regimes where leukaemia will be controlled. *Journal of the Royal Society, Interface*, *10*(81), 20120968.

Mahadevan, D., & List, A. F. (2004). Targeting the multidrug resistance-1 transporter in AML: Molecular regulation and therapeutic strategies. *Blood*, *104*(7), 1940–1951.

Mendez-Ferrer, S., Lucas, D., Battista, M., & Frenette, P. S. (2008). Haematopoietic stem cell release is regulated by circadian oscillations. *Nature*, *452*(7186), 442–447.

Mendez-Ferrer, S., Michurina, T. V., Ferraro, F., Mazloom, A. R., Macarthur, B. D., Lira, S. A., et al. (2010). Mesenchymal and haematopoietic stem cells form a unique bone marrow niche. *Nature, 466*(7308), 829–834.

Minami, Y., Stuart, S. A., Ikawa, T., Jiang, Y., Banno, A., Hunton, I. C., et al. (2008). BCR-ABL-transformed GMP as myeloid leukemic stem cells. *Proceedings of the National Academy of Sciences of the United States of America, 105*(46), 17967–17972.

Morrison, S. J., & Scadden, D. T. (2014). The bone marrow niche for haematopoietic stem cells. *Nature, 505*(7483), 327–334.

Nair, R. R., Tolentino, J. H., Argilagos, R. F., Zhang, L., Pinilla-Ibarz, J., & Hazlehurst, L. A. (2012). Potentiation of Nilotinib-mediated cell death in the context of the bone marrow microenvironment requires a promiscuous JAK inhibitor in CML. *Leukemia Research, 36*(6), 756–763.

Nemeth, M. J., Mak, K. K., Yang, Y., & Bodine, D. M. (2009). beta-Catenin expression in the bone marrow microenvironment is required for long-term maintenance of primitive hematopoietic cells. *Stem Cells, 27*(5), 1109–1119.

Nervi, B., Ramirez, P., Rettig, M. P., Uy, G. L., Holt, M. S., Ritchey, J. K., et al. (2009). Chemosensitization of acute myeloid leukemia (AML) following mobilization by the CXCR4 antagonist AMD3100. *Blood, 113*(24), 6206–6214.

Nilsson, S. K., Johnston, H. M., & Coverdale, J. A. (2001). Spatial localization of transplanted hemopoietic stem cells: Inferences for the localization of stem cell niches. *Blood, 97*(8), 2293–2299.

Nombela-Arrieta, C., Pivarnik, G., Winkel, B., Canty, K. J., Harley, B., Mahoney, J. E., et al. (2013). Quantitative imaging of haematopoietic stem and progenitor cell localization and hypoxic status in the bone marrow microenvironment. *Nature Cell Biology, 15*(5), 533–543.

Nusslein-Volhard, C., & Wieschaus, E. (1980). Mutations affecting segment number and polarity in Drosophila. *Nature, 287*(5785), 795–801.

Omatsu, Y., Sugiyama, T., Kohara, H., Kondoh, G., Fujii, N., Kohno, K., et al. (2010). The essential functions of adipo-osteogenic progenitors as the hematopoietic stem and progenitor cell niche. *Immunity, 33*(3), 387–399.

Osawa, M., Hanada, K., Hamada, H., & Nakauchi, H. (1996). Long-term lymphohematopoietic reconstitution by a single CD34-low/negative hematopoietic stem cell. *Science, 273*(5272), 242–245.

Ossenkoppele, G. J., Stussi, G., Maertens, J., van Montfort, K., Biemond, B. J., Breems, D., et al. (2012). Addition of bevacizumab to chemotherapy in acute myeloid leukemia at older age: A randomized phase 2 trial of the Dutch-Belgian cooperative trial group for hemato-oncology (HOVON) and the Swiss group for clinical cancer research (SAKK). *Blood, 120*(24), 4706–4711.

Passegue, E., Wagers, A. J., Giuriato, S., Anderson, W. C., & Weissman, I. L. (2005). Global analysis of proliferation and cell cycle gene expression in the regulation of hematopoietic stem and progenitor cell fates. *Journal of Experimental Medicine, 202*(11), 1599–1611.

Perry, J. M., & Li, L. (2007). Disrupting the stem cell niche: Good seeds in bad soil. *Cell, 129*(6), 1045–1047.

Pinho, S., Lacombe, J., Hanoun, M., Mizoguchi, T., Bruns, I., Kunisaki, Y., et al. (2013). PDGFRalpha and CD51 mark human nestin+ sphere-forming mesenchymal stem cells capable of hematopoietic progenitor cell expansion. *Journal of Experimental Medicine, 210*(7), 1351–1367.

Politi, K., & Pao, W. (2011). How genetically engineered mouse tumor models provide insights into human cancers. *Journal of Clinical Oncology, 29*(16), 2273–2281.

Quintarelli, C., De Angelis, B., Errichiello, S., Caruso, S., Esposito, N., Colavita, I., et al. (2014). Selective strong synergism of Ruxolitinib and second generation tyrosine kinase inhibitors to overcome bone marrow stroma related drug resistance in chronic myelogenous leukemia. *Leukemia Research, 38*(2), 236–242.

Raaijmakers, M. H., Mukherjee, S., Guo, S., Zhang, S., Kobayashi, T., Schoonmaker, J. A., et al. (2010). Bone progenitor dysfunction induces myelodysplasia and secondary leukaemia. *Nature, 464*(7290), 852–857.

Reya, T., Morrison, S. J., Clarke, M. F., & Weissman, I. L. (2001). Stem cells, cancer, and cancer stem cells. *Nature, 414*(6859), 105–111.

Roboz, G. J. (2011). Novel approaches to the treatment of acute myeloid leukemia. *Hematology/the Education Program of the American Society of Hematology, 2011*, 43–50.

Rowley, J. D. (1973). Letter: A new consistent chromosomal abnormality in chronic myelogenous leukaemia identified by quinacrine fluorescence and Giemsa staining. *Nature, 243*(5405), 290–293.

Samanta, A. K., Lin, H., Sun, T., Kantarjian, H., & Arlinghaus, R. B. (2006). Janus kinase 2: A critical target in chronic myelogenous leukemia. *Cancer Research, 66*(13), 6468–6472.

Santaguida, M., Schepers, K., King, B., Sabnis, A. J., Forsberg, E. C., Attema, J. L., et al. (2009). JunB protects against myeloid malignancies by limiting hematopoietic stem cell proliferation and differentiation without affecting self-renewal. *Cancer Cell, 15*(4), 341–352.

Scadden, D. T. (2006). The stem-cell niche as an entity of action. *Nature, 441*(7097), 1075–1079.

Schemionek, M., Spieker, T., Kerstiens, L., Elling, C., Essers, M., Trumpp, A., et al. (2012). Leukemic spleen cells are more potent than bone marrow-derived cells in a transgenic mouse model of CML. *Leukemia, 26*(5), 1030–1037.

Schepers, K., Pietras, E. M., Reynaud, D., Flach, J., Binnewies, M., Garg, T., et al. (2013). Myeloproliferative neoplasia remodels the endosteal bone marrow niche into a self-reinforcing leukemic niche. *Cell Stem Cell, 13*(3), 285–299.

Schofield, R. (1978). The relationship between the spleen colony-forming cell and the haemopoietic stem cell. *Blood Cells, 4*(1-2), 7–25.

Schreck, C., Bock, F., Grziwok, S., Oostendorp, R. A., & Istvanffy, R. (2014). Regulation of hematopoiesis by activators and inhibitors of Wnt signaling from the niche. *Annals of the New York Academy of Sciences, 1310*, 32–43.

Sennino, B., & McDonald, D. M. (2012). Controlling escape from angiogenesis inhibitors. *Nature Reviews. Cancer, 12*(10), 699–709.

Shiozawa, Y., & Taichman, R. S. (2010). Dysfunctional niches as a root of hematopoietic malignancy. *Cell Stem Cell, 6*(5), 399–400.

Showel, M. M., & Levis, M. (2014). Advances in treating acute myeloid leukemia. *F1000Prime Reports, 6*, 96.

Spencer, J. A., Ferraro, F., Roussakis, E., Klein, A., Wu, J., Runnels, J. M., et al. (2014). Direct measurement of local oxygen concentration in the bone marrow of live animals. *Nature, 508*(7495), 269–273.

Suda, T., Arai, F., & Hirao, A. (2005). Hematopoietic stem cells and their niche. *Trends in Immunology, 26*(8), 426–433.

Sugimura, R., He, X. C., Venkatraman, A., Arai, F., Box, A., Semerad, C., et al. (2012). Noncanonical Wnt signaling maintains hematopoietic stem cells in the niche. *Cell, 150*(2), 351–365.

Sugiyama, T., Kohara, H., Noda, M., & Nagasawa, T. (2006). Maintenance of the hematopoietic stem cell pool by CXCL12-CXCR4 chemokine signaling in bone marrow stromal cell niches. *Immunity, 25*(6), 977–988.

Taussig, D. C., Miraki-Moud, F., Anjos-Afonso, F., Pearce, D. J., Allen, K., Ridler, C., et al. (2008). Anti-CD38 antibody-mediated clearance of human repopulating cells masks the heterogeneity of leukemia-initiating cells. *Blood, 112*(3), 568–575.

Taussig, D. C., Vargaftig, J., Miraki-Moud, F., Griessinger, E., Sharrock, K., Luke, T., et al. (2010). Leukemia-initiating cells from some acute myeloid leukemia patients with mutated nucleophosmin reside in the CD34(-) fraction. *Blood, 115*(10), 1976–1984.

Tavian, M., & Peault, B. (2005). Embryonic development of the human hematopoietic system. *International Journal of Developmental Biology, 49*(2-3), 243–250.

Tavor, S., Petit, I., Porozov, S., Avigdor, A., Dar, A., Leider-Trejo, L., et al. (2004). CXCR4 regulates migration and development of human acute myelogenous leukemia stem cells in transplanted NOD/SCID mice. *Cancer Research, 64*(8), 2817–2824.

Visnjic, D., Kalajzic, Z., Rowe, D. W., Katavic, V., Lorenzo, J., & Aguila, H. L. (2004). Hematopoiesis is severely altered in mice with an induced osteoblast deficiency. *Blood, 103*(9), 3258–3264.

Wagers, A. J., Sherwood, R. I., Christensen, J. L., & Weissman, I. L. (2002). Little evidence for developmental plasticity of adult hematopoietic stem cells. *Science, 297*(5590), 2256–2259.

Walkley, C. R., Olsen, G. H., Dworkin, S., Fabb, S. A., Swann, J., McArthur, G. A., et al. (2007a). A microenvironment-induced myeloproliferative syndrome caused by retinoic acid receptor gamma deficiency. *Cell, 129*(6), 1097–1110.

Walkley, C. R., Shea, J. M., Sims, N. A., Purton, L. E., & Orkin, S. H. (2007b). Rb regulates interactions between hematopoietic stem cells and their bone marrow microenvironment. *Cell, 129*(6), 1081–1095.

Wang, Y., Krivtsov, A. V., Sinha, A. U., North, T. E., Goessling, W., Feng, Z., et al. (2010). The Wnt/beta-catenin pathway is required for the development of leukemia stem cells in AML. *Science, 327*(5973), 1650–1653.

Warner, J. K., Wang, J. C., Hope, K. J., Jin, L., & Dick, J. E. (2004). Concepts of human leukemic development. *Oncogene, 23*(43), 7164–7177.

Wei, J., Wunderlich, M., Fox, C., Alvarez, S., Cigudosa, J. C., Wilhelm, J. S., et al. (2008). Microenvironment determines lineage fate in a human model of MLL-AF9 leukemia. *Cancer Cell, 13*(6), 483–495.

Weisberg, E., Azab, A. K., Manley, P. W., Kung, A. L., Christie, A. L., Bronson, R., et al. (2012). Inhibition of CXCR4 in CML cells disrupts their interaction with the bone marrow microenvironment and sensitizes them to nilotinib. *Leukemia, 26*(5), 985–990.

Weisberg, E., Liu, Q., Nelson, E., Kung, A. L., Christie, A. L., Bronson, R., et al. (2012). Using combination therapy to override stromal-mediated chemoresistance in mutant FLT3-positive AML: Synergism between FLT3 inhibitors, dasatinib/multi-targeted inhibitors and JAK inhibitors. *Leukemia, 26*(10), 2233–2244.

Weisberg, E., Liu, Q., Zhang, X., Nelson, E., Sattler, M., Liu, F., et al. (2013). Selective Akt inhibitors synergize with tyrosine kinase inhibitors and effectively override stroma-associated cytoprotection of mutant FLT3-positive AML cells. *PLoS One, 8*(2), e56473.

Weissman, I. L. (2000). Stem cells: Units of development, units of regeneration, and units in evolution. *Cell, 100*(1), 157–168.

Williams, S. A., Anderson, W. C., Santaguida, M. T., & Dylla, S. J. (2013). Patient-derived xenografts, the cancer stem cell paradigm, and cancer pathobiology in the 21st century. *Laboratory Investigation, 93*(9), 970–982.

Wilson, A., & Trumpp, A. (2006). Bone-marrow haematopoietic-stem-cell niches. *Nature Reviews. Immunology, 6*(2), 93–106.

Winkler, I. G., Barbier, V., Nowlan, B., Jacobsen, R. N., Forristal, C. E., Patton, J. T., et al. (2012). Vascular niche E-selectin regulates hematopoietic stem cell dormancy, self renewal and chemoresistance. *Nature Medicine, 18*(11), 1651–1657.

Winkler, I. G., Sims, N. A., Pettit, A. R., Barbier, V., Nowlan, B., Helwani, F., et al. (2010). Bone marrow macrophages maintain hematopoietic stem cell (HSC) niches and their depletion mobilizes HSCs. *Blood, 116*(23), 4815–4828.

Yamamoto-Sugitani, M., Kuroda, J., Ashihara, E., Nagoshi, H., Kobayashi, T., Matsumoto, Y., et al. (2011). Galectin-3 (Gal-3) induced by leukemia microenvironment promotes drug resistance and bone marrow lodgment in chronic myelogenous leukemia. *Proceedings of the National Academy of Sciences of the United States of America, 108*(42), 17468–17473.

Yamazaki, S., Ema, H., Karlsson, G., Yamaguchi, T., Miyoshi, H., Shioda, S., et al. (2011). Nonmyelinating Schwann cells maintain hematopoietic stem cell hibernation in the bone marrow niche. *Cell, 147*(5), 1146–1158.

Yang, X., Sexauer, A., & Levis, M. (2014). Bone marrow stroma-mediated resistance to FLT3 inhibitors in FLT3-ITD AML is mediated by persistent activation of extracellular regulated kinase. *British Journal of Haematology, 164*(1), 61–72.

Yin, T., & Li, L. (2006). The stem cell niches in bone. *Journal of Clinical Investigation, 116*(5), 1195–1201.

Zahiragic, L., Schliemann, C., Bieker, R., Thoennissen, N. H., Burow, K., Kramer, C., et al. (2007). Bevacizumab reduces VEGF expression in patients with relapsed and refractory acute myeloid leukemia without clinical antileukemic activity. *Leukemia, 21*(6), 1310–1312.

Zhang, B., Ho, Y. W., Huang, Q., Maeda, T., Lin, A., Lee, S. U., et al. (2012). Altered microenvironmental regulation of leukemic and normal stem cells in chronic myelogenous leukemia. *Cancer Cell, 21*(4), 577–592.

Zhang, B., Li, M., McDonald, T., Holyoake, T. L., Moon, R. T., Campana, D., et al. (2013). Microenvironmental protection of CML stem and progenitor cells from tyrosine kinase inhibitors through N-cadherin and Wnt-beta-catenin signaling. *Blood, 121*(10), 1824–1838.

Zhao, C., Blum, J., Chen, A., Kwon, H. Y., Jung, S. H., Cook, J. M., et al. (2007). Loss of beta-catenin impairs the renewal of normal and CML stem cells in vivo. *Cancer Cell, 12*(6), 528–541.

Zhao, C., Chen, A., Jamieson, C. H., Fereshteh, M., Abrahamsson, A., Blum, J., et al. (2009). Hedgehog signalling is essential for maintenance of cancer stem cells in myeloid leukaemia. *Nature, 458*(7239), 776–779.

Zhou, B. O., Yue, R., Murphy, M. M., Peyer, J. G., & Morrison, S. J. (2014). Leptin-receptor-expressing mesenchymal stromal cells represent the main source of bone formed by adult bone marrow. *Cell Stem Cell, 15*(2), 154–168.

CHAPTER SEVEN

Perspectives on Epidermal Growth Factor Receptor Regulation in Triple-Negative Breast Cancer: Ligand-Mediated Mechanisms of Receptor Regulation and Potential for Clinical Targeting

Carly Bess Williams*, Adam C. Soloff[†], Stephen P. Ethier[‡], Elizabeth S. Yeh*,[1]

*Department of Cell and Molecular Pharmacology and Experimental Therapeutics, Medical University of South Carolina, Charleston, South Carolina, USA
[†]Department of Microbiology and Immunology, Medical University of South Carolina, Charleston, South Carolina, USA
[‡]Department of Pathology and Laboratory Medicine, Medical University of South Carolina, Charleston, South Carolina, USA
[1]Corresponding author: e-mail address: yeh@musc.edu

Contents

1. Introduction — 254
2. Regulation of EGFR Turnover and Signaling Outcomes — 257
3. Common Molecular Characteristics of TNBC and Their Relationship to EGFR — 264
   3.1 *BRCA1* and *BRCA2* — 264
   3.2 *TP53* — 265
   3.3 *PI3K/PTEN/AKT* — 266
4. Tumor Associated Macrophages in Breast Cancer Metastasis and Their Relationship with EGFR Signaling — 267
5. Future Perspectives on Therapy — 269
   5.1 Combination Therapy — 269
   5.2 Tumor-Associated Macrophage-Based Therapy — 271
6. Conclusions — 273
Acknowledgments — 273
References — 273

Abstract

Currently, there are no effective targeted therapies for triple-negative breast cancer (TNBC) indicating a critical unmet need for breast cancer patients. Tumors that fall into the triple-negative category of breast cancers do not respond to the targeted therapies currently approved for breast cancer treatment, such as endocrine therapy (tamoxifen, aromatase inhibitors) or human epidermal growth factor receptor-2 (HER2) inhibitors (trastuzumab, lapatinib), because these tumors lack the most common breast cancer markers: estrogen receptor, progesterone receptor, and HER2. While many patients with TNBC respond to chemotherapy, subsets of patients fare poorly and relapse very quickly. Studies indicate that epidermal growth factor receptor (EGFR) is frequently overrepresented in TNBC (>50%), suggesting EGFR could be used as a biomarker and target in breast cancer. While it is clear that this growth factor receptor plays an integral role in TNBC, little is known about the mechanisms of sustained EGFR activation and how to target this protein despite availability of EGFR-targeted inhibitors, suggesting that our understanding of EGFR deregulation in TNBC is incomplete.

1. INTRODUCTION

Breast cancer is a heterogeneous disease composed of a growing number of clinically and scientifically recognized subtypes. The heterogeneous nature of breast cancer has posed difficult implications for physicians and their patients, and despite continuing advances in prevention and treatment, breast cancer remains the second most common cause of mortality among women in the United States and the leading cause of cancer mortality among women worldwide (Parkin, Bray, Ferlay, & Pisani, 2005; Schmadeka, Harmon, & Singh, 2014; Valentin, da Silva, Privat, Alaoui-Jamali, & Bignon, 2012).

Current treatments for breast cancer are targeted toward particular subtype and molecular markers (Dent et al., 2007). Clinically, three main receptors: two hormones—estrogen receptor (ER) and progesterone receptors (PR), and one growth factor—human epidermal growth factor receptor-2 (HER2), are used for diagnostic classification. Breast cancer patients with tumors that are ER+/PR+ have lower risks of mortality after their diagnosis compared to women with ER−/PR− disease (Colditz, Rosner, Chen, Holmes, & Hankinson, 2004). Clinical trials have also shown that the survival advantage for women with hormone receptor-positive tumors is enhanced by treatment with adjuvant chemotherapeutic regimens that are supplemented with hormone-specific treatments (Goldhirsch et al., 2007). These hormone-directed therapies consist of two classes of drugs: aromatase inhibitors and selective estrogen receptor modulators (SERMs), so-called because they are targeted inhibitors that either block estrogen production or the effects of estrogen on its target cell. While more aggressive than ER+ breast cancers, those patients with tumors that are HER2+, in addition to chemotherapy, receive adjuvant or neoadjuvant targeted inhibitor to HER2, mainly trastuzumab and lapatinib (Goldhirsch et al., 2007). Although drug resistance is an ongoing issue in ER+ and HER2+ patients, reliable-targeted inhibitors are currently being employed in terms of first-line treatments.

Conversely, patients that present with ER−, PR−, and HER2− disease are considered triple-negative (TN), as defined by their lack of immunohistochemical staining for these receptors (Schmadeka et al., 2014), and therefore lack obvious avenues for targeted inhibition. Approximately 15% of breast cancer cases fall into the triple-negative breast cancer (TNBC) category and these cases tend to behave more aggressively than non-TNBCs (Schmadeka et al., 2014). Along with the aggressive nature, patients with this subtype of breast cancer tend to relapse more quickly and have a higher likelihood of developing central nervous system and visceral metastases than those non-TNBC patients (Rakha & Chan, 2011; Schmadeka et al., 2014). While TNBC patients tend to respond to chemotherapy for a subset of these patients, the effects are often short lived, with a general time to relapse occurring between 1 and 3 years and a median survival from the time of recurrence of $\sim$9 months (Dent et al., 2007).

Beyond the intrinsic classification of breast cancers discussed earlier, recent advances in gene expression analysis has further identified several distinct breast cancer subtypes by differentiating breast cancers into separate groups based only on gene expression patterns (Perou et al., 2000). Recently, Lehmann et al. reported there are seven subtypes of TNBC. These molecular subtypes are divided into two basal-like (BL1 and BL2), an immunomodulatory (IM), a mesenchymal (M), a mesenchymal stem-like (MSL), luminal androgen receptor (LAR), and a relatively undefined "unstable" (UNS) subclass (Lehmann et al., 2011). While TNBCs clearly display distinct gene expression profiles that provide hints at biology beyond the initial clinical immunohistochemical classifications, those profiles are highly dependent on how qualitative the experimental sample normalization and analysis is performed, and thus consistent utilization of data for clinical application can be considered as being in an early stage (Lehmann & Pietenpol, 2014).

One major hurdle for TNBC is the absence of targeted therapies such as those present for hormone sensitive and HER2+ patients. This represents a key challenge, as clinicians try to find curative solutions, palliate disease progression, and extend life when possible (Hudis & Gianni, 2011). With the current lack of effective targeted therapies for TNBC, treatment regimens often fail to slow tumor progression in a subset of patients (Schmadeka et al., 2014). Presumably, more accurate molecular classification of TNBCs and adequately powered prospective trials are necessary to validate predictive biomarkers in order to establish effective treatments for all patients (Schmadeka et al., 2014). To date several important molecules are identified as common regions of mutation, deletion, or overrepresentation in TNBC. These include the breast cancer genes 1 and 2, BRCA1 and BRCA2, and the gene encoding the p53 tumor suppressor protein, *TP53*. These molecules act to preserve genomic integrity and their loss of function is consistent with loss of genomic stability as a key feature of this breast cancer subtype. Growth factor signaling is also a common alteration in TNBC. Some of the most common areas for overactivity are alterations in the PI3K/AKT signaling pathway and the epidermal growth factor receptor tyrosine kinase (EGFR).

Though deregulation of EGFR is observed in all subtypes of breast cancer, EGFR is more frequently overexpressed in TNBC, including inflammatory breast cancer (Masuda et al., 2012), with overexpression or activation found in over 50% of TNBC cases and some reports as high as ~75% (Dent et al., 2007). Interestingly, *EGFR* gene amplifications and mutations are rarely found in breast cancer (Shapira, Lee, Vora, & Budman, 2013), yet the overexpression or overactivation of EGFR protein seen in TNBC is associated with large tumor size, poor differentiation, and poor clinical outcomes (Sainsbury, Farndon, Needham, Malcolm, & Harris, 1987; Salomon, Brandt, Ciardiello, & Normanno, 1995). A number of early studies demonstrated that an inverse relationship between EGFR and ER expression existed in breast cancer tumor samples highlighting that EGFR expression corresponds with the ER− status of the TN subtype (Harris, Nicholson, Sainsbury, Farndon, & Wright, 1989; Nicholson et al., 1990; O'Sullivan, Lewis, Harris, & McGee, 1993; Sainsbury et al., 1987; Sainsbury, Farndon, Sherbet, & Harris, 1985; Sainsbury et al., 1988).

Understanding that TNBCs overexpress EGFR, and that EGFR signaling has a central role in human cancers, led to the testing of anti-EGFR therapies including tyrosine kinase inhibitors gefitinib and erlotinib (Masuda et al., 2012), and the monoclonal antibody cetuximab (Shapira et al., 2013), in preclinical models and some clinical trials. Although a few clinical trial studies that combined cetuximab with certain chemotherapeutic agents showed promise in overall response rate, those studies tested only small patient groups and in general, EGFR targeting in clinical trials for TNBC have been disappointing because they do not show a significant effect on progression-free or overall survival (Baselga et al., 2013; Clark, Botrel, Paladini, & Ferreira, 2014; Nechushtan et al., 2014). This leads us to the question, why do anti-EGFR therapies tend to be unsuccessful despite the high propensity for EGFR overactivation/overrepresentation in TNBC?

In this chapter, we will provide an overview of EGFR in TNBC. Specifically, how regulation of this growth factor receptor in TNBC could circumvent treatment and the relationship of EGFR to the other common molecular changes found in this breast cancer subtype. Additionally, we will provide an overview of the emerging signaling relationship that is dependent on EGFR, between tumor cells and immune cells found in the tumor microenvironment, in particular tumor-associated macrophages (TAMs), and how this knowledge could be applied for future diagnostics or treatment.

2. REGULATION OF EGFR TURNOVER AND SIGNALING OUTCOMES

EGFR is a transmembrane glycoprotein encoded by a gene on chromosome 7p12 (Shapira et al., 2013). EGFR-mediated signaling is critical for the growth and development of multicellular organisms (Moghal & Sternberg, 1999) and has been shown to play a critical role in mammary gland development (Higashiyama et al., 2008; Lee et al., 2003; Luetteke et al., 1999). There are over 20 ligands known to bind and activate EGFR (Shapira et al., 2013), of which there are at least seven that regulate canonical EGFR signaling: EGF, transforming growth factor-α (TGFA), heparin-binding EGF-like growth factor (HB-EGF), amphiregulin (AREG), betacellulin (BTC), epiregulin (EREG), and epigen (EPGN) (Harris, Chung, & Coffey, 2003). Upon ligand binding to EGFR, the receptor either forms homodimers or heterodimerizes with other EGFR family members (HER2, HER3, and HER4) resulting in transphosphorylation at multiple tyrosine residues along the carboxyl-terminal tails of the receptor pair (Pennock & Wang, 2008). The phosphorylated tyrosine residues serve as sites of initiation for protein complexes that transduce downstream signals, ultimately resulting in cell biological outcomes including proliferation, survival, and key metabolic changes (Burgering & Coffer, 1995). As shown in Fig. 1, some of the major signaling pathways activated by EGFR receptors are mediated by proteins including PI3 kinase–AKT, Ras–Raf–MAPK, JNK, FAK, and PLCγ, key factors which are implicated in breast cancer cell proliferation, survival and stress response, and migration (Schulze, Deng, & Mann, 2005). Consequently, the deregulation of EGFR activity and

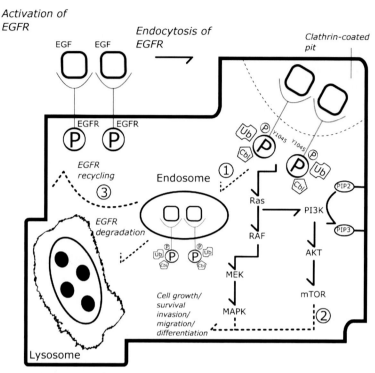

Figure 1 The activation of EGFR occurs via the binding of EGFR ligands, such as EGF, which promotes EGFR homodimerization or heterodimerization with one of the other EGFR family members; HER2, HER2, and HER4. This dimerization results in the receptor-mediated cross-phosphorylation of the cytoplasmic tails of the receptors, which recruits molecules that initiate endocytosis of the receptor complex. Once the receptor is internalized, one of three things can happen: (1) The E3-ubiquitin ligase, Cbl, binds to a phosphorylated tyrosine residue (Y1045) on the receptor. This event results in ubiquitination of the receptor and tags the receptor for lysosomal degradation. (2) The phosphorylated receptors can activate downstream molecules (e.g., Ras, Akt) that will in turn create a signaling cascade to promote cell invasion, migration, differentiation, growth, and/or survival. (3) The phosphorylated receptors are recycled back to the cell surface.

downstream signaling is considered a major factor in breast epithelial cell transformation and breast cancer progression.

The EGFR ligands not only induce EGFR signaling, but they also mediate internalization of this receptor via endocytosis; a process that removes parts of plasma membranes associated with receptor tyrosine kinases, other plasma membrane receptors, and glucose transporters (Shapira et al., 2013). EGFR receptors are internalized mainly via clathrin-dependent endocytosis and individual ligands elicit differential effects on signaling and endocytic

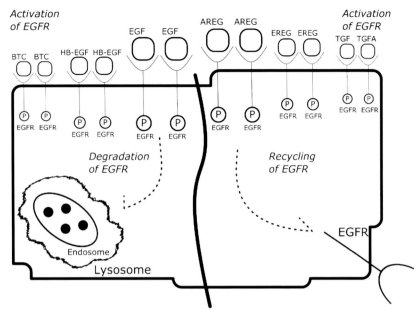

Figure 2 The different EGFR ligands prompt the receptor to follow different endosome-mediated outcomes. EGF, HB-EGF, and BTC have been shown to induce degradation of EGFR via the lysosome, whereas AREG, EREG, and TGFA have been shown to promote recycling of EGFR back to the cell surface.

sorting (Sorkin & Goh, 2008). As illustrated in Fig. 2, it was previously shown that three of the seven canonical ligands: HB-EGF, BTC, and EGF will promote receptor degradation whereas AREG, EREG, and TGFA binding will preferentially result in recycling (Baillo, Giroux, & Ethier, 2011). Interestingly, ligands including AREG and TGFA that preferentially drive EGFR recycling have also been implicated in autocrine/paracrine signaling in TN and inflammatory breast cancers (LeJeune et al., 1993; Ma et al., 2001; Panico et al., 1996; Streicher et al., 2007).

The most well-studied ligand–receptor interaction for EGFR across cell types is likely EGF–EGFR and observations, resulting from the study of EGF–EGFR initiated interaction serves as a general model of how EGFR is regulated in response to ligand interaction. Upon ligand binding, the EGF–EGFR complex is initially internalized into the clathrin-coated pit and transported through the cytoplasm (Sorkina, Huang, Beguinot, & Sorkin, 2002). Figure 1 illustrates how EGFR, as internalized cargo, can follow one of two cellular destinations: (1) either fuse with the lysosome leading to degradation of EGFR thus terminating the signaling of this protein or (2) recycle back to the plasma membrane and continue to signal (Sorkin & Goh, 2008). With EGFR ligands other than EGF, signaling may be continuous or temporally shifted based on whether a specific ligand preferentially directs EGFR to recycling or degradation pathways. Proper maintenance of the balance of available EGFR at the cell surface and the attenuation of EGFR signaling through degradation are therefore critical in maintaining intracellular homeostasis (Pennock & Wang, 2008).

A major signaling component that regulates EGFR degradation is the Casitas B-lineage Lymphoma (Cbl) family of proteins (Fig. 1). As E3 ubiquitin ligases, Cbl proteins play a prominent role in EGFR downregulation by ubiquitinating the receptor, prompting its degradation through the lysosomal degradation pathway (Joazeiro et al., 1999; Pennock & Wang, 2008). Polyubiquitination of EGFR is considered a major determinant in EGFR downregulation at both early and late stages of receptor sorting, when the receptor is taken into clathrin-coated pits as well as into the lysosome for degradation (Marmor & Yarden, 2004). EGFR recruits Cbl by binding to ligand, which causes subsequent phosphorylation of the receptor on tyrosine 1045 (Y1045). This leads to Cbl association with EGFR through its tyrosine kinase-binding domain at Y1045 of EGFR, when this residue is phosphorylated (Levkowitz et al., 1999). This contact between Cbl and EGFR then allows EGFR to phosphorylate Cbl, thereby recruiting adaptor proteins and Ub-loaded E2 proteins to the RING finger domain of Cbl (Zheng, Wang, Jeffrey, & Pavletich, 2000).

Cbl homologues, Cbl-b and c-Cbl, play distinct roles in EGFR regulation (Davies et al., 2004). Ubiquitination mediated by either Cbl-b or c-Cbl is thought to be sufficient to cause significant EGFR degradation (Pennock & Wang, 2008). However, temporal differences in their regulatory activities are apparent since c-Cbl interaction with EGFR is strongest at an early stage of EGFR trafficking when the receptor is rapidly ubiquitinated due to EGF ligand binding, while Cbl-b's peak association occurs later during a secondary wave of regulation (Pennock & Wang, 2008). Following peak c-Cbl association and the initial immediate rise in receptor ubiquitination, EGFR becomes partially deubiquitinated (Pennock & Wang, 2008). This is followed by c-Cbl dissociation from EGFR, specific dephosphorylation at Y1045, and a strong rise in Cbl-b association with EGFR (Pennock & Wang, 2008). It is possible that following EGF stimulation, each Cbl isoform acts on a separate spatiotemporal pool of EGFR to effect its degradation, and these processes may kinetically differ

(Pennock & Wang, 2008). However, the purpose of this biphasic ubiquitin-mediated regulation of EGFR is still relatively unclear. Regardless, in the absence of both c-Cbl and Cbl-b, EGFR is neither ubiquitinated nor degraded (Ahmad et al., 2014; Pennock & Wang, 2008).

The consequences of Cbl regulatory activity on EGFR signaling is inhibition of EGF-mediated growth and a decrease in signaling through PI3K–AKT and Ras–MAPK (ERK1/2) pathways (Ettenberg et al., 1999). Mutation of EGFR that renders the receptor insensitive to Cbl regulation, or directly interfering with Cbl-containing complexes by mutation or genetic deletion of Cbl, leads to disruption of EGFR internalization, degradation, and signaling (Ahmad et al., 2014; Pennock & Wang, 2008). However, while EGFR downregulation was severely inhibited due to impaired Cbl function, the receptor was still observed to localize to internal vesicles, indicating that internalization is not effected by Cbl dysfunction (Lill et al., 2000; Pennock & Wang, 2008). Since internalization occurs prior to both recycling and degradation, it is easy to speculate how Cbl dysfunction could sustain EGFR activation if a loss of EGFR degradation were to occur.

Evidence supporting this assertion shows that mutation of EGFR at tyrosine residue Y1045 (Pennock & Wang, 2008), which serves as docking sites for Cbl, sustains EGF-stimulated activation of its downstream effectors, ERK1/2 and AKT (Hartman, Zhao, & Agazie, 2013). In line with observations with Cbl, the EGFR phospho-tyrosine mutants were also resistant to EGF-induced degradation (Hartman et al., 2013). Furthermore, the interaction of c-Cbl with EGFR was reduced in cells expressing mutant EGFR (Hartman et al., 2013). This was accompanied by a reduction in ubiquitination and an increased proliferative potential as well as an increased ability to form colonies in soft agar (Hartman et al., 2013). These results suggest that the interaction between EGFR and its adaptor proteins, including Cbl, are critical for normal EGFR function. Therefore loss of this regulatory aspect for EGFR promotes its transforming potential and mechanistically, this could be due to a reduced ability of EGFR to be degraded.

It is necessary to highlight that the regulation of EGFR by Cbl has been almost exclusively examined in the context of EGF. The differential expression of EGFR ligands, such as AREG, that promote receptor recycling will most likely dramatically alter the cellular signaling responses downstream of ligand binding. Gene expression and protein signaling analysis confirm this assertion. Recent studies show that SUM 149PT TNBC cells exposed to AREG have differential gene and protein expression profiles than those exposed to EGF (Kappler et al., 2015). Moreover, it is essential to note the importance of EGFR protein regulation by these ligands. Certainly with the case of AREG, it is clear that ligand activation maintains the overexpression of EGFR protein by sustaining cell surface expression in TNBC cells (Willmarth et al., 2009). This observation and the concept that posttranslational protein regulation of EGFR is key in TNBC is supported by the lack of evidence to support EGFR amplification events in breast cancer (Bhargava et al., 2005; Nakajima et al., 2014; Shapira et al., 2013; Shawarby, Al-Tamimi, & Ahmed, 2011).

The EGFR ligand, AREG, has taken on an emerging role in TNBC as a key factor that maintains cell surface EGFR protein expression. AREG was initially identified by Shoyab and colleagues who isolated it from conditioned medium from MCF-7 breast cancer cells that had been treated with phorbol 12-myristate 13-acetate and found AREG could act as a growth factor (Shoyab, McDonald, Bradley, & Todaro, 1988). Additional studies went on to characterize AREG as an EGFR ligand due to its ability to partially compete with ^{125}I-EGF for binding to EGFR and AREG has since been classified to act exclusively through this receptor (Miyamoto et al., 2004; Salomon et al., 1995). Since its discovery, AREG has been implicated in several human cancers including breast cancer with some reports indicating that AREG is overexpressed in $\sim 50\%$ of human breast carcinomas (LeJeune et al., 1993). Studies show that AREG is expressed in breast tumor epithelial cells, rather than adjacent stroma, with many breast cancers displaying evidence that EGFR and AREG are coexpressed, suggesting that AREG autocrine loops may exist in a subset of patients (LeJeune et al., 1993).

Further investigation confirmed the existence of an AREG autocrine loop using the EGFR-positive SUM-149PT human breast cancer cell line (Berquin, Dziubinski, Nolan, & Ethier, 2001). SUM-149PT cells were originally isolated from a patient with TNBC and found to have constitutively active EGFR, while being EGF independent for growth (Baillo et al., 2011). It was determined that the EGFR activity in these cells was due to overexpression of AREG mRNA and protein (Berquin et al., 2001). The SUM-149PT cells display a type of self-sustaining autocrine loop between AREG and EGFR (Willmarth & Ethier, 2006), where the binding of AREG to EGFR causes increased transcription and secretion of AREG itself, allowing for more AREG to bind to other EGFR receptors, thus causing EGFR to be constitutively active (Baillo et al., 2011).

As shown in Fig. 2, even though EGF and AREG both serve as ligands for EGFR their signaling effects vary (Streicher et al., 2007). With regard to receptor turnover, Baldys et al. showed that the presence of AREG promotes the recycling of EGFR back to the cell surface preventing receptor degradation in the lysosome, whereas EGF induces rapid receptor downregulation (Baldys et al., 2009; Fig. 2). Differences between EGF and AREG functional activity are also observed. AREG interaction with EGFR is thought to promote cell motility and consequently tumor invasiveness (Willmarth et al., 2009). In the MCF10A human mammary epithelial cell model, cell motility and invasion was increased after stimulation with AREG. However, these effects were not seen when cells were stimulated with EGF (Willmarth & Ethier, 2006). Likewise, impairing AREG by shRNA knockdown in the SUM-149PT cell line inhibited invasion (Baillo et al., 2011).

Consistent with a proinvasive and metastatic role for AREG, activation of EGFR by AREG contributes to the synthesis, secretion, and activation of proteins that are involved with these processes (Silvy, Giusti, Martin, & Berthois, 2001). AREG was shown to have a significant relationship with the matrix metalloprotease-9 (MMP-9) protein (Kondapaka, Fridman, & Reddy, 1997), which is involved in the breakdown of the basement membrane creating an environment conducive for metastasis. When MMP-9 expression is increased, it modulates the expression of AREG, whereas AREG inhibition prohibits the expression of MMP-9 (Ma et al., 2010). Additional evidence stemming from AREG-induced gene expression network analysis implicates interleukin-1α (IL-1α) and IL-1β as key factors that regulates AREG-directed functional effects. Further investigation elucidated an AREG-mediated interleukin-1 (IL-1)/nuclear factor-κB (NF-κB)-positive feedback loop that is driven by EGFR activation (Streicher et al., 2007). Not only is IL-1 required for AR-dependent cell proliferation, but also it has been linked to the expression and regulation of MMPs, a relationship that is consistent with the metastatic role for AREG in breast cancer progression (Karin & Greten, 2005).

Although much less is known about how EGFR ligands regulate the metastatic functions of EGFR, these functions are driven in part by AREG/EGFR-mediated deregulation of the signaling described earlier. Additional evidence suggests that AREG/EGFR regulates fibronectin-dependent signaling as well (Kappler et al., 2015). Other evidence indicates a role for EGFR in the regulation of signaling molecules that mediate cell polarization or other features of epithelial differentiation (Wang et al., 1998).

Recently, EGFR has been shown to promote epithelial–mesenchymal transition (EMT), a process by which cells undergo a morphologic switch from a polarized epithelial phenotype to a mesenchymal fibroblastoid phenotype (Masuda et al., 2012; Thiery, 2002). EMT has been described as a key process for promoting tumor migration and invasion (Bernards & Weinberg, 2002; Mani et al., 2008, 2007; Radisky, 2005; Yang et al., 2004). Evidence suggests that the EGFR-tyrosine kinase inhibitor, erlotinib, inhibits migration and potentially reverses EMT (Zhang et al., 2009). Erlotinib treatment enhanced expression of E-cadherin but lowered expression of vimentin, implying a transition from the mesenchymal phenotype back to epithelial, a process referred to MET (Zhang et al., 2009). Arguably, mesenchymal cells could be more sensitive to erlotinib, rather than erlotinib-inducing MET. Consequently, a shift in the population of cells remaining, mainly epithelial in structure, would allow epithelial cells to take on a more evident role in breast cancer cell cultures or tumors. However, EMT is thought to result in transcriptional reprogramming of the tumor cell and its transition to a mesenchymal phenotype is thought to be promoted by abnormal survival signals through proteins: cMET, ERK1/2, and/or AKT (Buck et al., 2007). Consistent with this assertion, the Ras-ERK1/2 pathway, which becomes activated downstream of EGFR, has been shown to also regulate EMT, tumor invasion, and metastasis (Doehn et al., 2009; Masuda et al., 2012).

The major enigma in EGFR targeting in TNBC is that clinically approved EGFR inhibitors have not been as effective as predicted despite their proven ability to interfere with EGFR signaling. It has been suggested that this failure can be attributed to the ability of breast cancer cells to sustain receptor activity through continuously recycling and activation of EGFR (Lurje & Lenz, 2009; Martinazzi, Crivelli, Zampatti, & Martinazzi, 1993; Salomon et al., 1995). Given the connection between propagation of EGFR signaling and its relationship to endocytic trafficking of the receptor, it is not hard to imagine that disruption of EGFR's natural endocytic cycle could prompt intracellular changes associated with the cellular transformation process. Thus theoretically, better patient outcomes could potentially be seen in combining anti-EGFR therapies with medications that inhibit endosomal trafficking or more specifically promote EGFR turnover (Shapira et al., 2013).

3. COMMON MOLECULAR CHARACTERISTICS OF TNBC AND THEIR RELATIONSHIP TO EGFR

3.1 *BRCA1* and *BRCA2*

Several molecular characteristics are common to TNBC. BRCA1 and BRCA2 are among the most well known of the molecular changes associated with TNBC. Mutations in *BRCA1* and *BRCA2* are characterized by cellular defects in the process of homologous recombination in double-stranded DNA break repair (Schmadeka et al., 2014). Compared to the general population, patients with germline mutations in either of these genes have a 20- to 30-fold increased risk for breast cancer (Murphy & Moynahan, 2010). More than 75% of tumors arising in women carrying a germline mutation in *BRCA1* have a TN phenotype (Foulkes, Smith, & Reis-Filho, 2010) and mutation of *BRCA1* or *BRCA2* additionally correlates with development of ovarian cancer (Brose et al., 2002; Howlader et al., 2014). Moreover, *BRCA1* mutations and any resulting protein deficiency may be associated with a higher rate of TNBC in younger women diagnosed with breast cancer (Mahamodhossen, Liu, & Rong-Rong, 2013).

The high incidence of EGFR overexpression in TNBC suggests that the probability that *BRCA1* mutant breast cancers harbor EGFR overexpression is high. Recent evidence from both ovarian and breast cancer research shows that *BRCA1* loss in human breast and ovarian cancer samples and BRCA1-deficient breast cancer models leads to upregulation of EGFR (Burga et al., 2011; Li et al., 2013; van Diest, van der Groep, & van der Wall, 2006), suggesting that the activity of these signaling molecules are linked, presenting opportunity for targeting EGFR in BRCA1-deficient patients. Experimental modeling using *Brca1*-deficient mice confirms this notion as targeting EGFR was effective in increasing the time of disease-free survival of the *Brca1*-deficient animals treated with erlotinib compared to placebo (Burga et al., 2011). Consistent with these findings, evidence shows that BRCA1 transcriptionally suppresses AREG expression and thus, loss of BRCA1 derepresses AREG levels and activity leading to EGFR stabilization (Lamber, Horwitz, & Parvin, 2010). Thus, targeting elements of the EGFR pathway, including AREG, is a potential area of intervention in BRCA1-deficient patients.

3.2 *TP53*

Mutation or loss of *TP53* resulting in loss of p53 protein function is a common alteration found in TNBC with some reports as high as ∼90% (Lehmann & Pietenpol, 2014; Stefansson et al., 2011). The majority of missense mutations in p53 cluster into "hotspot" codons within the DNA-binding domain of p53 (R248Q, R248W, R175H, R273H, R273C, and G245S) and there is evidence that these mutations can provide oncogenic potential beyond the simple loss of p53 function (Liedtke et al.,

2008; Olivier et al., 2002, 2006). Because of the high rate of mutations in *TP53*, considerable interest lies in identifying small molecules that restore p53 activity (Lehmann & Pietenpol, 2014). Experimental evidence also shows that *TP53* mutation can also lead to a gain of function in p53 endowing tumor cells with metastatic and invasive abilities (Brosh & Rotter, 2009). Consequently, proper maintenance of p53 is critical for normal breast epithelial cell function.

Interestingly, the expression of mutant p53 is able to drive invasion and metastasis via an endosomal trafficking pathway by recycling integrins and EGFR back to the plasma membrane of tumor cells (Muller et al., 2009). Thus, the overexpression of EGFR in TNBC could be in part attributed to endosomal recycling powered by oncogenic *TP53* mutations (Manie et al., 2009). Targeting these gain-of-function mutations, restoring wild-type p53 function, or targeting cell cycle checkpoint vulnerabilities regulated by p53 have the potential to make a large clinical impact, as *TP53* alteration is a common event in TNBC (Lehmann & Pietenpol, 2014).

3.3 PI3K/PTEN/AKT

Deregulation of AKT signaling is frequently found in TNBC and largely occurs through the mutation of the AKT upstream regulatory protein PI3K (encoded by *PI3KCA*) or loss of function in *PTEN*. There are three isoforms of AKT: AKT-1, -2, and -3 (Hennessy, Smith, Ram, Lu, & Mills, 2005), and while gene mutations in *AKT* are rare, one of these isoforms, *AKT3*, is present in ER− tumors (Chin et al., 2014) and may be mutated in a subset of TNBC (O'Hurley et al., 2014). Mutations in *PI3KCA* have been found in all subtypes of breast cancer (Bader, Kang, Zhao, & Vogt, 2005) and are frequently associated with loss of *PTEN* in breast tumors (Saal et al., 2005). Although most commonly connected with activation of AKT, mutation of the gene encoding PI3K, *PI3KCA*, can aid in tumor progression via AKT-dependent and AKT-independent mechanisms (Vasudevan et al., 2009).

As a prominent survival pathway, PI3K-AKT signaling circumvents apoptotic cell death (Ilieva et al., 2003) and feedback signaling that enforces continuous signaling through this modality has been highlighted as a major mechanism for drug resistance (Hennessy et al., 2005). Consequently, significant effort has been made to focus on targeting this pathway in breast cancer and clinical trial analysis of AKT inhibition in breast cancer is ongoing.

EGFR is one of the main growth factor receptors that activate AKT signaling and much of the knowledge gained by studies focused on AKT signaling downstream of EGFR are specific to AKT survival signaling. However, recent evidence has also implicated AKT in regulating EGFR trafficking and degradation (Er, Mendoza, Mackey, Rameh, & Blenis, 2013), which could have significant impact on treatment modalities. Continued future investigation in this area may bring to light unexpected functions for AKT signaling and its relationship with endocytosis that could be applied to TNBC.

4. TUMOR ASSOCIATED MACROPHAGES IN BREAST CANCER METASTASIS AND THEIR RELATIONSHIP WITH EGFR SIGNALING

Emerging evidence strongly argues for immune cells as playing a significant role in breast cancer pathogenesis resulting in inflammation being proposed as the seventh hallmark of cancer (Mantovani & Sica, 2010). A distinct genetic signature enriched for immune cell signaling and transduction pathways has been identified in the IM subtype of TNBC, but the impact of such a genetic analysis on clinical outcome has yet to be determined (Lehmann et al., 2011; Masuda et al., 2013). Although multiple immunosuppressive cell types have been identified, such as myeloid-derived suppressor cells and regulatory T cells (T-reg), Tumor Associated Macrophages (TAMs) comprise the most abundant population in mammary tumors and exhibit a robust and unique influence upon disease. As such, infiltration of macrophages in human mammary tumors is strongly associated with high vascular grade, reduced relapse-free survival, and reduced overall survival and serves as an independent prognostic indicator of breast cancer (Leek et al., 1996). Thus, the balance between pro- and antitumor immunity is a critical factor in breast cancer with TAMs representing a major contributor to pathology.

Early studies using cytokine release assays suggested that cytokines are released from primary breast carcinomas cell lines (O'Sullivan et al., 1993). In turn, analysis of primary tumors showed that cells resembling activated macrophages released EGF but this was not seen in the normal or malignant epithelial cells (O'Sullivan et al., 1993). More recently, elegant intravital imaging studies have demonstrated that direct interactions between malignant cells and TAMs are required for migration and invasion in models of breast cancer. In line with the earlier observations, this phenomenon appears to be dependent upon TAM-derived EGF, and subsequent positive feedback loop of paracrine EGF and colony-stimulated factor-1 (CSF1) signaling with tumor cells (Goswami et al., 2005;

Wyckoff et al., 2007). Secretion of CSF1 from breast cancer cells recruits monocytes, macrophage precursors, from circulation and upregulates their expression of EGF upon conversion to TAMs (Goswami et al., 2005). In turn, within the tumor, stroma-activated macrophages but not normal or malignant epithelial cells, are the predominant contributors of EGF in primary breast carcinomas (O'Sullivan et al., 1993). EGF and CSF1 were shown to induce the formation of invadopodia in mammary adenocarcinoma cells and podosomes in TAMs, respectively, involved in extracellular matrix degradation and remodeling (Condeelis & Pollard, 2006). Multiphoton microscopy has illustrated that tumor cell intravasation occurs in association with perivascular macrophages in animal models of mammary tumors and that intravasation may occur in the absence of local angiogenesis (Wyckoff et al., 2007). The coordinated movement of cancer cells and perivascular macrophages is dependent upon this positive feedback loop established via CSF1 produced by cancer cells and EGF produced by perivascular TAMs (Goswami et al., 2005; Wyckoff et al., 2007). Furthermore, local secretion of EGF preferentially stimulates EGFR-expressing breast cancer cells, inducing the pluripotency gene *SOX-2* through activation of the transcription factor, signal transducer, and activator of transcription 3, enhancing their survival and proliferation and is negatively correlated with clinical outcome (Leek et al., 2000; Yang et al., 2013).

Additional EGFR ligands have likewise been implicated in paracrine signaling between TAMs and mammary tumor cells. However, the exact range of EGFR agonists secreted by TAMs is unknown. Recent evidence demonstrates that TAMs secrete HB-EGF and prominent HB-EGF staining was observed in human breast cancer samples (Hoerger et al., 2013). Within the same study, HB-EGF plasma levels were found to correlate with primary tumor size and lymph node dissemination of mammary breast carcinomas (Hoerger et al., 2013). Further evidence suggests that EGFR ligands are additionally produced by breast cancer cells, which participate in feedback loops that regulate expression of macrophage chemoattractants. AREG, TGFA, and HB-EGF were found to be highly expressed in the MDA MB 231 TNBC cell line. Knockdown of AREG or TGFA in MDA MB 231 cells decreased motility and expression of the macrophage chemoattractant, CSF1, and the monocyte chemoattractant, chemokine (C-C motif) ligand 2 (CCL2) (Nickerson, Mill, Wu, Riese, & Foley, 2013). *In vivo*, AREG and TGFA knockdown reduced tumor growth, angiogenesis, and macrophage attraction (Nickerson et al., 2013).

With TAMs being the most thoroughly studied innate immune component, the impact of TAM precursor tumor-associated monocytes on tumor progression is poorly understood. Interestingly, it was demonstrated that infiltrating monocytes are able to promote angiogenesis in a xenograft-based HER2/neu-derived mouse mammary tumor model (De Palma et al., 2005). Furthermore, it was shown that tumor-associated monocytes secrete EREG (Hoerger et al., 2013). However, subsequent analysis revealed that insignificant levels of EREG were found in human breast cancer patient plasma samples (Hoerger et al., 2013). As follows the significance of this observation remains unknown. As a whole, current evidence points to a prominent role for TAMs and their interaction with EGFR signaling in tumor cells as a major proponent of breast cancer metastasis.

5. FUTURE PERSPECTIVES ON THERAPY
5.1 Combination Therapy

Paradoxically, TNBCs show increased sensitivity to chemotherapy, such as anthracyclines and taxanes, compared to non-TNBC (Liedtke et al., 2008), which may be attributed to the highly proliferative nature of tumors in this subclass of breast cancer (Hudis & Gianni, 2011; Hugh et al., 2009; Loesch et al., 2010). While a large proportion of patients achieve pathological complete response from chemotherapy given in the adjuvant setting, for certain subsets the response to chemotherapy is short lived and the rate of disease relapse is high. As DNA damage and repair pathways are commonly deregulated in TNBC due to frequent mutation of *TP53*, *BRCA1*, and *BRCA2*, addition of platinum salts, which are DNA crosslinking agents, to chemotherapeutic regimens have shown added benefit (Hudis & Gianni, 2011). Furthermore, the defect in homologous recombination that results from loss-of-function mutations in *BRCA1* and *BRCA2* can cause TNBC cells to be dependent on DNA base excision repair, largely mediated by PARP, as an alternative DNA repair pathway. This idea led to the testing of PARP inhibitors for TNBC treatment and recent clinical trial data suggest that PARP inhibition shows potential. However, while early phase I and II clinical trials were encouraging (O'Shaughnessy et al., 2011; Tutt et al., 2010), later phase III trials showed no improvement in overall or progression-free survival after providing PARP inhibitor with platinum agent to metastatic TNBC patients enrolled in this study (O'Shaughnessy et al., 2011). As it currently stands, an appreciable lack of therapeutic options for TNBC patients that do not initially respond robustly to adjuvant chemotherapy is apparent. Moreover, no specific targeted therapies are currently applied as a part of a standard treatment regimen for breast cancer patients with the TN subtype.

Since TNBC is comprised of multiple distinct subgroups with distinct clinical characteristics and variable responses to therapy, in order to realize the ultimate goal of personalized medicine, patients will need to be screened for distinct biomarkers that will likely be developed alongside their associated targeted drug(s). As TNBC are heterogeneous, it may also need to be taken into consideration that multiple cell types could be present within a single tumor, such as those that are characteristically mesenchymal and those that are epithelial. As such, effective treatment could require agents tailored to specific cell types combined with a more detailed understanding of the mechanistic regulation of tangible targets, such as EGFR. As discussed in this chapter, previous trials showed that many patients with EGFR-expressing tumors did not respond to EGFR-targeted therapy (Masuda et al., 2012), which suggests that EGFR expression alone does not indicate tumor cell dependence on EGFR signaling but rather a codependence of multiple regulatory pathways that likely influence one another due to cross talk between pathways.

Combined targeting of EGFR and its associated downstream signaling components could alter chemosensitivity due to rewiring of apoptotic signaling networks (Masuda et al., 2012) and as already discussed, considerable effort is taking place in this arena. In addition to the signaling modalities, we have discussed presently, there is also evidence for significant interactions of EGFR with other receptor tyrosine kinases, and it is possible that such alternative signaling pathways are linked to therapeutic resistance (Buck et al., 2007). Indeed, most TNBC ($\sim$70%) overexpress genes involved in invasion and metastasis as well as genes involved in proliferation and resistance to apoptosis (Sorlie et al., 2001). Thus, combining EGFR-targeted therapy with drugs targeting alternative signaling pathways could improve efficacy (Masuda et al., 2012; Fig. 3).

It is certainly possible that the failure of EGFR-targeted therapy to produce a large effect on reducing breast tumor size could be pointing to a lack of therapeutic effect on the proliferative or survival functions of EGFR. One possible hypothesis is that this failure could indicate that EGFR's proliferative, metastatic, and cell survival functions are uncoupled in TNBC. Consequently, EGFR inhibitors could be effective toward EGFR-directed proliferation but not prosurvival or metastatic functions. Emerging evidence has also connected endosome protein networks to driving breast cancer cell

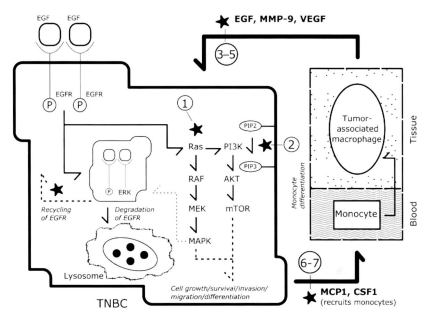

Figure 3 The relationship between tumor cells and tumor-associated macrophages (TAMs) in the tumor microenvironment. Tumor cells can secrete cytokines, like CCL2 and/or CSF1, which will recruit monocytes from the circulating blood to the tumor environment. These monocytes can differentiate into TAMs that secrete factors including EGF, MMP-9, and/or VEGF, promoting tumor cell migration and metastasis. Potential points of intervention are indicated (see numbered labels corresponding with Table 1).

migration through control of focal adhesion disassembly (Mendoza et al., 2013), likely through a Rac1-dependent mechanism (Balaji & Colicelli, 2013; Lichius et al., 2014). Consequently, the incorporation of multiple-targeted therapies directed at distinct domains of this transmembrane receptor, a better understanding of the specific functions of critical downstream effectors involved in signal transduction pathways driven by EGFR, determining the role of EGFR in specific cancer cell types, and understanding the intricate cross talk between molecules that drive TNBC needs to be a major focus for future development of therapeutic options for patients with early and metastatic breast cancer of the TN subtype (Kelly & Buzdar, 2013; Fig. 3).

5.2 Tumor-Associated Macrophage-Based Therapy

Macrophages have emerged as critical factors in breast cancer metastasis and thus represent an attractive cancer specific but tumor cell autonomous target for combinational therapy (Panni, Linehan, & DeNardo, 2013). Given the strong dependence of TNBC on EGFR signaling, inhibition of TAM-derived paracrine signaling between tumor cells and macrophages represents a novel mechanism for treatment. TAM-dependent EGF-mediated metastasis and cancer stem cell support provide specific targets for this mechanism (Fig. 3). Current intervention strategies have focused on inhibition of recruitment, reprogramming function, and depletion of TAMs and their progenitors. As indicated in Fig. 3 and Table 1, clinical strategies include targeting the prominent CSF1–CSF1 receptor and CCL2–CCL2 receptor signaling axis, which impairs monocyte recruitment to mammary tumors and consequently macrophage differentiation within mammary tumors (Abraham et al., 2010; Qian et al., 2011). Evidence shows that genetic manipulation, administration of neutralizing antibodies, or antisense RNA to ablate either CSF1 or CCL2 signaling inhibit the development of primary tumors as well as bone marrow and lung metastasis in the polyoma middle T mouse models of spontaneous breast cancer and xenotransplants of mammary tumor cells (DeNardo et al., 2011; Lin, Nguyen, Russell, & Pollard, 2001; Lu & Kang, 2009; Qian et al., 2011). Notably, inhibition of mammary tumor metastasis following inhibition of CSF1 signaling likely results from downregulation of TAM-derived EGF production (Lin et al., 2001; Patsialou et al., 2009), providing additional

Table 1 Potential Strategies for Targeting in TNBC

Numbers Correspond to Fig. 3	Molecule	Inhibitor(s)
1	Ras	Reolysin
2	PI3K–AKT–mTOR	MK-2206, GSK2141795, AZD2014, Everolimus, BYL719, BKM120
3	EGF	Gefitinib, erlotinib, cetuximab
4	MMP-9	(Not yet developed for clinical use)
5	VEGF	Bevacizumab, Pazopanib, Lucatinib
6	CCL2	Carlumab (CNT0888), MLN1202, PF-04136309
7	CSF1	PLX3397, PLX7486, IMC-CS4 (LY3022855), AMG820

Each number corresponds to a molecule in Fig. 3 that can be inhibited for therapeutic purposes. Potential agents are listed for each molecule. Therapeutic agents currently in clinical trials are listed.

support for the concept that EGFR signaling in tumor cells is critical for breast cancer progression and points toward a possible metastatic function for EGFR. Consequently, inhibition of TAM populations or their signaling properties represents an attractive, highly specific treatment option when considering combination interventions.

6. CONCLUSIONS

Significant research is directed toward identifying effective targeted therapies for TNBC. As an emerging biomarker, EGFR is an exciting target but a clear role for EGFR in TNBC is not fully defined. Future efforts should be geared toward delineating the molecular mechanisms that deregulate EGFR in TNBC, with a focus on EGFR ligands and posttranslational protein regulation within the endosome pathway. Additional focus on how to target EGFR effectively in conjunction with other targeted agents should take into account not only the heterogeneous populations of cells within the tumor but also the immune cells within the tumor microenvironment.

ACKNOWLEDGMENTS

The authors would like to thank Bryan Granger and acknowledge him for his assistance with the artwork associated with this chapter. The Yeh lab is supported by research funding from an American Cancer Society Institutional Research Grant (IRG-97-219-14) awarded to the Hollings Cancer Center at MUSC, by research funding from a Department of Defense grant (W81XWH-11-2-0229) at MUSC, and by an award from the Concern Foundation.

REFERENCES

Abraham, D., Zins, K., Sioud, M., Lucas, T., Schafer, R., Stanley, E. R., et al. (2010). Stromal cell-derived CSF-1 blockade prolongs xenograft survival of CSF-1-negative neuroblastoma. *International Journal of Cancer, 126*(6), 1339–1352. http://dx.doi.org/10.1002/ijc.24859.

Ahmad, G., Mohapatra, B. C., Schulte, N. A., Nadeau, S. A., Luan, H., Zutshi, N., et al. (2014). Cbl-family ubiquitin ligases and their recruitment of CIN85 are largely dispensable for epidermal growth factor receptor endocytosis. *The International Journal of Biochemistry & Cell Biology, 57*, 123–134. http://dx.doi.org/10.1016/j.biocel.2014.10.019.

Bader, A. G., Kang, S., Zhao, L., & Vogt, P. K. (2005). Oncogenic PI3K deregulates transcription and translation. *Nature Reviews. Cancer, 5*, 921–929. http://dx.doi.org/10.1038/nrc1753.

Baillo, A., Giroux, C., & Ethier, S. P. (2011). Knock-down of amphiregulin inhibits cellular invasion in inflammatory breast cancer. *Journal of Cellular Physiology, 226*(10), 2691–2701. http://dx.doi.org/10.1002/jcp.22620.

Balaji, K., & Colicelli, J. (2013). RIN1 regulates cell migration through RAB5 GTPases and ABL tyrosine kinases. *Communicative & Integrative Biology, 6*(5), e25421. http://dx.doi.org/10.4161/cib.25421.

Baldys, A., Gooz, M., Morinelli, T. A., Lee, M. H., Raymond, J. R., Jr., Luttrell, L. M., et al. (2009). Essential role of c-Cbl in amphiregulin-induced recycling and signaling of the endogenous epidermal growth factor receptor. *Biochemistry*, *48*(7), 1462–1473. http://dx.doi.org/10.1021/bi801771g.

Baselga, J., Gomez, P., Greil, R., Braga, S., Climent, M. A., Wardley, A. M., et al. (2013). Randomized phase II study of the anti-epidermal growth factor receptor monoclonal antibody cetuximab with cisplatin versus cisplatin alone in patients with metastatic triple-negative breast cancer. *Journal of Clinical Oncology*, *31*(20), 2586–2592. http://dx.doi.org/10.1200/JCO.2012.46.2408.

Bernards, R., & Weinberg, R. A. (2002). A progression puzzle. *Nature*, *418*(6900), 823. http://dx.doi.org/10.1038/418823a.

Berquin, I. M., Dziubinski, M. L., Nolan, G. P., & Ethier, S. P. (2001). A functional screen for genes inducing epidermal growth factor autonomy of human mammary epithelial cells confirms the role of amphiregulin. *Oncogene*, *20*(30), 4019–4028. http://dx.doi.org/10.1038/sj.onc.1204537.

Bhargava, R., Gerald, W. L., Li, A. R., Pan, Q., Lal, P., Ladanyi, M., et al. (2005). EGFR gene amplification in breast cancer: Correlation with epidermal growth factor receptor mRNA and protein expression and HER-2 status and absence of EGFR-activating mutations. *Modern Pathology*, *18*(8), 1027–1033. http://dx.doi.org/10.1038/modpathol.3800438.

Brose, M. S., Rebbeck, T. R., Calzone, K. A., Stopfer, J. E., Nathanson, K. L., & Weber, B. L. (2002). Cancer risk estimates for BRCA1 mutation carriers identified in a risk evaluation program. *Journal of the National Cancer Institute*, *94*(18), 1365–1372.

Brosh, R., & Rotter, V. (2009). When mutants gain new powers: News from the mutant p53 field. *Nature Reviews. Cancer*, *9*(10), 701–713. http://dx.doi.org/10.1038/nrc2693.

Buck, E., Eyzaguirre, A., Barr, S., Thompson, S., Sennello, R., Young, D., et al. (2007). Loss of homotypic cell adhesion by epithelial-mesenchymal transition or mutation limits sensitivity to epidermal growth factor receptor inhibition. *Molecular Cancer Therapeutics*, *6*(2), 532–541. http://dx.doi.org/10.1158/1535-7163.MCT-06-0462.

Burga, L. N., Hu, H., Juvekar, A., Tung, N. M., Troyan, S. L., Hofstatter, E. W., et al. (2011). Loss of BRCA1 leads to an increase in epidermal growth factor receptor expression in mammary epithelial cells, and epidermal growth factor receptor inhibition prevents estrogen receptor-negative cancers in BRCA1-mutant mice. *Breast Cancer Research*, *13*(2), R30. http://dx.doi.org/10.1186/bcr2850.

Burgering, B. M., & Coffer, P. J. (1995). Protein kinase B (c-Akt) in phosphatidylinositol-3-OH kinase signal transduction. *Nature*, *376*(6541), 599–602. http://dx.doi.org/10.1038/376599a0.

Chin, Y. R., Yoshida, T., Marusyk, A., Beck, A. H., Polyak, K., & Toker, A. (2014). Targeting Akt3 signaling in triple-negative breast cancer. *Cancer Research*, *74*(3), 964–973. http://dx.doi.org/10.1158/0008-5472.CAN-13-2175.

Clark, O., Botrel, T. E., Paladini, L., & Ferreira, M. B. (2014). Targeted therapy in triple-negative metastatic breast cancer: A systematic review and meta-analysis. *Core Evidence*, *9*, 1–11. http://dx.doi.org/10.2147/CE.S52197.

Colditz, G. A., Rosner, B. A., Chen, W. Y., Holmes, M. D., & Hankinson, S. E. (2004). Risk factors for breast cancer according to estrogen and progesterone receptor status. *Journal of the National Cancer Institute*, *96*(3), 218–228.

Condeelis, J., & Pollard, J. W. (2006). Macrophages: Obligate partners for tumor cell migration, invasion, and metastasis. *Cell*, *124*(2), 263–266. http://dx.doi.org/10.1016/j.cell.2006.01.007.

Davies, G. C., Ettenberg, S. A., Coats, A. O., Mussante, M., Ravichandran, S., Collins, J., et al. (2004). Cbl-b interacts with ubiquitinated proteins; differential functions of the

UBA domains of c-Cbl and Cbl-b. *Oncogene, 23*(42), 7104–7115. http://dx.doi.org/10.1038/sj.onc.1207952.

De Palma, M., Venneri, M. A., Galli, R., Sergi Sergi, L., Politi, L. S., Sampaolesi, M., et al. (2005). Tie2 identifies a hematopoietic lineage of proangiogenic monocytes required for tumor vessel formation and a mesenchymal population of pericyte progenitors. *Cancer Cell, 8*(3), 211–226. http://dx.doi.org/10.1016/j.ccr.2005.08.002.

DeNardo, D. G., Brennan, D. J., Rexhepaj, E., Ruffell, B., Shiao, S. L., Madden, S. F., et al. (2011). Leukocyte complexity predicts breast cancer survival and functionally regulates response to chemotherapy. *Cancer Discovery, 1*(1), 54–67. http://dx.doi.org/10.1158/2159-8274.CD-10-0028.

Dent, R., Trudeau, M., Pritchard, K. I., Hanna, W. M., Kahn, H. K., Sawka, C. A., et al. (2007). Triple-negative breast cancer: Clinical features and patterns of recurrence. *Clinical Cancer Research, 13*(15 Pt. 1), 4429–4434. http://dx.doi.org/10.1158/1078-0432.CCR-06-3045.

Doehn, U., Hauge, C., Frank, S. R., Jensen, C. J., Duda, K., Nielsen, J. V., et al. (2009). RSK is a principal effector of the RAS-ERK pathway for eliciting a coordinate promotile/invasive gene program and phenotype in epithelial cells. *Molecular Cell, 35*(4), 511–522. http://dx.doi.org/10.1016/j.molcel.2009.08.002.

Er, E. E., Mendoza, M. C., Mackey, A. M., Rameh, L. E., & Blenis, J. (2013). AKT facilitates EGFR trafficking and degradation by phosphorylating and activating PIKfyve. *Science Signaling, 6*(279), ra45. http://dx.doi.org/10.1126/scisignal.2004015.

Ettenberg, S. A., Keane, M. M., Nau, M. M., Frankel, M., Wang, L. M., Pierce, J. H., et al. (1999). Cbl-b inhibits epidermal growth factor receptor signaling. *Oncogene, 18*(10), 1855–1866. http://dx.doi.org/10.1038/sj.onc.1202499.

Foulkes, W. D., Smith, I. E., & Reis-Filho, J. S. (2010). Triple-negative breast cancer. *The New England Journal of Medicine, 363*(20), 1938–1948. http://dx.doi.org/10.1056/NEJMra1001389.

Goldhirsch, A., Wood, W. C., Gelber, R. D., Coates, A. S., Thurlimann, B., Senn, H. J., et al. (2007). Progress and promise: Highlights of the international expert consensus on the primary therapy of early breast cancer 2007. *Annals of Oncology, 18*(7), 1133–1144. http://dx.doi.org/10.1093/annonc/mdm271.

Goswami, S., Sahai, E., Wyckoff, J. B., Cammer, M., Cox, D., Pixley, F. J., et al. (2005). Macrophages promote the invasion of breast carcinoma cells via a colony-stimulating factor-1/epidermal growth factor paracrine loop. *Cancer Research, 65*(12), 5278–5283. http://dx.doi.org/10.1158/0008-5472.CAN-04-1853.

Harris, R. C., Chung, E., & Coffey, R. J. (2003). EGF receptor ligands. *Experimental Cell Research, 284*(1), 2–13.

Harris, A. L., Nicholson, S., Sainsbury, J. R., Farndon, J., & Wright, C. (1989). Epidermal growth factor receptors in breast cancer: Association with early relapse and death, poor response to hormones and interactions with neu. *Journal of Steroid Biochemistry, 34*(1-6), 123–131.

Hartman, Z., Zhao, H., & Agazie, Y. M. (2013). HER2 stabilizes EGFR and itself by altering autophosphorylation patterns in a manner that overcomes regulatory mechanisms and promotes proliferative and transformation signaling. *Oncogene, 32*(35), 4169–4180. http://dx.doi.org/10.1038/onc.2012.418.

Hennessy, B. T., Smith, D. L., Ram, P. T., Lu, Y., & Mills, G. B. (2005). Exploiting the PI3K/AKT pathway for cancer drug discovery. *Nature Reviews. Drug Discovery, 4*(12), 988–1004. http://dx.doi.org/10.1038/nrd1902.

Higashiyama, S., Iwabuki, H., Morimoto, C., Hieda, M., Inoue, H., & Matsushita, N. (2008). Membrane-anchored growth factors, the epidermal growth factor family: Beyond receptor ligands. *Cancer Science, 99*(2), 214–220. http://dx.doi.org/10.1111/j.1349-7006.2007.00676.x.

Hoerger, M., Epstein, R. M., Winters, P. C., Fiscella, K., Duberstein, P. R., Gramling, R., et al. (2013). Values and options in cancer care (VOICE): Study design and rationale for a patient-centered communication and decision-making intervention for physicians, patients with advanced cancer, and their caregivers. *BMC Cancer, 13*, 188. http://dx.doi.org/10.1186/1471-2407-13-188.

Howlader, N., Chen, V. W., Ries, L. A., Loch, M. M., Lee, R., DeSantis, C., et al. (2014). Overview of breast cancer collaborative stage data items–their definitions, quality, usage, and clinical implications: A review of SEER data for 2004–2010. *Cancer, 120*(Suppl. 23), 3771–3780. http://dx.doi.org/10.1002/cncr.29059.

Hudis, C. A., & Gianni, L. (2011). Triple-negative breast cancer: An unmet medical need. *The Oncologist, 16*(Suppl. 1), 1–11. http://dx.doi.org/10.1634/theoncologist.2011-S1-01.

Hugh, J., Hanson, J., Cheang, M. C., Nielsen, T. O., Perou, C. M., Dumontet, C., et al. (2009). Breast cancer subtypes and response to docetaxel in node-positive breast cancer: Use of an immunohistochemical definition in the BCIRG 001 trial. *Journal of Clinical Oncology, 27*(8), 1168–1176. http://dx.doi.org/10.1200/JCO.2008.18.1024.

Ilieva, H., Nagano, I., Murakami, T., Shiote, M., Shoji, M., & Abe, K. (2003). Sustained induction of survival p-AKT and p-ERK signals after transient hypoxia in mice spinal cord with G93A mutant human SOD1 protein. *Journal of the Neurological Sciences, 215*(1-2), 57–62.

Joazeiro, C. A., Wing, S. S., Huang, H., Leverson, J. D., Hunter, T., & Liu, Y. C. (1999). The tyrosine kinase negative regulator c-Cbl as a RING-type, E2-dependent ubiquitin-protein ligase. *Science, 286*(5438), 309–312.

Kappler, C. S., Guest, S. T., Irish, J. C., Garrett-Mayer, E., Kratche, Z., Wilson, R. C., et al. (2015). Oncogenic signaling in amphiregulin and EGFR-expressing PTEN-null human breast cancer. *Molecular Oncology, 9*(2), 527–543. http://dx.doi.org/10.1016/j.molonc.2014.10.006.

Karin, M., & Greten, F. R. (2005). NF-kappaB: Linking inflammation and immunity to cancer development and progression. *Nature Reviews. Immunology, 5*(10), 749–759. http://dx.doi.org/10.1038/nri1703.

Kelly, C. M., & Buzdar, A. U. (2013). Using multiple targeted therapies in oncology: Considerations for use, and progress to date in breast cancer. *Drugs, 73*(6), 505–515. http://dx.doi.org/10.1007/s40265-013-0044-0.

Kondapaka, S. B., Fridman, R., & Reddy, K. B. (1997). Epidermal growth factor and amphiregulin up-regulate matrix metalloproteinase-9 (MMP-9) in human breast cancer cells. *International Journal of Cancer, 70*(6), 722–726.

Lamber, E. P., Horwitz, A. A., & Parvin, J. D. (2010). BRCA1 represses amphiregulin gene expression. *Cancer Research, 70*(3), 996–1005. http://dx.doi.org/10.1158/0008-5472.CAN-09-2842.

Lee, D. C., Sunnarborg, S. W., Hinkle, C. L., Myers, T. J., Stevenson, M. Y., Russell, W. E., et al. (2003). TACE/ADAM17 processing of EGFR ligands indicates a role as a physiological convertase. *Annals of the New York Academy of Sciences, 995*, 22–38.

Leek, R. D., Hunt, N. C., Landers, R. J., Lewis, C. E., Royds, J. A., & Harris, A. L. (2000). Macrophage infiltration is associated with VEGF and EGFR expression in breast cancer. *The Journal of Pathology, 190*(4), 430–436. http://dx.doi.org/10.1002/(SICI)1096-9896(200003)190:4<430::AID-PATH538>3.0.CO;2-6.

Leek, R. D., Lewis, C. E., Whitehouse, R., Greenall, M., Clarke, J., & Harris, A. L. (1996). Association of macrophage infiltration with angiogenesis and prognosis in invasive breast carcinoma. *Cancer Research, 56*(20), 4625–4629.

Lehmann, B. D., Bauer, J. A., Chen, X., Sanders, M. E., Chakravarthy, A. B., Shyr, Y., et al. (2011). Identification of human triple-negative breast cancer subtypes and preclinical models for selection of targeted therapies. *The Journal of Clinical Investigation, 121*(7), 2750–2767. http://dx.doi.org/10.1172/JCI45014.

Lehmann, B. D., & Pietenpol, J. A. (2014). Identification and use of biomarkers in treatment strategies for triple-negative breast cancer subtypes. *The Journal of Pathology, 232*(2), 142–150. http://dx.doi.org/10.1002/path.4280.

LeJeune, S., Leek, R., Horak, E., Plowman, G., Greenall, M., & Harris, A. L. (1993). Amphiregulin, epidermal growth factor receptor, and estrogen receptor expression in human primary breast cancer. *Cancer Research, 53*(15), 3597–3602.

Levkowitz, G., Waterman, H., Ettenberg, S. A., Katz, M., Tsygankov, A. Y., Alroy, I., et al. (1999). Ubiquitin ligase activity and tyrosine phosphorylation underlie suppression of growth factor signaling by c-Cbl/Sli-1. *Molecular Cell, 4*(6), 1029–1040.

Li, D., Bi, F. F., Cao, J. M., Cao, C., Li, C. Y., & Yang, Q. (2013). Effect of BRCA1 on epidermal growth factor receptor in ovarian cancer. *Journal of Experimental & Clinical Cancer Research, 32*, 102. http://dx.doi.org/10.1186/1756-9966-32-102.

Lichius, A., Goryachev, A. B., Fricker, M. D., Obara, B., Castro-Longoria, E., & Read, N. D. (2014). CDC-42 and RAC-1 regulate opposite chemotropisms in Neurospora crassa. *Journal of Cell Science, 127*(Pt. 9), 1953–1965. http://dx.doi.org/10.1242/jcs.141630.

Liedtke, C., Mazouni, C., Hess, K. R., Andre, F., Tordai, A., Mejia, J. A., et al. (2008). Response to neoadjuvant therapy and long-term survival in patients with triple-negative breast cancer. *Journal of Clinical Oncology, 26*(8), 1275–1281. http://dx.doi.org/10.1200/JCO.2007.14.4147.

Lill, N. L., Douillard, P., Awwad, R. A., Ota, S., Lupher, M. L., Jr., Miyake, S., et al. (2000). The evolutionarily conserved N-terminal region of Cbl is sufficient to enhance downregulation of the epidermal growth factor receptor. *The Journal of Biological Chemistry, 275*(1), 367–377.

Lin, E. Y., Nguyen, A. V., Russell, R. G., & Pollard, J. W. (2001). Colony-stimulating factor 1 promotes progression of mammary tumors to malignancy. *The Journal of Experimental Medicine, 193*(6), 727–740.

Loesch, D., Greco, F. A., Senzer, N. N., Burris, H. A., Hainsworth, J. D., Jones, S., et al. (2010). Phase III multicenter trial of doxorubicin plus cyclophosphamide followed by paclitaxel compared with doxorubicin plus paclitaxel followed by weekly paclitaxel as adjuvant therapy for women with high-risk breast cancer. *Journal of Clinical Oncology, 28*(18), 2958–2965. http://dx.doi.org/10.1200/JCO.2009.24.1000.

Lu, X., & Kang, Y. (2009). Chemokine (C-C motif) ligand 2 engages CCR2+ stromal cells of monocytic origin to promote breast cancer metastasis to lung and bone. *The Journal of Biological Chemistry, 284*(42), 29087–29096. http://dx.doi.org/10.1074/jbc.M109.035899.

Luetteke, N. C., Qiu, T. H., Fenton, S. E., Troyer, K. L., Riedel, R. F., Chang, A., et al. (1999). Targeted inactivation of the EGF and amphiregulin genes reveals distinct roles for EGF receptor ligands in mouse mammary gland development. *Development, 126*(12), 2739–2750.

Lurje, G., & Lenz, H. J. (2009). EGFR signaling and drug discovery. *Oncology, 77*(6), 400–410. http://dx.doi.org/10.1159/000279388.

Ma, L., de Roquancourt, A., Bertheau, P., Chevret, S., Millot, G., Sastre-Garau, X., et al. (2001). Expression of amphiregulin and epidermal growth factor receptor in human breast cancer: Analysis of autocriny and stromal-epithelial interactions. *The Journal of Pathology, 194*(4), 413–419.

Ma, L., Huet, E., Serova, M., Berthois, Y., Calvo, F., Mourah, S., et al. (2010). Antisense inhibition of amphiregulin expression reduces EGFR phosphorylation in transformed human breast epithelial cells. *Anticancer Research, 30*(6), 2101–2106.

Mahamodhossen, Y. A., Liu, W., & Rong-Rong, Z. (2013). Triple-negative breast cancer: New perspectives for novel therapies. *Medical Oncology, 30*(3), 653. http://dx.doi.org/10.1007/s12032-013-0653-1.

Mani, S. A., Guo, W., Liao, M. J., Eaton, E. N., Ayyanan, A., Zhou, A. Y., et al. (2008). The epithelial-mesenchymal transition generates cells with properties of stem cells. *Cell*, *133*(4), 704–715. http://dx.doi.org/10.1016/j.cell.2008.03.027.

Mani, S. A., Yang, J., Brooks, M., Schwaninger, G., Zhou, A., Miura, N., et al. (2007). Mesenchyme Forkhead 1 (FOXC2) plays a key role in metastasis and is associated with aggressive basal-like breast cancers. *Proceedings of the National Academy of Sciences of the United States of America*, *104*(24), 10069–10074. http://dx.doi.org/10.1073/pnas.0703900104.

Manie, E., Vincent-Salomon, A., Lehmann-Che, J., Pierron, G., Turpin, E., Warcoin, M., et al. (2009). High frequency of TP53 mutation in BRCA1 and sporadic basal-like carcinomas but not in BRCA1 luminal breast tumors. *Cancer Research*, *69*(2), 663–671. http://dx.doi.org/10.1158/0008-5472.CAN-08-1560.

Mantovani, A., & Sica, A. (2010). Macrophages, innate immunity and cancer: Balance, tolerance, and diversity. *Current Opinion in Immunology*, *22*(2), 231–237. http://dx.doi.org/10.1016/j.coi.2010.01.009.

Marmor, M. D., & Yarden, Y. (2004). Role of protein ubiquitylation in regulating endocytosis of receptor tyrosine kinases. *Oncogene*, *23*(11), 2057–2070. http://dx.doi.org/10.1038/sj.onc.1207390.

Martinazzi, M., Crivelli, F., Zampatti, C., & Martinazzi, S. (1993). Relationships between epidermal growth factor receptor (EGF-R) and other predictors of prognosis in breast carcinomas. An immunohistochemical study. *Pathologica*, *85*(1100), 637–644.

Masuda, H., Baggerly, K. A., Wang, Y., Iwamoto, T., Brewer, T., Pusztai, L., et al. (2013). Comparison of molecular subtype distribution in triple-negative inflammatory and non-inflammatory breast cancers. *Breast Cancer Research*, *15*(6), R112. http://dx.doi.org/10.1186/bcr3579.

Masuda, H., Zhang, D., Bartholomeusz, C., Doihara, H., Hortobagyi, G. N., & Ueno, N. T. (2012). Role of epidermal growth factor receptor in breast cancer. *Breast Cancer Research and Treatment*, *136*(2), 331–345. http://dx.doi.org/10.1007/s10549-012-2289-9.

Mendoza, P., Ortiz, R., Diaz, J., Quest, A. F., Leyton, L., Stupack, D., et al. (2013). Rab5 activation promotes focal adhesion disassembly, migration and invasiveness in tumor cells. *Journal of Cell Science*, *126*(Pt. 17), 3835–3847. http://dx.doi.org/10.1242/jcs.119727.

Miyamoto, S., Hirata, M., Yamazaki, A., Kageyama, T., Hasuwa, H., Mizushima, H., et al. (2004). Heparin-binding EGF-like growth factor is a promising target for ovarian cancer therapy. *Cancer Research*, *64*(16), 5720–5727. http://dx.doi.org/10.1158/0008-5472.CAN-04-0811.

Moghal, N., & Sternberg, P. W. (1999). Multiple positive and negative regulators of signaling by the EGF-receptor. *Current Opinion in Cell Biology*, *11*(2), 190–196.

Muller, P. A., Caswell, P. T., Doyle, B., Iwanicki, M. P., Tan, E. H., Karim, S., et al. (2009). Mutant p53 drives invasion by promoting integrin recycling. *Cell*, *139*(7), 1327–1341. http://dx.doi.org/10.1016/j.cell.2009.11.026.

Murphy, C. G., & Moynahan, M. E. (2010). BRCA gene structure and function in tumor suppression: A repair-centric perspective. *Cancer Journal*, *16*(1), 39–47. http://dx.doi.org/10.1097/PPO.0b013e3181cf0204.

Nakajima, H., Ishikawa, Y., Furuya, M., Sano, T., Ohno, Y., Horiguchi, J., et al. (2014). Protein expression, gene amplification, and mutational analysis of EGFR in triple-negative breast cancer. *Breast Cancer*, *21*(1), 66–74. http://dx.doi.org/10.1007/s12282-012-0354-1.

Nechushtan, H., Vainer, G., Stainberg, H., Salmon, A. Y., Hamburger, T., & Peretz, T. (2014). A phase 1/2 of a combination of Cetuximab and Taxane for "triple negative" breast cancer patients. *Breast*, *23*(4), 435–438. http://dx.doi.org/10.1016/j.breast.2014.03.003.

Nicholson, S., Wright, C., Sainsbury, J. R., Halcrow, P., Kelly, P., Angus, B., et al. (1990). Epidermal growth factor receptor (EGFr) as a marker for poor prognosis in node-negative breast cancer patients: Neu and tamoxifen failure. *The Journal of Steroid Biochemistry and Molecular Biology*, *37*(6), 811–814.

Nickerson, N. K., Mill, C. P., Wu, H. J., Riese, D. J., 2nd., & Foley, J. (2013). Autocrine-derived epidermal growth factor receptor ligands contribute to recruitment of tumor-associated macrophage and growth of basal breast cancer cells in vivo. *Oncology Research*, *20*(7), 303–317.

O'Hurley, G., Daly, E., O'Grady, A., Cummins, R., Quinn, C., Flanagan, L., et al. (2014). Investigation of molecular alterations of AKT-3 in triple-negative breast cancer. *Histopathology*, *64*(5), 660–670. http://dx.doi.org/10.1111/his.12313.

Olivier, M., Eeles, R., Hollstein, M., Khan, M. A., Harris, C. C., & Hainaut, P. (2002). The IARC TP53 database: New online mutation analysis and recommendations to users. *Human Mutation*, *19*(6), 607–614. http://dx.doi.org/10.1002/humu.10081.

Olivier, M., Langerod, A., Carrieri, P., Bergh, J., Klaar, S., Eyfjord, J., et al. (2006). The clinical value of somatic TP53 gene mutations in 1,794 patients with breast cancer. *Clinical Cancer Research*, *12*(4), 1157–1167. http://dx.doi.org/10.1158/1078-0432.CCR-05-1029.

O'Shaughnessy, J., Osborne, C., Pippen, J. E., Yoffe, M., Patt, D., Rocha, C., et al. (2011). Iniparib plus chemotherapy in metastatic triple-negative breast cancer. *The New England Journal of Medicine*, *364*(3), 205–214. http://dx.doi.org/10.1056/NEJMoa1011418.

O'Sullivan, C., Lewis, C. E., Harris, A. L., & McGee, J. O. (1993). Secretion of epidermal growth factor by macrophages associated with breast carcinoma. *Lancet*, *342*(8864), 148–149.

Panico, L., D'Antonio, A., Salvatore, G., Mezza, E., Tortora, G., De Laurentiis, M., et al. (1996). Differential immunohistochemical detection of transforming growth factor alpha, amphiregulin and CRIPTO in human normal and malignant breast tissues. *International Journal of Cancer*, *65*(1), 51–56. http://dx.doi.org/10.1002/(SICI)1097-0215(19960103)65:1<51::AID-IJC9>3.0.CO;2-0.

Panni, R. Z., Linehan, D. C., & DeNardo, D. G. (2013). Targeting tumor-infiltrating macrophages to combat cancer. *Immunotherapy*, *5*(10), 1075–1087. http://dx.doi.org/10.2217/imt.13.102.

Parkin, D. M., Bray, F., Ferlay, J., & Pisani, P. (2005). Global cancer statistics, 2002. *CA: A Cancer Journal for Clinicians*, *55*(2), 74–108.

Patsialou, A., Wyckoff, J., Wang, Y., Goswami, S., Stanley, E. R., & Condeelis, J. S. (2009). Invasion of human breast cancer cells in vivo requires both paracrine and autocrine loops involving the colony-stimulating factor-1 receptor. *Cancer Research*, *69*(24), 9498–9506. http://dx.doi.org/10.1158/0008-5472.CAN-09-1868.

Pennock, S., & Wang, Z. (2008). A tale of two Cbls: Interplay of c-Cbl and Cbl-b in epidermal growth factor receptor downregulation. *Molecular and Cellular Biology*, *28*(9), 3020–3037. http://dx.doi.org/10.1128/MCB.01809-07.

Perou, C. M., Sorlie, T., Eisen, M. B., van de Rijn, M., Jeffrey, S. S., Rees, C. A., et al. (2000). Molecular portraits of human breast tumours. *Nature*, *406*(6797), 747–752. http://dx.doi.org/10.1038/35021093.

Qian, B. Z., Li, J., Zhang, H., Kitamura, T., Zhang, J., Campion, L. R., et al. (2011). CCL2 recruits inflammatory monocytes to facilitate breast-tumour metastasis. *Nature*, *475*(7355), 222–225. http://dx.doi.org/10.1038/nature10138.

Radisky, D. C. (2005). Epithelial-mesenchymal transition. *Journal of Cell Science*, *118*(Pt. 19), 4325–4326. http://dx.doi.org/10.1242/jcs.02552.

Rakha, E. A., & Chan, S. (2011). Metastatic triple-negative breast cancer. *Clinical Oncology (Royal College of Radiologists)*, *23*(9), 587–600. http://dx.doi.org/10.1016/j.clon.2011.03.013.

Saal, L. H., Holm, K., Maurer, M., Memeo, L., Su, T., Wang, X., et al. (2005). PIK3CA mutations correlate with hormone receptors, node metastasis, and ERBB2, and are mutually exclusive with PTEN loss in human breast carcinoma. *Cancer Research*, 65(7), 2554–2559. http://dx.doi.org/10.1158/0008-5472-CAN-04-3913.

Sainsbury, J. R., Farndon, J. R., Needham, G. K., Malcolm, A. J., & Harris, A. L. (1987). Epidermal-growth-factor receptor status as predictor of early recurrence of and death from breast cancer. *Lancet*, 1(8547), 1398–1402.

Sainsbury, J. R., Farndon, J. R., Sherbet, G. V., & Harris, A. L. (1985). Epidermal-growth-factor receptors and oestrogen receptors in human breast cancer. *Lancet*, 1(8425), 364–366.

Sainsbury, J. R., Nicholson, S., Angus, B., Farndon, J. R., Malcolm, A. J., & Harris, A. L. (1988). Epidermal growth factor receptor status of histological sub-types of breast cancer. *British Journal of Cancer*, 58(4), 458–460.

Salomon, D. S., Brandt, R., Ciardiello, F., & Normanno, N. (1995). Epidermal growth factor-related peptides and their receptors in human malignancies. *Critical Reviews in Oncology/Hematology*, 19(3), 183–232.

Schmadeka, R., Harmon, B. E., & Singh, M. (2014). Triple-negative breast carcinoma: Current and emerging concepts. *American Journal of Clinical Pathology*, 141(4), 462–477. http://dx.doi.org/10.1309/AJCPQN8GZ8SILKGN.

Schulze, W. X., Deng, L., & Mann, M. (2005). Phosphotyrosine interactome of the ErbB-receptor kinase family. *Molecular Systems Biology*, 1(2005), 0008. http://dx.doi.org/10.1038/msb4100012.

Shapira, I., Lee, A., Vora, R., & Budman, D. R. (2013). P53 mutations in triple negative breast cancer upregulate endosomal recycling of epidermal growth factor receptor (EGFR) increasing its oncogenic potency. *Critical Reviews in Oncology/Hematology*, 88(2), 284–292. http://dx.doi.org/10.1016/j.critrevonc.2013.05.003.

Shawarby, M. A., Al-Tamimi, D. M., & Ahmed, A. (2011). Very low prevalence of epidermal growth factor receptor (EGFR) protein expression and gene amplification in Saudi breast cancer patients. *Diagnostic Pathology*, 6, 57. http://dx.doi.org/10.1186/1746-1596-6-57.

Shoyab, M., McDonald, V. L., Bradley, J. G., & Todaro, G. J. (1988). Amphiregulin: A bifunctional growth-modulating glycoprotein produced by the phorbol 12-myristate 13-acetate-treated human breast adenocarcinoma cell line MCF-7. *Proceedings of the National Academy of Sciences of the United States of America*, 85(17), 6528–6532.

Silvy, M., Giusti, C., Martin, P. M., & Berthois, Y. (2001). Differential regulation of cell proliferation and protease secretion by epidermal growth factor and amphiregulin in tumoral versus normal breast epithelial cells. *British Journal of Cancer*, 84(7), 936–945. http://dx.doi.org/10.1054/bjoc.2000.1678.

Sorkin, A., & Goh, L. K. (2008). Endocytosis and intracellular trafficking of ErbBs. *Experimental Cell Research*, 314(17), 3093–3106. http://dx.doi.org/10.1016/j.yexcr.2008.08.013.

Sorkina, T., Huang, F., Beguinot, L., & Sorkin, A. (2002). Effect of tyrosine kinase inhibitors on clathrin-coated pit recruitment and internalization of epidermal growth factor receptor. *The Journal of Biological Chemistry*, 277(30), 27433–27441. http://dx.doi.org/10.1074/jbc.M201595200.

Sorlie, T., Perou, C. M., Tibshirani, R., Aas, T., Geisler, S., Johnsen, H., et al. (2001). Gene expression patterns of breast carcinomas distinguish tumor subclasses with clinical implications. *Proceedings of the National Academy of Sciences of the United States of America*, 98(19), 10869–10874. http://dx.doi.org/10.1073/pnas.191367098.

Stefansson, O. A., Jonasson, J. G., Olafsdottir, K., Hilmarsdottir, H., Olafsdottir, G., Esteller, M., et al. (2011). CpG island hypermethylation of BRCA1 and loss of pRb as co-occurring events in basal/triple-negative breast cancer. *Epigenetics*, 6(5), 638–649. http://dx.doi.org/10.4161/epi.6.5.15667.

Streicher, K. L., Willmarth, N. E., Garcia, J., Boerner, J. L., Dewey, T. G., & Ethier, S. P. (2007). Activation of a nuclear factor kappaB/interleukin-1 positive feedback loop by amphiregulin in human breast cancer cells. *Molecular Cancer Research, 5*(8), 847–861. http://dx.doi.org/10.1158/1541-7786.MCR-06-0427.

Thiery, J. P. (2002). Epithelial-mesenchymal transitions in tumour progression. *Nature Reviews. Cancer, 2*(6), 442–454. http://dx.doi.org/10.1038/nrc822.

Tutt, A., Robson, M., Garber, J. E., Domchek, S. M., Audeh, M. W., Weitzel, J. N., et al. (2010). Oral poly(ADP-ribose) polymerase inhibitor olaparib in patients with BRCA1 or BRCA2 mutations and advanced breast cancer: A proof-of-concept trial. *Lancet, 376*(9737), 235–244. http://dx.doi.org/10.1016/S0140-6736(10)60892-6.

Valentin, M. D., da Silva, S. D., Privat, M., Alaoui-Jamali, M., & Bignon, Y. J. (2012). Molecular insights on basal-like breast cancer. *Breast Cancer Research and Treatment, 134*(1), 21–30. http://dx.doi.org/10.1007/s10549-011-1934-z.

van Diest, P. J., van der Groep, P., & van der Wall, E. (2006). EGFR expression predicts BRCA1 status in patients with breast cancer. *Clinical Cancer Research, 12*(2), 670. http://dx.doi.org/10.1158/1078-0432.CCR-05-2098, author reply 671.

Vasudevan, K. M., Barbie, D. A., Davies, M. A., Rabinovsky, R., McNear, C. J., Kim, J. J., et al. (2009). AKT-independent signaling downstream of oncogenic PIK3CA mutations in human cancer. *Cancer Cell, 16*(1), 21–32. http://dx.doi.org/10.1016/j.ccr.2009.04.012.

Wang, F., Weaver, V. M., Petersen, O. W., Larabell, C. A., Dedhar, S., Briand, P., et al. (1998). Reciprocal interactions between beta1-integrin and epidermal growth factor receptor in three-dimensional basement membrane breast cultures: A different perspective in epithelial biology. *Proceedings of the National Academy of Sciences of the United States of America, 95*(25), 14821–14826.

Willmarth, N. E., Baillo, A., Dziubinski, M. L., Wilson, K., Riese, D. J., 2nd., & Ethier, S. P. (2009). Altered EGFR localization and degradation in human breast cancer cells with an amphiregulin/EGFR autocrine loop. *Cellular Signalling, 21*(2), 212–219. http://dx.doi.org/10.1016/j.cellsig.2008.10.003.

Willmarth, N. E., & Ethier, S. P. (2006). Autocrine and juxtacrine effects of amphiregulin on the proliferative, invasive, and migratory properties of normal and neoplastic human mammary epithelial cells. *The Journal of Biological Chemistry, 281*(49), 37728–37737. http://dx.doi.org/10.1074/jbc.M606532200.

Wyckoff, J. B., Wang, Y., Lin, E. Y., Li, J. F., Goswami, S., Stanley, E. R., et al. (2007). Direct visualization of macrophage-assisted tumor cell intravasation in mammary tumors. *Cancer Research, 67*(6), 2649–2656. http://dx.doi.org/10.1158/0008-5472.CAN-06-1823.

Yang, J., Liao, D., Chen, C., Liu, Y., Chuang, T. H., Xiang, R., et al. (2013). Tumor-associated macrophages regulate murine breast cancer stem cells through a novel paracrine EGFR/Stat3/Sox-2 signaling pathway. *Stem Cells, 31*(2), 248–258. http://dx.doi.org/10.1002/stem.1281.

Yang, J., Mani, S. A., Donaher, J. L., Ramaswamy, S., Itzykson, R. A., Come, C., et al. (2004). Twist, a master regulator of morphogenesis, plays an essential role in tumor metastasis. *Cell, 117*(7), 927–939. http://dx.doi.org/10.1016/j.cell.2004.06.006.

Zhang, D., LaFortune, T. A., Krishnamurthy, S., Esteva, F. J., Cristofanilli, M., Liu, P., et al. (2009). Epidermal growth factor receptor tyrosine kinase inhibitor reverses mesenchymal to epithelial phenotype and inhibits metastasis in inflammatory breast cancer. *Clinical Cancer Research, 15*(21), 6639–6648. http://dx.doi.org/10.1158/1078-0432.CCR-09-0951.

Zheng, N., Wang, P., Jeffrey, P. D., & Pavletich, N. P. (2000). Structure of a c-Cbl-UbcH7 complex: RING domain function in ubiquitin-protein ligases. *Cell, 102*(4), 533–539.

CHAPTER EIGHT

The Quest for an Effective Treatment for an Intractable Cancer: Established and Novel Therapies for Pancreatic Adenocarcinoma

Bridget A. Quinn[*], Nathaniel A. Lee[*,†], Timothy P. Kegelman[*], Praveen Bhoopathi[*], Luni Emdad[*,‡,§], Swadesh K. Das[*,‡,§], Maurizio Pellecchia[¶], Devanand Sarkar[*,‡,§], Paul B. Fisher[*,‡,§,1]

[*]Department of Human and Molecular Genetics, Virginia Commonwealth University, School of Medicine, Richmond, Virginia, USA
[†]Department of Surgery, Virginia Commonwealth University, School of Medicine, Richmond, Virginia, USA
[‡]VCU Institute of Molecular Medicine, Virginia Commonwealth University, School of Medicine, Richmond, Virginia, USA
[§]VCU Massey Cancer Center, Virginia Commonwealth University, School of Medicine, Richmond, Virginia, USA
[¶]Sanford-Burnham Medical Research Institute, La Jolla, California, USA
[1]Corresponding author: e-mail address: pbfisher@vcu.edu

Contents

1. Pancreatic Cancer	284
2. Current Pancreatic Cancer Therapies	285
2.1 Standards of Care—An Overview	285
2.2 5-Fluorouracil	287
2.3 Gemcitabine	288
2.4 Gemcitabine Combinations	289
3. Novel Therapeutic Strategies	292
3.1 Ephrin Receptor Targeting	292
3.2 Sabutoclax and Minocycline	295
3.3 Poly I:C	297
4. Future Perspectives	301
Acknowledgments	302
References	302

Abstract

With therapies that date back to the 1950s, and few newly approved treatments in the last 20 years, pancreatic cancer remains a significant challenge for the development of novel therapeutics. Current regimens have successfully extended patient survival, in uracil uptake into DNA compared to normal tissue (Rutman, Cantarow, & Paschkis, 1954), and early nucleic acid analogues showed promising but limited antitumor activity (Jaffe, Handschumacher, & Welch, 1957; Stock, 1954). Decades after its introduction, 5-FU remains widely used for numerous cancer indications, especially intractable pancreatic cancer and metastatic colorectal cancer, but also for breast, head and neck, ovarian, and gastric cancers (Sargent et al., 2009; Wilson et al., 2014).

Orally administered derivatives of 5-FU, including capecitabine, rely on metabolic conversion via enzymes overexpressed in tumor cells (Liu, Cao, Russell, Handschumacher, & Pizzorno, 1998; Mori et al., 2000). Breakdown of 5-FU itself results in the production of the toxic metabolite fluoro-β-alanine, which can lead to deleterious cardiac and neurological side effects (Kato et al., 2001). Formulations that include 5-chloro-2,4-dihydroxypyridine (also known as gimeracil) slow the metabolism of 5-FU, leading to both higher availability to tumor cells and lower levels of fluoro-β-alanine (Kato et al., 2001). 5-FU and other inhibitors of TS remain an integral part of combination therapies, with expanding applications and new combinations being tested. For example, inhibitors of DNA repair enzymes, especially those involved in base excision repair such as APE-1, UDG, and PARP1, potentiate the effects of TS inhibitors (Al-Safi, Odde, Shabaik, & Neamati, 2012; Bulgar et al., 2012; Geng, Huehls, Wagner, Huntoon, & Karnitz, 2011; Huehls et al., 2011; Simeonov et al., 2009; Weeks, Fu, & Gerson, 2013; Wilson et al., 2014).

2.3 Gemcitabine

In 1997, Burris et al. published a clinical study comparing gemcitabine to 5-fluorouracil for the treatment of pancreatic cancer. In this study, 126 patients were enrolled with 63 per treatment group. 23.8% of patients showed clinical benefit with gemcitabine, as compared to only 4.8% of 5-FU-treated patients. The median survival was shown to be 5.65 months for gemcitabine and 4.41 months for 5-FU. Finally, 18% of patients treated with gemcitabine were alive at 12-month time point, while survival at this time point for patients treated with 5-FU was only 2% (Burris et al., 1997). This trial helped encourage the FDA to approve gemcitabine for the treatment of pancreatic cancer in 1998. Gemcitabine is currently a standard treatment used for patients with pancreatic cancer. Despite this, the drug only provides minimal benefit to patients.

Similar to 5-FU, gemcitabine's cytotoxic effects are due to its ability to act as a pyrimidine analog, specifically for deoxycytidine triphosphate, which allows it to be incorporated into DNA during replication. After gemcitabine is incorporated, another nucleotide may be added to the chain, but inhibition of chain elongation subsequently occurs. DNA damage repair is not able to remove the drug and, consequently, apoptosis occurs (Moysan, Bastiat, & Benoit, 2013).

Gemcitabine enters the cell through multiple cell membrane transporters, though the sodium-independent transporter, hENT1, has been shown to preferentially transport gemcitabine (Moysan et al., 2013). Although there are multiple mechanisms of gemcitabine resistance, one important mechanism associates with expression of this protein. Giovannetti et al. showed that patients with tumors that express high amounts of hENT1 have a greater survival advantage with gemcitabine treatment as compared to those with lower hENT1 expression (Giovannetti et al., 2006). Patients with higher hENT1 expression have tumors that can more readily take up gemcitabine, leading to an increased clinical benefit. However, in many tumors, low expression of gemcitabine transporters translates to a need for the drug to be administered frequently and at high doses, two things that can add significantly to drug toxicity.

2.4 Gemcitabine Combinations

Since gemcitabine is currently the standard of care in pancreatic cancer treatment, many preclinical and clinical studies focus on the combination of gemcitabine with another chemotherapeutic or targeted agent (Cunningham et al., 2009). The use of combination therapy in cancer treatment is quickly becoming the mainstay and a strategy that has the greatest chance of success in treating patients. This approach offers promise in treating patients with pancreatic cancers that may be resistant to one of the two or more agents used in therapy, but sensitive to other components of the combination therapy. While multiple combinations with gemcitabine have been evaluated, only a few have yielded encouraging data or resulted in a change in approved pancreatic cancer therapy.

2.4.1 Gemcitabine–Erlotinib

Human epidermal growth factor receptor, HER1/EGFR, is often overexpressed in pancreatic cancer. Erlotinib, a HER1/EGFR inhibitor, has been shown to block the intrinsic tyrosine kinase activity and thus downstream signaling through the RAF–ERK pathway *in vitro* and in xenograft tumor models. When used in combination with other chemotherapeutics, erlotinib potentiates their effects, enhancing cell death (Ng, Tsao,

Nicklee, & Hedley, 2002). The efficacy of erlotinib in combination with gemcitabine was evaluated in an international stage III clinical trial enrolling patients with locally advanced and metastatic pancreatic adenocarcinoma. Patients received gemcitabine plus erlotinib or gemcitabine and placebo. The median overall survival reached significance with 6.24 months in the combination therapy arm and 5.91 months with monotherapy. One-year survival was 23% and 17%, respectively. Progression-free survival was also improved with combination therapy, 3.75 months versus 3.55 months. Toxicity was similar between the two arms, though rash was significantly more common in the erlotinib arm (Moore et al., 2007). Rash development was correlated with improved outcomes, though subsequent studies showed this effect was lost with escalating dosages until rash induction (Van Cutsem et al., 2014). While these endpoints reached statistical significance, their clinical relevance was limited suggesting efficacy of this regimen may only be applicable to subgroups of patients.

2.4.2 Gemcitabine-FOLFIRINOX

FOLFIRINOX, a combination therapy including leucovorin, 5-FU, irinotecan, and oxaliplatin, combines drugs with distinct toxicities and known synergistic effects. 5-FU, discussed above, is a pyrimidine analog that irreversibly inhibits TS leading to cell death in rapidly replicating cells. Leucovorin, also known as folinic acid, potentiates the effect of 5-FU by also inhibiting the activity of TS (Mullany, Svingen, Kaufmann, & Erlichman, 1998). Irinotecan inhibits topoisomerase 1, limiting DNA replication (Ueno et al., 2007), while oxaliplatin causes cross-link formation between DNA strands leading to decreased replication and cell death (Zeghari-Squalli, Raymond, Cvitkovic, & Goldwasser, 1999).

FOLFIRINOX was first reported as a therapy for pancreatic adenocarcinoma in 2010 and was subsequently evaluated in the ACCORD-11 trial that found it to have significant promise compared to the previous standard of care, gemcitabine monotherapy. Overall survival in the FOLFIRINOX group was 11.1 months compared to 6.8 months in the gemcitabine group. This was at the cost of increased side effects, including neutropenia, thrombocytopenia, anemia, sensory neuropathy, diarrhea, and transaminitis (Conroy et al., 2011). Despite increased toxicity, global health status and time to deterioration were significantly improved in the treatment arm (Gourgou-Bourgade et al., 2013). Considering these side effects, FOLFIRINOX is only recommended as first line treatment for patients with good performance status.

Although not all patients are considered appropriate for systemic therapy with FOLFIRINOX, it is the first major addition to the chemotherapy arsenal for pancreatic cancer in over a decade. Numerous studies are underway to minimize toxicities, maximize efficacy, and to clarify the role of FOLFIRINOX in neoadjuvant and adjuvant settings. A combination therapy using gemcitabine with abraxane followed by FOLFIRINOX has been reported to show some efficacy, and trials are underway to further evaluate this therapeutic option (Kunzmann et al., 2014; Marsh, Talamonti, Katz, & Herman, 2015).

2.4.3 Gemcitabine–Abraxane

Paclitaxel is a commonly used chemotherapeutic drug most often used in breast, lung, and ovarian cancer, and AIDS-related sarcomas. As a microtubule inhibitor, paclitaxel acts to stabilize polymerized microtubules during mitosis, thus leading to cell cycle arrest in the G2 and M phases. Solubility is a major issue with this drug, as it must be administered in a solution with Cremophor and dehydrated ethanol. These vehicles can have toxic effects on their own and have led to the inability to use higher doses of this chemotherapeutic drug in patients. Furthermore, Cremophor-bound paclitaxel has a tendency to form micelles that trap the drug in the center, limiting its efficacy (Hoy, 2014).

Recently, a novel formulation of paclitaxel has been developed in which the drug is bound to albumin (nab-paclitaxel or abraxane). High-pressure homogenization is used to combine albumin and paclitaxel. Once injected, the drug is free to bind/unbind albumin or other molecules in the blood. This results in greater amounts of unbound drug in the circulation as compared to the Cremophor-bound formulation. However, bound nab-paclitaxel tends to bind albumin, which helps to facilitate its entry into tumor cells, thus nab-paclitaxel can use normal albumin transport mechanisms to gain entry. Additionally, some albumin-binding proteins, such as SPARC, show high prevalence in the tumor microenvironment, although the clinical significance of this is still unclear (Hoy, 2014). In mouse studies, pancreatic xenograft tumors were found to have a 2.8-fold increase in intratumoral gemcitabine concentration when the drug was administered in combination with nab-paclitaxel, which may be due to nab-paclitaxel-induced disruption of the stroma (Alvarez et al., 2013).

A Phase III clinical study in patients with metastatic pancreatic cancer evaluated nab-paclitaxel + gemcitabine versus gemcitabine alone in a total of 861 patients. The median survival was 8.5 months in the nab-paclitaxel + gemcitabine group versus 6.7 months in the gemcitabine group. The survival rate at 1 year was 35% versus 22%, and 9% versus 4% at 2 years. The response rate was 23% for combination group and 7% for gemcitabine alone. Toxicities included neutropenia, fatigue, and neuropathy (Von Hoff et al., 2013). These results encouraged the FDA to approve the combination of gemcitabine and nab-paclitaxel (gem-abraxane) for metastatic pancreatic cancer.

3. NOVEL THERAPEUTIC STRATEGIES

Toxicity is a major issue with many chemotherapeutic agents, and as a result, the creation of more targeted therapies with lower risks of toxicity has become an attractive strategy in developing cancer therapeutics. Many conventional chemotherapeutic drugs work well in killing cells, but because they also target normal cells, often lead to high levels of toxicity. Targeted therapies focus on specifically attacking cancer cells while sparing normal cells, thereby reducing side effects. One specific strategy of targeted therapy involves modifying currently used drugs to make them cancer specific. This often involves identifying an extracellular cancer cell biomarker that the modified drug can target. One such target is the ephrin receptor.

3.1 Ephrin Receptor Targeting

Ephrin receptors are a family of tyrosine kinase receptors involved in neuronal connectivity, blood vessel development, and cell–cell interactions. EphA2 was identified in 1990 and is expressed in the majority of epithelial cells. In cancer cells, EphA2 is highly overexpressed and encourages communication not only between individual cancer cells but also between cancer cells and surrounding stromal or vascular cells. EphA2 overexpression also correlates with poor prognosis in patients. Despite EphA2 overexpression, expression of its ligand, EphrinA1, often remains normal even in a cancerous state. This can lead to accumulation of inactivated EphA2 and subsequent oncogenic activity (Tandon, Vemula, & Mittal, 2011).

EphA2 is being actively studied as a potential target for developing cancer therapies and for tumor diagnosis. Targeting peptides (YSAYPDSVPMMS) were produced that, similar to the natural ligand for this receptor, selectively bind to EphA2 and cause receptor activation and internalization (Wang et al., 2012). These peptides when linked to commonly used chemotherapeutic drugs provide a specific delivery strategy for these drugs to tumor cells. Once the receptor is activated, the peptide and its attached drug are internalized into a lysosome, where the peptide is degraded and the drug is free to exert its toxic effects on the cell (Wang et al., 2012). Previous studies have shown that paclitaxel conjugated with these peptides shows increased efficacy in prostate and renal cancers (Wang et al., 2013).

To improve the EphA2-targeting peptide (YSA) as a delivery agent for chemotherapeutic agents to prostate cancer and other EpA2-overexpressing cancer cells (such as renal cancer) and to make this novel delivery system more drug-like, nonnatural amino acids were introduced into the targeting peptide, referred to as the YNH peptide. In the modified YNH peptide, norleucine and homoserine replace the two methionine residues of YSA, and D-tyrosine replaces the L-tyrosine in the first position of the YNH peptide, resulting in dYNH (Wang et al., 2013). Both YNH and dYNH were complexed with paclitaxel (YNH-PTX and dYNH-PTX, respectively) and evaluated both *in vitro* and *in vivo* for activity in human prostate and mouse renal cancer cells. dYNH-PTX displayed enhanced stability versus YSA-PTX or YNH-PTX in mouse serum with enhanced antitumor and antivascular activity as compared with vehicle or paclitaxel treatments (Wang et al., 2013). Moreover, these proof-of-principle studies support the concept of structurally modifying the EphA2 ligand to develop drug conjugates with potential to treat a variety of cancer types that display elevated EphA2 expression.

Recent studies by Barile et al. have scrutinized the chemical determinants responsible for the stability and degradation of the EphA2-targeting peptides in serum (Barile et al., 2014). These analyses were approached by modifying both the peptide and the linker between the peptide and paclitaxel in the drug conjugate. Modifications were identified that created drug conjugates with enhanced attributes, including enhanced stability in rat plasma and an ability to reduce tumor size in a human prostate cancer xenograft model in comparison with animals treated with paclitaxel alone. Critical rate-limiting degradation sites were identified in the peptide–drug conjugates, providing a path forward for developing next-generation targeting molecules that display further enhanced stability and potentially therapeutic efficacy (Barile et al., 2014).

Although short peptides represent potentially efficacious agents with high specificity, these positive traits are often a trade-off since stability and degradation are significant issues when they are used *in vivo*. In these contexts, the EphA2–drug conjugates previously produced, although quite active *in vivo*, do not have long half-lives in this setting. The "Holy Grail" for these delivery agents would be a molecule that is appropriately optimized to display robust and specific tumor targeting with exceptional stability and the capacity to be conjugated with a diverse range of therapeutic drugs. Progress toward achieving this objective has recently been obtained (Wu et al., 2015). A unique EphA2 tumor-targeting agent has been developed, 123B9, through a combination of NMR-guided structure activity relationships with appropriate biochemical and cell culture studies. Complexing 123B9 with paclitaxel results in enhanced bioactivity versus paclitaxel alone in both a human pancreatic cancer xenograft model and a lung colonization and metastasis model. The current data suggest that the enhanced activity in this latter model may relate to its ability to target the tumor vasculature. Further experiments are required to confirm this hypothesis. Additionally, the approaches used in these newer studies provide a means of deriving and optimizing additional tumor-selective homing agents.

Although outcomes remain poor overall, chemotherapy remains the mainstay of pancreatic cancer treatment regimens. There is a vital need to develop novel therapies that provide greater clinical benefit to patients. A recent study evaluated modifications of both gemcitabine and paclitaxel against pancreatic cancer. Gemcitabine, the current first-line treatment for pancreatic cancer, does not offer a great therapeutic benefit to patients. Recent work aimed to investigate a modified version of gemcitabine as an alternate to the traditional drug (unpublished data: B.A. Quinn, E. Barile, S. Wang, B. Wu, M. Pellecchia, P.B. Fisher). This modified drug, YNH-gemcitabine, consists of the drug with the first-generation peptide attached (YNH). This peptide was designed to specifically bind to EphA2 receptor, which is overexpressed on pancreatic cancer cells (Van den Broeck, Vankelecom, Van Eijsden, Govaere, & Topal, 2012). The result is that the attached gemcitabine is internalized into the cell via EphA2, bypassing its normal mechanism of cellular entry. In principle, this allows a greater amount of gemcitabine to enter the pancreatic cancer cell. This first-generation modified drug showed greater tumor growth inhibition and prolonged survival than gemcitabine in a xenograft model of pancreatic cancer (unpublished data: B.A. Quinn, B. Wu, S.K.Das, M. Pellecchia, P.B. Fisher).

However, despite these encouraging initial results in pancreatic cancer, the YNH peptide was found to have less than optimal plasma stability (Barile et al., 2014). The terminal tyrosine of this peptide, which is essential for specific EphA2 binding, was shown to be susceptible to aminopeptidases in the blood, leading to degradation. This resulted in a half-life of only a few minutes. To improve the half-life, a newer derivative of the previously developed YNH family of compounds, 123B9, described above, was created. 123B9 contains a synthetic tyrosine that is resistant to aminopeptidase degradation, leading to a significantly longer half-life in blood (approximately 4 h) and, hopefully, greater efficacy as well (Wu et al., 2015). Due to the recent approval of gemcitabine and abraxane as a first-line therapy for metastatic pancreatic cancer, 123B9-paclitaxel and gemcitabine were also evaluated with promising results (unpublished data: B.A. Quinn, B. Wu, S.K. Das, M. Pellecchia, P.B. Fisher). Additional studies focusing on comparing 123B9-gemcitabine to gemcitabine in pancreatic cancer are currently in progress.

3.2 Sabutoclax and Minocycline

Sabutoclax, a small-molecule BH3 mimetic, binds to and inhibits the function of the antiapoptotic Bcl-2 proteins (Azab et al., 2012; Dash et al., 2011; Goff et al., 2013; Hedvat et al., 2012; Jackson et al., 2012; Placzek et al., 2011; Quinn et al., 2011; Thomas et al., 2013; Varadarajan et al., 2013; Wei et al., 2010). In recent studies, the efficacy of a novel combination was evaluated employing sabutoclax and minocycline, a synthetic tetracycline and understudied potential anticancer agent in the treatment of pancreatic cancer (Quinn et al., 2015). In addition to testing this combination in multiple pancreatic cancer cell lines *in vitro*, several mouse models were also used to scrutinize these drugs *in vivo*, including the commonly used KPC mouse model (Fig. 1). Sabutoclax induced growth arrest and apoptosis in pancreatic cancer cells and synergized with minocycline. Together, these two drugs showed profound cytotoxicity that was caspase dependent and occurred through the mitochondrial pathway of apoptosis (Quinn et al., 2015). Furthermore, the toxicity induced by sabutoclax and minocycline was reliant upon loss of phosphorylated STAT3, with reintroduction of activated Stat3 capable of rescuing cells from toxicity. *In vivo* work showed that this combination inhibited tumor growth in immune-deficient and immune-competent models and prolonged survival in the KPC transgenic mouse (Fig. 2). Sabutoclax and minocycline promoted profound cytotoxicity in pancreatic cancer, both *in vitro* and in multiple *in vivo* animal models providing significant survival benefits (Quinn et al., 2015). These drugs offer a novel and exciting direction for developing potential effective therapeutic options for patients with this devastating disease (Fig. 3).

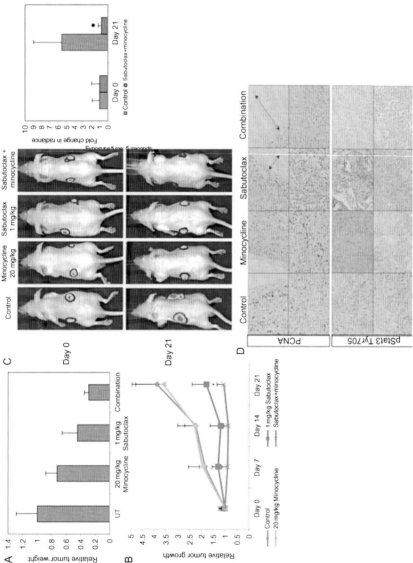

Figure 1 Sabutoclax decreases tumor growth in a subcutaneous human pancreatic tumor xenograft model and this effect is enhanced by the addition of minocycline. $n=5$ mice/group. (A) Relative tumor weight as normalized to the control animals at the end of the experiment. (B) Growth kinetics of MIA PaCa-2-luc subcutaneous tumors on the flanks of athymic mice. $*p < 0.04$ as compared to all other groups. (C) Bioluminescence imaging (BLI) of tumors and image quantification. (D) Formalin-fixed paraffin-embedded tumor sections were stained with p-Stat3 Y705 and PCNA. Arrows in PCNA images show margin of negatively stained tumor area at the periphery of each tumor. *Adapted from Quinn et al. (2015)*. (See the color plate.)

3.3 Poly I:C

Immune modulation continues to be a focus of therapeutic development, as shown through the use of polyinosine–polycytidylic acid (pIC) to induce antitumoral responses. pIC is a synthetic dsRNA, which stimulates toll-like receptors (TLRs) to induce dendritic and natural killer (NK) cell activity

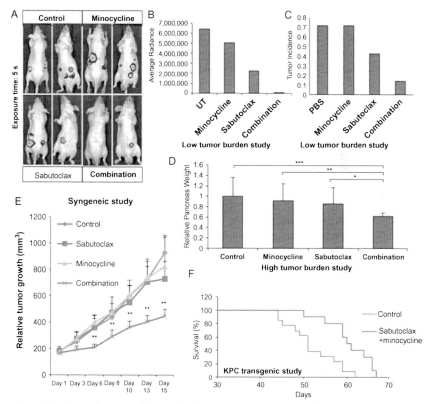

Figure 2 Effects of minocycline and sabutoclax on tumor growth in a quasi-orthotopic xenograft model, a syngeneic KPC-derived tumor model, and on the survival of the KPC transgenic mouse model. (A and B) Low tumor burden study (1×10^6 cells injected i.p.), BLI images of mice and image quantification. (C) Tumor incidence in pancreas—measured gross examination of animals at necropsy and by imaging. (D) High tumor burden study (5×10^6 cells injected i.p.), pancreas weight at necropsy. Relative tumor weight normalized to control. $*p=0.03$, $**p=0.01$, $***p=0.004$. (E) Syngeneic study: tumor growth kinetics: Pdx-1-Cre/K-$ras^{LSL-G12D}$/$p53^{flox/wt}$-derived tumor cells were implanted subcutaneously in control Pdx-1-Cre negative/K-$ras^{LSL-G12D}$/$p53^{flox/wt}$ mice. $**p<0.01$ as compared to all other groups. (F) Kaplan–Meier survival curve of Pdx-1-Cre/K-$ras^{LSL-G12D}$/$p53^{flox/flox}$ mice treated with minocycline and sabutoclax. $*p=0.001$. $n=12$ mice (control group); $n=10$ mice (sabutoclax+minocycline group). *Adapted from Quinn et al. (2015).* (See the color plate.)

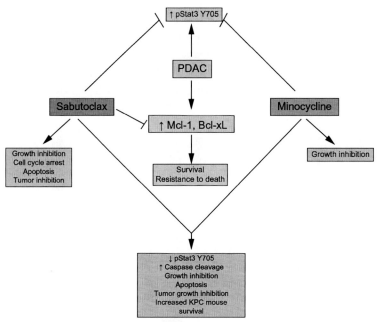

Figure 3 Schematic model showing therapeutic effects of sabutoclax and minocycline in PDAC. *Adapted from Quinn et al. (2015)*.

(Ammi et al., 2015; Bhoopathi et al., 2014). TLRs act as primary sensors, recognizing innate patterns of bacterial or viral infection (Barral et al., 2009). Activation can lead to decreases in the immunosuppressive functions of T_{reg} cells, often aberrantly overactive in tumor microenvironments (Liu, Zhang, & Zhao, 2010). Through these pathways, pIC can directly activate dendritic cells and promote NK cells to attack tumors by mimicking viral RNA (Perrot et al., 2010). While pIC has been utilized extensively in clinical trials (Gnjatic, Sawhney, & Bhardwaj, 2010), only recently has the mechanism of action in pancreatic cells been uncovered (Bhoopathi et al., 2014). pIC exposure induces elevated levels of proteins involved in viral and tumoral host defenses, including type I IFNs, OAS, RIG-I helicase, and MDA-5 (Barral et al., 2009; Gnjatic et al., 2010). Early trials with pIC were not promising due to lack of stability and poor IFN induction. By complexing with polyethylenimine (PEI), which is known to increase transfection efficiency of DNA, siRNA, and RNA *in vivo* (Bhang, Gabrielson, Laterra, Fisher, & Pomper, 2011), [pIC]^{PEI} demonstrated dramatic reduction in cancer growth, increased toxic autophagy, and apoptosis

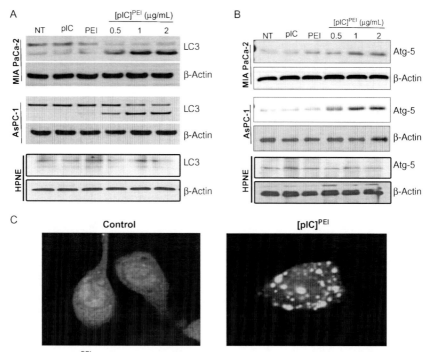

Figure 4 [pIC]^{PEI} induces autophagy in PDAC cells. (A and B) HPNE (h-TERT-immortalized normal pancreatic ductal epithelial cells) and PDAC cells were treated with either [pIC], PEI, or indicated doses of [pIC]^{PEI} for 48 h, and cell lysates were subjected to Western blotting to detect LC3 (A) and Atg5 (B). β-Actin served as a loading control. (C) AsPC-1 cells were treated with 1 microg/mL of [pIC]^{PEI} for 48 h and stained for LC3 localization. Results are representative of three independent experiments. *Used with permission from Bhoopathi et al. (2014).*

(Fig. 4), and promoted an antitumoral immune response (Besch et al., 2009; Inao et al., 2012; Tormo et al., 2009). Recent experiments in pancreatic cancer showed that [pIC]^{PEI} repressed XIAP and survivin expression, as well as promoting an immune response through induction of MDA-5, RIG-I, and NOXA. Furthermore, Akt activation was inhibited by [pIC]^{PEI} in pancreatic cancer cells, proving to be a crucial step in apoptosis through XIAP and survivin degradation (Bhoopathi et al., 2014). [pIC]^{PEI} administered *in vivo* to quasi-orthotopic models of pancreatic cancer showed significant inhibition of growth and progression (Fig. 5). Since [pIC]^{PEI} does not show evidence of toxicity, this could be a promising novel therapy to be used alone or most likely in combination with established protocols for treating pancreatic cancer.

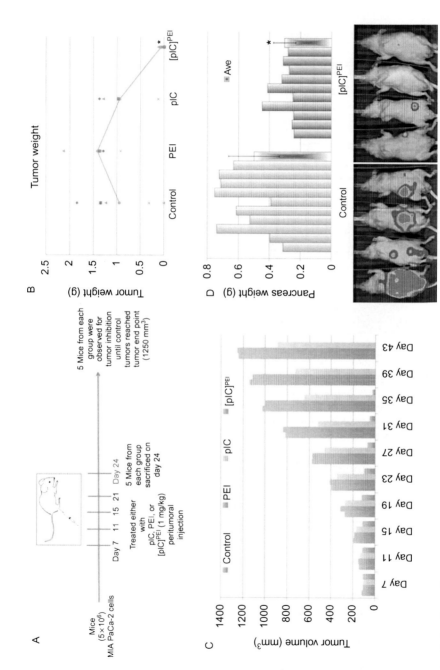

Figure 5 See legend on next page.

4. FUTURE PERSPECTIVES

The particularly poor prognosis associated with pancreatic cancer likely derives from a combination of its typically late presentation, anatomical location that lends to widespread invasion, and the disparate genetic mutations found in these tumors. This heterogeneity in mutations drives a principally invasive and resistant phenotype, leading to an extremely aggressive cancer. A number of models are available to test prospective therapies, but each has advantages and drawbacks. Like many transgenic models of human cancer, pancreatic cancer models show significant variability in time to tumor development as well as kinetics of tumor growth. While this mimics the variation observed in human cancers, it makes particular therapeutic approaches more challenging, such as early detection, cancer vaccine development, and preventative therapies. To address this problem, bioluminescent markers have been utilized to enable tumor monitoring through noninvasive imaging (Minn et al., 2014; Pomper & Fisher, 2014). Moreover, metastases can be identified through these imaging techniques, opening the door to more thoroughly test antimetastatic therapies (Minn et al., 2014; Pomper & Fisher, 2014).

In the development of future therapies, targeting the Bcl-2 family of proteins could provide useful ways to combat the resistance of pancreatic tumor cells to current chemotherapies and radiation. The antiapoptotic proteins in

---

Figure 5 *In vivo* cytoplasmic delivery of [pIC] using *in vivo* jetPEI decreases human pancreatic tumor growth in subcutaneous and quasi-orthotopic models. (A) Mice were injected subcutaneously with MIA PaCa-2 cells (5×10^6) and once tumors reached approximately 75 mm^3, they were divided into four groups. Each group was treated with four doses of [pIC], PEI, or [pIC]^{PEI} (1 mg/kg), as indicated in the schematic. One set of mice (5 mice from each group) was sacrificed 3 days after the last dose of [pIC]^{PEI}. The other group was maintained until the control tumors reached the IACUC endpoint (1250 mm^3). (B) Tumor weights were measured and are presented graphically. Mean of the tumor weights of each group of mice is shown in line graph. $*p < 0.01$ versus control. (C) Tumor volumes were measured periodically using a vernier caliper and are presented graphically. (D) MIA PaCa-2-luc cells (5×10^6) were injected intraperitoneally into nude mice. BLI was performed every week after tumor cell implantation. After 2 weeks following cell implantation, mice were divided into two groups of 10 mice each. One group was used as control, without treatment, and the other group was injected twice weekly with [pIC]^{PEI} (1 mg/kg) i.p. (total four doses). Control and treated mice were observed for tumor progression using BLI. Once the mice were sacrificed, the pancreas was weighed and the data presented graphically. $*p < 0.01$ versus control. *Used with permission from Bhoopathi et al. (2014).* (See the color plate.)

this family, particularly Mcl-1, are overexpressed in resistant tumors, and inhibiting their actions can sensitize cancer cells to other therapeutic agents (Quinn et al., 2011). New therapies will have to demonstrate significant gains over current regimens, while avoiding unacceptable toxicities. Small-molecule development is expanding at an accelerated pace and may be equipped to reach the goals of a physiologically stable compound that can target tumor-initiating cells specifically. While it is unlikely one molecule will be up to the task, combination therapies have strong potential to make clinically significant gains.

ACKNOWLEDGMENTS

This work was supported by National Institutes of Health, National Cancer Institute Grants, 1R01 CA134721 (P.B.F.) and 1R01 CA168517 (M.P. and P.B.F.), a VCU Massey Cancer Center Developmental Grant (P.B.F.), and the James S. McDonnell Foundation (D.S.). D.S. is a Harrison Scholar in the MCC and P.B.F. holds the Thelma Newmeyer Corman Chair in Cancer Research in the MCC.

REFERENCES

Al-Safi, R., Odde, S., Shabaik, Y., & Neamati, N. (2012). Small-molecule inhibitors of APE1 DNA repair function: An overview. *Current Molecular Pharmacology*, 5(1), 14–35.

Alvarez, R., Musteanu, M., Garcia-Garcia, E., Lopez-Casas, P., Megias, D., Guerra, C., et al. (2013). Stromal disrupting effects of nab-paclitaxel in pancreatic cancer. *British Journal of Cancer*, 109(4), 926–933.

Ammi, R., De Waele, J., Willemen, Y., Van Brussel, I., Schrijvers, D. M., Lion, E., et al. (2015). Poly(I:C) as cancer vaccine adjuvant: Knocking on the door of medical breakthroughs. *Pharmacology & Therapeutics*, 146, 120–131.

Azab, B., Dash, R., Das, S. K., Bhutia, S. K., Shen, X. N., Quinn, B. A., et al. (2012). Enhanced delivery of mda-7/IL-24 using a serotype chimeric adenovirus (Ad.5/3) in combination with the Apogossypol derivative BI-97C1 (Sabutoclax) improves therapeutic efficacy in low CAR colorectal cancer cells. *Journal of Cellular Physiology*, 227(5), 2145–2153.

Barile, E., Wang, S., Das, S. K., Noberini, R., Dahl, R., Stebbins, J. L., et al. (2014). Design, synthesis and bioevaluation of an EphA2 receptor-based targeted delivery system. *ChemMedChem*, 9, 1403–1412.

Barral, P. M., Sarkar, D., Su, Z., Barber, G. N., DeSalle, R., Racaniello, V. R., et al. (2009). Functions of the cytoplasmic RNA sensors RIG-I and MDA-5: Key regulators of innate immunity. *Pharmacology & Therapeutics*, 124(2), 219–234.

Besch, R., Poeck, H., Hohenauer, T., Senft, D., Häcker, G., Berking, C., et al. (2009). Proapoptotic signaling induced by RIG-I and MDA-5 results in type I interferon-independent apoptosis in human melanoma cells. *The Journal of Clinical Investigation*, 119(8), 2399–2411.

Bhang, H. C., Gabrielson, K. L., Laterra, J., Fisher, P. B., & Pomper, M. G. (2011). Tumor-specific imaging through progression elevated gene-3 promoter-driven gene expression. *Nature Medicine*, 17(1), 123–129.

Bhoopathi, P., Quinn, B. A., Gui, Q., Shen, X. N., Grossman, S. R., Das, S. K., et al. (2014). Pancreatic cancer-specific cell death induced in vivo by cytoplasmic-delivered polyinosine-polycytidylic acid. *Cancer Research*, 74(21), 6224–6235.

Bulgar, A. D., Weeks, L. D., Miao, Y., Yang, S., Xu, Y., Guo, C., et al. (2012). Removal of uracil by uracil DNA glycosylase limits pemetrexed cytotoxicity: Overriding the limit with methoxyamine to inhibit base excision repair. *Cell Death and Disease*, *3*, e252.

Burris, H. A., Moore, M. J., Andersen, J., Green, M. R., Rothenberg, M. L., Modiano, M. R., et al. (1997). Improvements in survival and clinical benefit with gemcitabine as first-line therapy for patients with advanced pancreas cancer: A randomized trial. *Journal of Clinical Oncology: Official Journal of the American Society of Clinical Oncology*, *15*(6), 2403–2413.

Conroy, T., Desseigne, F., Ychou, M., Bouché, O., Guimbaud, R., Bécouarn, Y., et al. (2011). FOLFIRINOX versus gemcitabine for metastatic pancreatic cancer. *The New England Journal of Medicine*, *364*(19), 1817–1825.

Cunningham, D., Chau, I., Stocken, D. D., Valle, J. W., Smith, D., Steward, W., et al. (2009). Phase III randomized comparison of gemcitabine versus gemcitabine plus capecitabine in patients with advanced pancreatic cancer. *Journal of Clinical Oncology: Official Journal of the American Society of Clinical Oncology*, *27*(33), 5513–5518.

Dash, R., Azab, B., Quinn, B. A., Shen, X., Wang, X. Y., Das, S. K., et al. (2011). Apogossypol derivative BI-97C1 (Sabutoclax) targeting Mcl-1 sensitizes prostate cancer cells to mda-7/IL-24-mediated toxicity. *Proceedings of the National Academy of Sciences of the United States of America*, *108*(21), 8785–8790.

Geng, L., Huehls, A. M., Wagner, J. M., Huntoon, C. J., & Karnitz, L. M. (2011). Checkpoint signaling, base excision repair, and PARP promote survival of colon cancer cells treated with 5-fluorodeoxyuridine but not 5-fluorouracil. *PLoS One*, *6*(12), e28862.

Giovannetti, E., Del Tacca, M., Mey, V., Funel, N., Nannizzi, S., Ricci, S., et al. (2006). Transcription analysis of human equilibrative nucleoside transporter-1 predicts survival in pancreas cancer patients treated with gemcitabine. *Cancer Research*, *66*(7), 3928–3935.

Gnjatic, S., Sawhney, N. B., & Bhardwaj, N. (2010). Toll-like receptor agonists: Are they good adjuvants? *Cancer Journal*, *16*(4), 382–391.

Goff, D. J., Court Recart, A., Sadarangani, A., Chun, H. J., Barrett, C. L., Krajewska, M., et al. (2013). A pan-BCL2 inhibitor renders bone-marrow-resident human leukemia stem cells sensitive to tyrosine kinase inhibition. *Cell Stem Cell*, *12*(3), 316–328.

Gourgou-Bourgade, S., Bascoul-Mollevi, C., Desseigne, F., Ychou, M., Bouché, O., Guimbaud, R., et al. (2013). Impact of FOLFIRINOX compared with gemcitabine on quality of life in patients with metastatic pancreatic cancer: Results from the PRODIGE 4/ACCORD 11 randomized trial. *Journal of Clinical Oncology: Official Journal of the American Society of Clinical Oncology*, *31*(1), 23–29.

Hedvat, M., Emdad, L., Das, S. K., Kim, K., Dasgupta, S., Thomas, S., et al. (2012). Selected approaches for rational drug design and high throughput screening to identify anti-cancer molecules. *Anti-Cancer Agents in Medicinal Chemistry*, *12*(9), 1143–1155.

Heidelberger, C., Chaudhuri, N. K., Danneberg, P., Mooren, D., Griesbach, L., Duschinsky, R., et al. (1957). Fluorinated pyrimidines, a new class of tumour-inhibitory compounds. *Nature*, *179*(4561), 663–666.

Hoy, S. M. (2014). Albumin-bound paclitaxel: A review of its use for the first-line combination treatment of metastatic pancreatic cancer. *Drugs*, *74*(15), 1757–1768.

Huehls, A. M., Wagner, J. M., Huntoon, C. J., Geng, L., Erlichman, C., Patel, A. G., et al. (2011). Poly(ADP-ribose) polymerase inhibition synergizes with 5-fluorodeoxyuridine but not 5-fluorouracil in ovarian cancer cells. *Cancer Research*, *71*(14), 4944–4954.

Inao, T., Harashima, N., Monma, H., Okano, S., Itakura, M., Tanaka, T., et al. (2012). Antitumor effects of cytoplasmic delivery of an innate adjuvant receptor ligand, poly(I:C), on human breast cancer. *Breast Cancer Research and Treatment*, *134*(1), 89–100.

Jackson, R. S., 2nd., Placzek, W., Fernandez, A., Ziaee, S., Chu, C. Y., Wei, J., et al. (2012). Sabutoclax, a Mcl-1 antagonist, inhibits tumorigenesis in transgenic mouse and human xenograft models of prostate cancer. *Neoplasia, 14*(7), 656–665.

Jaffe, J. J., Handschumacher, R. E., & Welch, A. D. (1957). Studies on the carcinostatic activity in mice of 6-azauracil riboside (azauridine), in comparison with that of 6-azauracil. *The Yale Journal of Biology and Medicine, 30*(3), 168–175.

Kato, T., Shimamoto, Y., Uchida, J., Ohshimo, H., Abe, M., Shirasaka, T., et al. (2001). Possible regulation of 5-fluorouracil-induced neuro- and oral toxicities by two biochemical modulators consisting of S-1, a new oral formulation of 5-fluorouracil. *Anticancer Research, 21*(3), 1705–1712.

Kunzmann, V., Herrmann, K., Bluemel, C., Kapp, M., Hartlapp, I., & Steger, U. (2014). Intensified neoadjuvant chemotherapy with nab-paclitaxel plus gemcitabine followed by FOLFIRINOX in a patient with locally advanced unresectable pancreatic cancer. *Case Reports in Oncology, 7*(3), 648–655.

Liu, M., Cao, D., Russell, R., Handschumacher, R. E., & Pizzorno, G. (1998). Expression, characterization, and detection of human uridine phosphorylase and identification of variant uridine phosphorolytic activity in selected human tumors. *Cancer Research, 58*(23), 5418–5424.

Liu, G., Zhang, L., & Zhao, Y. (2010). Modulation of immune responses through direct activation of toll-like receptors to T cells. *Clinical and Experimental Immunology, 160*(2), 168–175.

Marsh, R. D. W., Talamonti, M. S., Katz, M. H., & Herman, J. M. (2015). Pancreatic cancer and FOLFIRINOX: A new era and new questions. *Cancer Medicine*, in press. [Epub ahead of print], PMID 25693729.

Miller, J. A., Miller, E. C., & Finger, G. C. (1953). On the enhancement of the carcinogenicity of 4-dimethylaminoazobenzene by fluoro-substitution. *Cancer Research, 13*(1), 93–97.

Minn, I., Menezes, M. E., Sarkar, S., Yarlagadda, K., Das, S. K., Emdad, L., et al. (2014). Molecular-genetic imaging of cancer. *Advances in Cancer Research, 124*, 131–169.

Moore, M. J., Goldstein, D., Hamm, J., Figer, A., Hecht, J. R., Gallinger, S., et al. (2007). Erlotinib plus gemcitabine compared with gemcitabine alone in patients with advanced pancreatic cancer: A phase III trial of the National Cancer Institute of Canada Clinical Trials Group. *Journal of Clinical Oncology: Official Journal of the American Society of Clinical Oncology, 25*(15), 1960–1966.

Mori, K., Hasegawa, M., Nishida, M., Toma, H., Fukuda, M., Kubota, T., et al. (2000). Expression levels of thymidine phosphorylase and dihydropyrimidine dehydrogenase in various human tumor tissues. *International Journal of Oncology, 17*(1), 33–38.

Moysan, E., Bastiat, G., & Benoit, J. (2013). Gemcitabine versus modified gemcitabine: A review of several promising chemical modifications. *Molecular Pharmaceutics, 10*(2), 430–444.

Mullany, S., Svingen, P. A., Kaufmann, S. H., & Erlichman, C. (1998). Effect of adding the topoisomerase I poison 7-ethyl-10-hydroxycamptothecin (SN-38) to 5-fluorouracil and folinic acid in HCT-8 cells: Elevated dTTP pools and enhanced cytotoxicity. *Cancer Chemotherapy and Pharmacology, 42*(5), 391–399.

National Cancer Institute. (2015). *Surveillance, epidemiology, and end results (SEER) program (www.seer.cancer.gov) research data (1973–2011)*. Retrieved March 17, 2015, from http://seer.cancer.gov/statfacts/html/pancreas.html.

Ng, S. S. W., Tsao, M., Nicklee, T., & Hedley, D. W. (2002). Effects of the epidermal growth factor receptor inhibitor OSI-774, Tarceva, on downstream signaling pathways and apoptosis in human pancreatic adenocarcinoma. *Molecular Cancer Therapeutics, 1*(10), 777–783.

Perrot, I., Deauvieau, F., Massacrier, C., Hughes, N., Garrone, P., Durand, I., et al. (2010). TLR3 and rig-like receptor on myeloid dendritic cells and rig-like receptor on human

NK cells are both mandatory for production of IFN-gamma in response to double-stranded RNA. *The Journal of Immunology, Virus Research and Experimental Chemotherapy, 185*(4), 2080–2088.
Placzek, W. J., Sturlese, M., Wu, B., Cellitti, J. F., Wei, J., & Pellecchia, M. (2011). Identification of a novel Mcl-1 protein binding motif. *The Journal of Biological Chemistry, 286*(46), 39829–39835.
Pomper, M. G., & Fisher, P. B. (2014). Emerging applications of molecular imaging to oncology. Preface. *Advances in Cancer Research, 124*, xiii.
Quinn, B. A., Dash, R., Azab, B., Sarkar, S., Das, S. K., Kumar, S., et al. (2011). Targeting Mcl-1 for the therapy of cancer. *Expert Opinion on Investigational Drugs, 20*(10), 1397–1411.
Quinn, B. A., Dash, R., Sarkar, S., Azab, B., Bhoopathi, P., Das, S. K., et al. (2015). Pancreatic cancer combination therapy using a BH3 mimetic and a synthetic tetracycline. *Cancer Research*, in press.
Rutman, R. J., Cantarow, A., & Paschkis, K. E. (1954). Studies in 2-acetylaminofluorene carcinogenesis. III. The utilization of uracil-2-C14 by preneoplastic rat liver and rat hepatoma. *Cancer Research, 14*(2), 119–123.
Ryan, D. P., Hong, T. S., & Bardeesy, N. (2014). Pancreatic adenocarcinoma. *The New England Journal of Medicine, 371*(22), 2140–2141.
Sargent, D., Sobrero, A., Grothey, A., O'Connell, M. J., Buyse, M., Andre, T., et al. (2009). Evidence for cure by adjuvant therapy in colon cancer: Observations based on individual patient data from 20,898 patients on 18 randomized trials. *Journal of Clinical Oncology: Official Journal of the American Society of Clinical Oncology, 27*(6), 872–877.
Siegel, R., Ma, J., Zou, Z., & Jemal, A. (2014). Cancer statistics, 2014. *CA: A Cancer Journal for Clinicians, 64*, 9–29.
Simeonov, A., Kulkarni, A., Dorjsuren, D., Jadhav, A., Shen, M., McNeill, D. R., et al. (2009). Identification and characterization of inhibitors of human apurinic/apyrimidinic endonuclease APE1. *PLoS One, 4*(6), e5740.
Stock, C. C. (1954). Experimental cancer chemotherapy. *Advances in Cancer Research, 2*, 425–492.
Tandon, M., Vemula, S. V., & Mittal, S. K. (2011). Emerging strategies for EphA2 receptor targeting for cancer therapeutics. *Expert Opinion on Therapeutic Targets, 15*(1), 31–51.
Thomas, S., Quinn, B. A., Das, S. K., Dash, R., Emdad, L., Dasgupta, S., et al. (2013). Targeting the Bcl-2 family for cancer therapy. *Expert Opinion on Therapeutic Targets, 17*(1), 61–75.
Tormo, D., Checińska, A., Alonso-Curbelo, D., Pérez-Guijarro, E., Cañón, E., Riveiro-Falkenbach, E., et al. (2009). Targeted activation of innate immunity for therapeutic induction of autophagy and apoptosis in melanoma cells. *Cancer Cell, 16*(2), 103–114.
Ueno, H., Okusaka, T., Funakoshi, A., Ishii, H., Yamao, K., Ishikawa, O., et al. (2007). A phase II study of weekly irinotecan as first-line therapy for patients with metastatic pancreatic cancer. *Cancer Chemotherapy and Pharmacology, 59*(4), 447–454.
Van Cutsem, E., Li, C., Nowara, E., Aprile, G., Moore, M., Federowicz, I., et al. (2014). Dose escalation to rash for erlotinib plus gemcitabine for metastatic pancreatic cancer: The phase II RACHEL study. *British Journal of Cancer, 111*(11), 2067–2075.
Van den Broeck, A., Vankelecom, H., Van Eijsden, R., Govaere, O., & Topal, B. (2012). Molecular markers associated with outcome and metastasis in human pancreatic cancer. *Journal of Experimental & Clinical Cancer Research, 31*, 68.
Varadarajan, S., Butterworth, M., Wei, J., Pellecchia, M., Dinsdale, D., & Cohen, G. M. (2013). Sabutoclax (BI97C1) and BI112D1, putative inhibitors of MCL-1, induce mitochondrial fragmentation either upstream of or independent of apoptosis. *Neoplasia, 15*(5), 568–578.

Varadhachary, G. R., Tamm, E. P., Abbruzzese, J. L., Xiong, H. Q., Crane, C. H., Wang, H., et al. (2006). Borderline resectable pancreatic cancer: Definitions, management, and role of preoperative therapy. *Annals of Surgical Oncology, 13*(8), 1035–1046.

Visser, B. C., Ma, Y., Zak, Y., Poultsides, G. A., Norton, J. A., & Rhoads, K. F. (2012). Failure to comply with NCCN guidelines for the management of pancreatic cancer compromises outcomes. *HPB, 14*(8), 539–547.

Von Hoff, D. D., Ervin, T., Arena, F. P., Chiorean, E. G., Infante, J., Moore, M., et al. (2013). Increased survival in pancreatic cancer with nab-paclitaxel plus gemcitabine. *The New England Journal of Medicine, 369*(18), 1691–1703.

Waddell, N., Pajic, M., Patch, A., Chang, D. K., Kassahn, K. S., Bailey, P., et al. (2015). Whole genomes redefine the mutational landscape of pancreatic cancer. *Nature, 518*(7540), 495–501.

Wang, S., Noberini, R., Stebbins, J. L., Das, S., Zhang, Z., Wu, B., et al. (2013). Targeted delivery of paclitaxel to EphA2-expressing cancer cells. *Clinical Cancer Research, 19*(1), 128–137.

Wang, S., Placzek, W. J., Stebbins, J. L., Mitra, S., Noberini, R., Koolpe, M., et al. (2012). Novel targeted system to deliver chemotherapeutic drugs to EphA2-expressing cancer cells. *Journal of Medicinal Chemistry, 55*(5), 2427–2436.

Weeks, L. D., Fu, P., & Gerson, S. L. (2013). Uracil-DNA glycosylase expression determines human lung cancer cell sensitivity to pemetrexed. *Molecular Cancer Therapeutics, 12*(10), 2248–2260.

Wei, J., Stebbins, J. L., Kitada, S., Dash, R., Placzek, W., Rega, M. F., et al. (2010). BI-97C1, an optically pure Apogossypol derivative as pan-active inhibitor of antiapoptotic B-cell lymphoma/leukemia-2 (Bcl-2) family proteins. *Journal of Medicinal Chemistry, 53*(10), 4166–4176.

Wilson, P. M., Danenberg, P. V., Johnston, P. G., Lenz, H. J., & Ladner, R. D. (2014). Standing the test of time: Targeting thymidylate biosynthesis in cancer therapy. *Nature Reviews. Clinical Oncology, 11*(5), 282–298.

Wu, B., Wang, S., De, S. K., Barile, E., Quinn, B. A., Zharkikh, I., et al. (2015). *Novel EphA2 agonists targeting its ligand binding domain*. Chemistry & Biology, submitted for publication.

Zeghari-Squalli, N., Raymond, E., Cvitkovic, E., & Goldwasser, F. (1999). Cellular pharmacology of the combination of the DNA topoisomerase I inhibitor SN-38 and the diaminocyclohexane platinum derivative oxaliplatin. *Clinical Cancer Research, 5*(5), 1189–1196.

INDEX

Note: Page numbers followed by "*f*" indicate figures and "*t*" indicate tables.

A

Abiraterone acetate (AA)
    AR, 128–130, 129*f*
    mCRPC, 144–145
Acute lymphoblastic leukemia (ALL), 243
Acute myeloid leukemia (AML), 228–229, 236, 244
    clinical trials, 242
    FLT3 inhibitors, 242
    genetic and cell type heterogeneity, 230–231
    Hedgehog signaling pathway, 239–240
    incidence, 230–231
    induction chemotherapy, 230–231
    inhibitors, therapeutic efficacy of, 243
    Notch signaling pathway, 240–241
    osteoblasts, 232–233
    VLA4-targeting antibodies, 233–234
    xenografted immunodeficient mouse model, 241–242
Amphiregulin (AREG)
    characterization, 262
    and EGF, 259*f*, 262–263
    fibronectin-dependent signaling, 263
    knockdown, 268
    MMP-9, 263
    SUM-149PT cells, 262
Androgen deprivation therapy (ADT), 125–126, 128
Androgen receptor (AR)
    AA, 128–130, 129*f*, 144–145
    ADT, 128
    ARE, 126–127
    coregulatory proteins, 127–128
    DHT, 126–127
    dynein motor proteins, 134–135
    enzalutamide (MDV3100), 130, 145
    EPI-002, 130–131
    nuclear translocation, 127
    testosterone, 126–127
Androgen responsive elements (ARE), 126–127

Aromatase inhibitors, 254–255

B

BCR-ABL protein expression, 230, 238–239, 242
Betacellulin (BTC), 257–258
Bivatuzumab, 242
Bone marrow microenvironment (BMM)
    CAR cells, 235
    cell types, 231–232
    HSCs, 228–232
    LSCs, 228–231
        Hedgehog signaling pathway, 239–240
        Notch signaling pathway, 240–241
        targeting therapies, 242–243
        in vivo models, 241–242
        Wnt/β-catenin signaling pathway, 237–239
    osteoblasts, 232–233
    perivascular MSCs, 234
    sympathetic neuronal cells, 235–236
    vascular endothelial cells, 233–234
Bortezomib (Btz)
    apoptotic resistance and autophagy, 197–199
    bromodomain, 210–213, 213*f*
    Il-6/STAT3 signaling axis, 206–207
    MDM2 inhibitors, 205–207
    next-generation proteasome inhibitors, 201–205
    pleiotropic anti-MM activity, 193, 194*f*
    PSMb5 gene, 195*f*, 196–199
    redox signaling, 201–205
    therapeutic monoclonal antibodies, 209–210
Bromodomain and extra terminal domain (BET) inhibitors, 192–193

C

Cabazitaxel (CBZ), 133, 136–139
    MCAK, 138*f*
    MCF-7, 139–140

Cabazitaxel (CBZ) (*Continued*)
  TROPIC Phase III clinical trial, 147
Carboplatin, 144
CAR cells. *See* CXCL12-abundant reticular (CAR) cells
Casitas B-lineage Lymphoma (Cbl), EGFR downregulation, 258f, 260–262
Castration-resistant prostate cancer (CRPC)
  AA, 144–145
  AR expression (*see* Androgen receptor (AR))
  Cabazitaxel, 147
  clusterin targeting (OGX-011), 146
  dynein motor proteins, 134–135
  enzalutamide, 145
  kinesin motor proteins, 135–139
  platinum-based therapy, 144
  taxanes (*see* Taxanes)
  therapeutic promises and value, 146–147
Cell encapsulation technology
  biomaterials, 176–177
  GBM treatment, 177–179
  physiologic environment, 176–177
  scaffold-based delivery system, 176–177
  sECM-encapsulated hMSC-S-TRAIL, 177–179, 178f
Cetuximab, 256–257
Chemotherapy
  PDAC, 286–287
  TNBC, 254–255, 269–270
Chromosomal instability (CIN), 136–139
Chronic myelogenous leukemia (CML), 193–196, 228–229, 236, 244
  BCR-ABL protein expression, 230
  Hedgehog signaling pathway, 239–240
  inhibitors, therapeutic efficacy of, 243
  Notch signaling pathway, 240
  osteoblasts, 232–233
  TKI, 230
  Wnt/β-catenin signaling pathway, 238–239
Clusterin, 146
CML. *See* Chronic myelogenous leukemia (CML)
c-Myc oncoprotein, 10–15, 204–205, 210–212
Combination therapy
  CRPC, 144–147
  PDAC, 286–287, 289
  TNBC, 269–271
CXCL12-abundant reticular (CAR) cells, 231–235

D

Dihydrotestosterone (DHT), 126–130
DNA binding domain (DBD), 127–128
Docetaxel (DOC), 131–133
  MCAK, 138f
  MCF-7, 139–140
Downstream effectors, MTA proteins, 13t
  nontranscription targets, 23
  transcriptional targets, 22–23
Dynein motor proteins, 134–135

E

EMT. *See* Epithelial-mesenchymal transition (EMT)
ENCyclopedia of DNA Elements (ENCODE), 53–54, 111
Endothelial cells (ECs), 233–234
Enzalutamide (MDV3100)
  AR, 130
  mCRPC, 145
Ephrin receptors, 292–295
Epidermal growth factor (EGF), 260–263, 267–268
Epidermal growth factor receptor tyrosine kinase (EGFR)
  activation, 257–258, 258f
  AKT signaling, 266–267
  anti-EGFR therapies, 256–257
  AREG, 259f, 261–263
  BRCA1 and BRCA2, 265
  Cbl, 258f, 260–262
  combination therapy, 270–271, 271f
  EMT, 263–264
  endocytic trafficking, 264
  ligands, 258–259, 259f
  mediate internalization, 258–259
  MET, 263–264
  TAM, 268–269, 271–273
  TP53, 266
Epigen (EPGN), 257–258
Epiregulin (EREG), 257–258
Epithelial-mesenchymal transition (EMT), 142–143, 263–264

Erlotinib, 256–257
Estrogen receptor (ER), 254–255
ETS-Related Gene (ERG), 140–141

F

5-Fluorouracil (5-FU), 287–288
FOLFIRINOX, 285–286, 290–291
Forkhead box protein 1 (FOXO1), 133–134

G

Gefitinib, 256–257
Gemcitabine, 288
   and abraxane, 291–292
   cytotoxic effects, 288–289
   and erlotinib, 289–290
   and FOLFIRINOX, 290–291
   hENT1, 289
Glioma, 58–59, 65–66, 74–99
Glutathione S-transferase-pi (GSTP), 202–206
Growth factor signaling, TNBC, 255–256

H

HBV-encoded X protein (HBx), 20–21
Hedgehog signaling pathway
   LSCs and BMM, 239–240
Hematopoietic stem cells (HSCs), 228–232
Histone deacetylase (HDAC) enzymes, 212–213
Homeobox (HOX) proteins, 76–99
HSCs. *See* Hematopoietic stem cells (HSCs)
Human epidermal growth factor receptor-2 (HER2), 254–255
Human kinesin-14 (HSET), 136–139
Human Protein Atlas Database, 54–57
Hypoxia, 21

I

Imatinib, 230
Immunomodulatory TNBC, 255
Immuno signaling pathways
   complement cascade, 107
   CTLA4 pathway, 106–107
   interferon, 105–106
   interleukins, 105
Intermittent chemotherapy, 143–144
In vivo murine models, 241–242

K

Kidney renal papillary carcinoma, 59–60
Kinesin
   Eg5, 135
   facilitate progression, 135
   HSET, 136–139
   MCAK, 135–136
   organelle transport, 135

L

Lapatinib, 254–255
Leukemia-initiating cells (LIC), 241–242
Leukemia-propagating cells (LPC), 243
Leukemia stem cells (LSCs)
   AML (*see* Acute myeloid leukemia (AML))
   BMM
      Hedgehog signaling pathway, 239–240
      Notch signaling pathway, 240–241
      in vivo models, 241–242
      Wnt/β-catenin signaling pathway, 237–239
   CML (*see* Chronic myelogenous leukemia (CML))
Ligand-binding domain (LBD), 127
Liver hepatocellular carcinoma, *mda-9*'s expression pattern, 59–60
LSCs. *See* Leukemia stem cells (LSCs)
Luminal androgen receptor (LAR), 255

M

Matrix metalloprotease-9 (MMP-9) protein, 263
mda-9/Syntenin. *See* Melanoma differentiation associated gene-9 (MDA-9)
Melanoma differentiation associated gene-9 (MDA-9)
   analytical tools, 57–58
   cancer progression, 51–52
   cellular processes, 51
   CpG methylation, 52–53, 72–74
   c-Src, 51–52
   DNA methyltransferases, 52–53
   exons, 51
   extracellular matrix, 51–52
   genomic datasets, 54–58

Melanoma differentiation associated gene-9 (MDA-9) (*Continued*)
   glioma
      at cg1719774, 111
      upregulated and downregulated genes, 76–99, 114
      VIGOR approach, 74–76, 74*f*
   GSEA analysis, 113
      collagens, 103–104
      ES and FDR q, 109
      IGF signaling, 107–108
      immuno signaling pathways, 105–107
      integrins and focal adhesion genes, 104
      metallopepties and proteins, 99–103
      neurite outgrowth, 104–105
      NF-κB activation, 108
      phospholipid metabolism, 108–109
      VEGF pathway, 108
   GSEA-identified gene sets, 109–111
   interferon β, 51–52
   invasion/metastasis stage, 51–52
   matrix metalloproteinases, 51–52
   mda-9's expression pattern, 58–59
      affymetrix SNP array, 61
      copy number and cg1719774 methylation, 68–69
      CpG sites, 62–65
      glioma, 58–59, 65–66
      hisones modification, 113
      Illumina HiSeq 2000 RNA Sequencing platform, 61
      illumina 450K array, 61–62
      kidney renal papillary carcinoma, 59–60, 67
      liver hepatocellular carcinoma, 59–60, 67
      in metastasis samples, 60
      PANCAN dataset, 68–69, 111
      prostate cancer, 59–60, 67
      TCGA, 69–71
   PDZ domains, 51
   subtraction hybridization approach, 51–52
   transcription factors, 52–53
   transcript variants, 51
Mesenchymal stem cells (MSCs), 161, 234
Mesenchymal stem-like TNBC, 255
Mesenchymal TNBC, 255

Metastasis-associated proteins (MTA proteins)
   BAH domain, 4–5
   curcumin, 31–32, 33*t*
   downstream effectors, 13*t*
      nontranscription targets, 23
      transcriptional targets, 22–23
   β-elemene, 31–32, 33*t*
   ELM2 domain functions, 4–5
   expression and clinical significance, 25*t*
      breast cancer, 30–31
      cervical cancer, 31
      digestive system cancer, 29–30
      head and neck cancer, 24
      prostate cancer, 31
      respiratory system cancer, 30
   genistein, 31–32, 33*t*
   MTA1, 2–3, 3*f*
   MTA2, 3, 3*f*
   MTA3, 3–4, 3*f*
   MTA1s, 2–3, 3*f*
   NuRD corepressor complex, 5–6
   NURF coactivator complex, 6–8, 8*f*
   portulacerebroside A, 33*t*
   pterostilbene, 31–32, 33*t*
   resveratrol, 31–32, 33*t*
   SANT domain, 4–5
   subcellular location, 9–10
   upstream regulators, 11*t*
      cell stressors, 20–21
      enzymes and protein kinases, 8*f*, 19–20
      growth factors and receptors, 17–18
      hormone receptors, 18–19
      miRNAs, 16–17
      transcription factors and coregulators, 10–16
   ZG29p, 2–3, 3*f*
Mi-2. *See* NuRD corepressor complex
MicroRNAs (miRNAs), 16–17
Minocycline, 297–298*f*
Mitotic centromere-associated kinesin (MCAK), 135–136, 137–138*f*
Monoclonal antibody
   Btz, 209–210
   cetuximab, 256–257
MSCs. *See* Mesenchymal stem cells (MSCs)
MTA proteins. *See* Metastasis-associated proteins (MTA proteins)

Murine double minute 2 (MDM2) inhibitors, 192–193
Myeloproliferative neoplasms (MPN), 235–236

N

Natalizumab, 242
Neural cell adhesion molecule (NCAM), 104–105
Notch signaling pathway, LSCs and BMM, 240–241
Nuclear localization signal (NLS), 127
Nucleosome remodeling factor (NURF) coactivator complex, 6–8, 8f
NuRD corepressor complex, 5–6

O

Oncolytic viruses (OVs), 169–171
Osteoblasts, 232–233

P

PANCAN dataset, 68–69, 111
Pancreatic ductal adenocarcinoma (PDAC)
 antiapoptotic proteins, 301–302
 bioluminescent markers, 301
 chemotherapy, 286–287
 classifications, 285, 285t
 combination therapies, 286–287, 286t
 ephrin receptors, 292–295
 5-fluorouracil, 287–288
 gemcitabine, 288–289
 genome, 284–285
 metastases, 301
 minocycline, 295–296, 297–298f
 poly 1:C, 297–300, 299–300f
 risk factors, 284
 sabutoclax, 295–296, 296–298f
 surgical resection, 285–286
 symptoms, 284
Perivascular MSCs, 234
Phosphatidylinositol 4,5-bisphosphate, 76–99
Polyinosine-polycytidylic acid (pIC), 297–300
Postate-specific antigen (PSA), 125–126
Progesterone receptors (PR), 254–255
Prostate cancer. *See also* Castration-resistant prostate cancer (CRPC)

mda-9's expression pattern, 59–60
Proteasome inhibitor (PI). *See* Bortezomib (Btz)
Proteasome subunit beta5 (PSMB5) gene, 195f, 196–199

R

Reactive oxygen species (ROS), 201–205
Redox signaling, 201–205
Retinoic acid receptor-gamma (RARγ), 236–237

S

Sabutoclax, 295–296, 296–298f
Selective estrogen receptor modulators (SERMs), 254–255
Sex-hormone-binding protein (SHBP), 126–127
Signal transducer and activator of transcription 3 (STAT3) signaling, 208–209
Siltuximab, 208–209
^{153}Sm-ethylenediaminetetramethylenephosphonate (^{153}Sm-EDTMP), 145
Specificity protein 1 (SP1), 15–16
ST8 alpha-*N*-acetyl-neuraminide alpha-2,8-sialyltransferase 4 (ST8SIA4), 104–105
Stathmin, 141–142
Stem cell factor (SCF), 232–233
Stem cell therapy
 antiangiogenic agents, 171–172
 cell encapsulation technology
  biomaterials, 176–177
  GBM treatment, 177–179
  physiologic environment, 176–177
  scaffold-based delivery system, 176–177
  sECM-encapsulated hMSC-S-TRAIL, 177–179, 178f
 IFN-α, 167–168
 IFN-β, 167
 interleukins (ILs), 165–166
 iPSCs, 161
 MSCs, 161
 NSCs, 161
 OVs, 169–171
 proapoptotic proteins, 172–173

Stem cell therapy (*Continued*)
  prodrugs, 168–169
  to promote tumor cell death, 164–165, 165f
  synergistic approaches, 173–176
  tumor homing and migratory properties, 161–164
SUM-149PT cells, 262
Sympathetic neuronal cells, 235–236

T

Taxanes, 126
  blocking AR translocation, 133–134
  Cabazitaxel, 133
  Docetaxel treatment, 131–133
  EMT, 142–143
  impairing centrosome clustering, 142
  intermittent chemotherapy, 143–144
  MCF-7, 139–140
  mechanism, 131, 132f, 137f
  microtubule binding dynamics, 140–142
  P-gp, 139–140
  tubulin mutations, 140
TCGA Genome Characterization (and Data Coordination) Centers, 54–57
TCGA Pan Cancer (PANCAN) dataset, 54–57
Thymidylate synthase (TS), 287
TKI. *See* Tyrosine kinase inhibitors (TKI)
Tocilizumab, 208–209
Trastuzumab, 254–255
Triple-negative breast cancer (TNBC)
  BRCA1 and BRCA2, 264–265
  chemotherapy, 255
  combination therapy, 269–271, 271f
  EGFR (*see* Epidermal growth factor receptor tyrosine kinase (EGFR))
  growth factor signaling, 255–256
  PI3K/AKT signaling pathway, 255–256, 266–267
  subtypes, 255
  TAM, 267–269, 271–273, 272t
  TP53, 265–266
β-Tubulin, 140
Tumor-associated macrophages (TAMs), 257, 267–269, 271–273, 272t
Tyrosine kinase inhibitors (TKI), 230, 256–257, 292

U

UCSC Genome Browser, 51, 54–58, 61–62, 68–69, 71–72, 113
Upstream regulators, MTA proteins, 11t
  cell stressors, 20–21
  enzymes and protein kinases, 8f, 19–20
  growth factors and receptors, 17–18
  hormone receptors, 18–19
  miRNAs, 16–17
  transcription factors and coregulators, 10–16

V

Vascular endothelial cells, 233–234
Virtual Gene Overexpression or Repression (VIGOR) approach, 74–76, 74f, 112, 114–115

W

Wnt/β-catenin signaling pathway, 237–239

X

Xenografted immunodeficient mouse model, 241–242

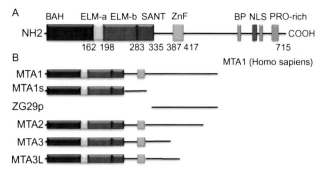

Figure 1, Da-Qiang Li and Rakesh Kumar (See Page 3 of this volume.)

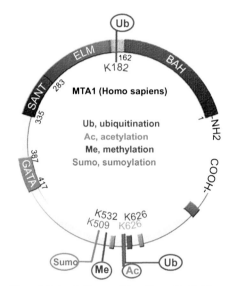

Figure 2, Da-Qiang Li and Rakesh Kumar (See Page 8 of this volume.)

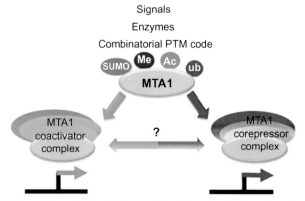

Figure 3, Da-Qiang Li and Rakesh Kumar (See Page 8 of this volume.)

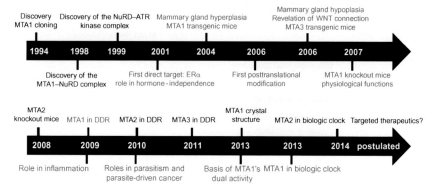

Figure 4, Da-Qiang Li and Rakesh Kumar (See Page 34 of this volume.)

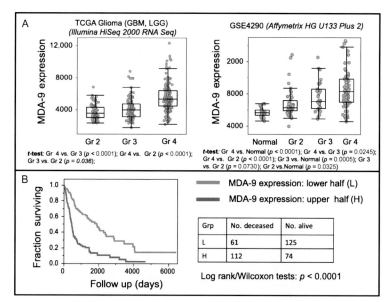

Figure 2, Manny D. Bacolod et al. (See Page 58 of this volume.)

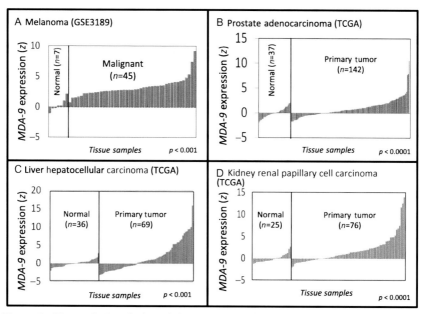

Figure 3, Manny D. Bacolod et al. (See Page 59 of this volume.)

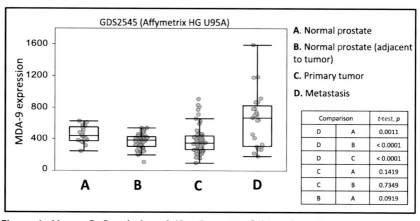

Figure 4, Manny D. Bacolod et al. (See Page 60 of this volume.)

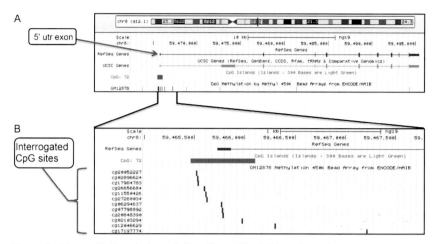

Figure 5, Manny D. Bacolod *et al.* (See Page 64 of this volume.)

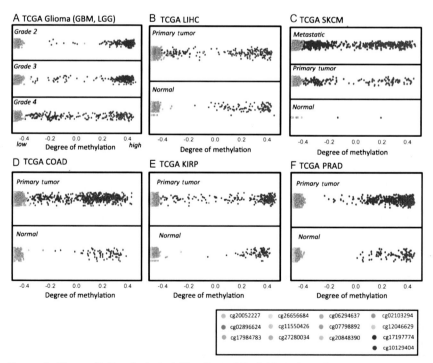

Figure 6, Manny D. Bacolod *et al.* (See Page 64 of this volume.)

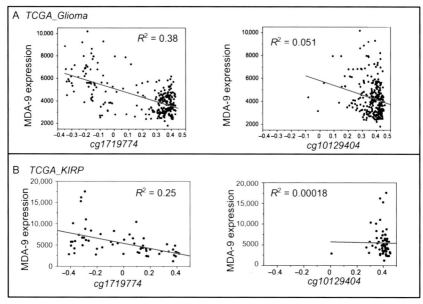

Figure 7, Manny D. Bacolod *et al.* (See Page 65 of this volume.)

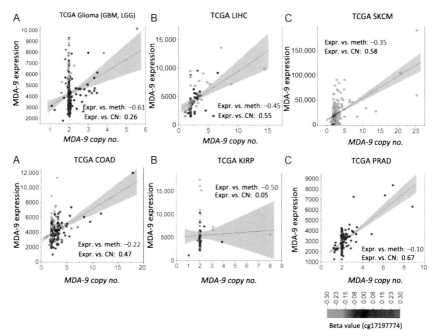

Figure 8, Manny D. Bacolod *et al.* (See Page 66 of this volume.)

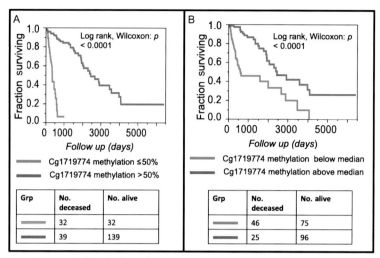

Figure 9, Manny D. Bacolod *et al.* (See Page 66 of this volume.)

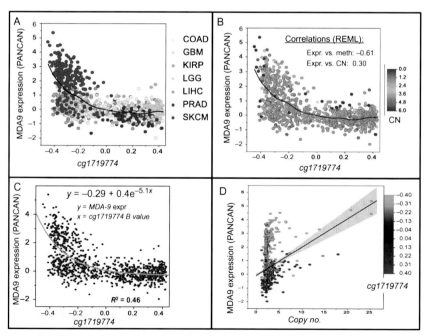

Figure 10, Manny D. Bacolod *et al.* (See Page 68 of this volume.)

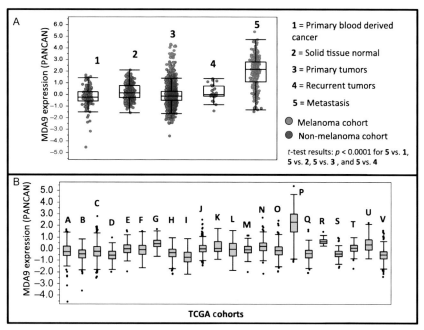

Figure 11, Manny D. Bacolod *et al.* (See Page 70 of this volume.)

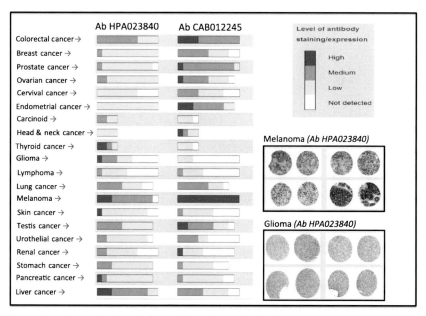

Figure 12, Manny D. Bacolod *et al.* (See Page 71 of this volume.)

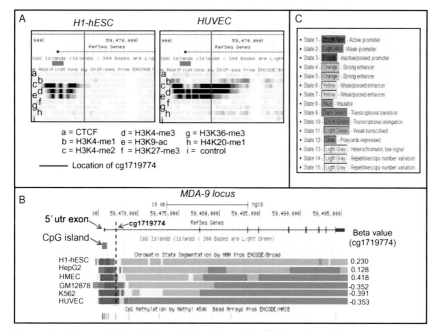

Figure 13, Manny D. Bacolod *et al.* (See Page 73 of this volume.)

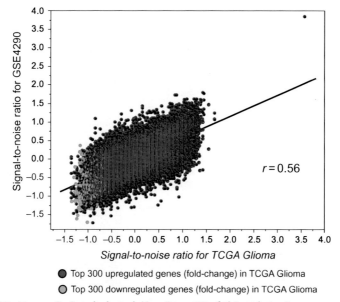

Figure 15, Manny D. Bacolod *et al.* (See Page 75 of this volume.)

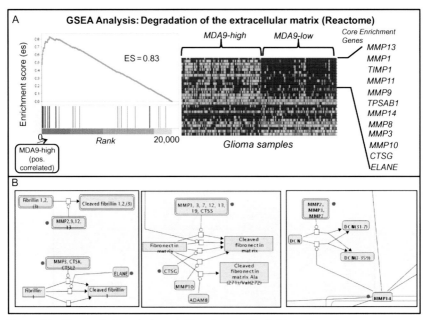

Figure 16, Manny D. Bacolod et al. (See Page 103 of this volume.)

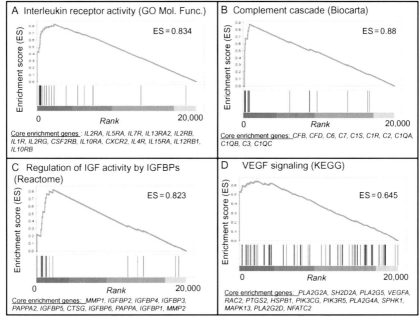

Figure 17, Manny D. Bacolod et al. (See Page 106 of this volume.)

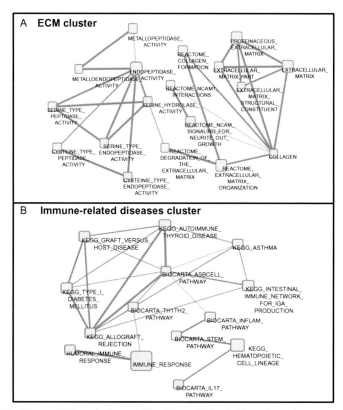

Figure 18, Manny D. Bacolod et al. (See Page 110 of this volume.)

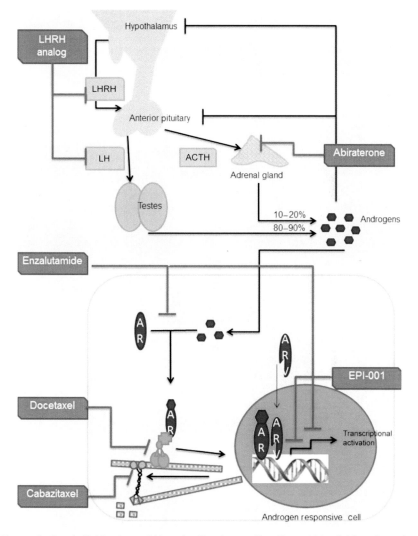

Figure 1, Sarah K. Martin and Natasha Kyprianou (See Page 129 of this volume.)

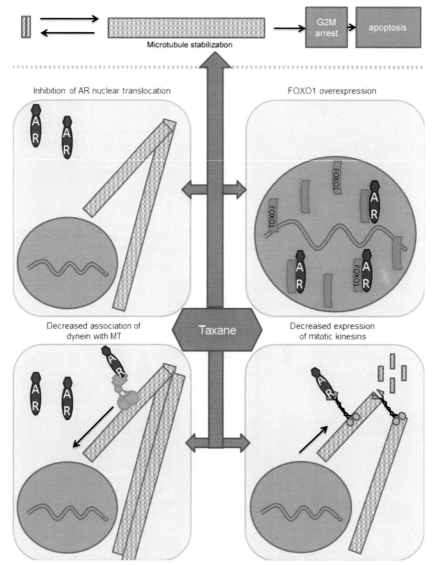

Figure 2, Sarah K. Martin and Natasha Kyprianou (See Page 132 of this volume.)

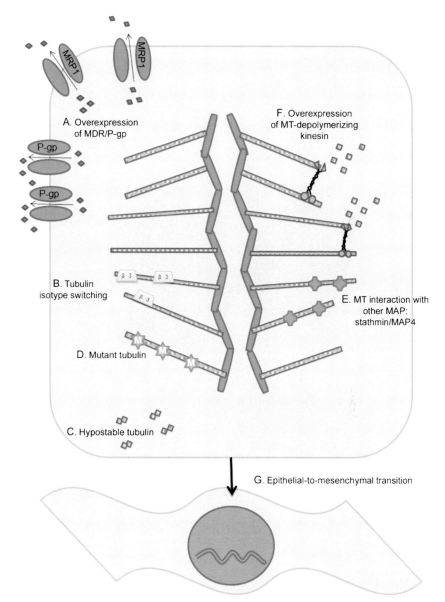

Figure 3, Sarah K. Martin and Natasha Kyprianou (See Page 137 of this volume.)

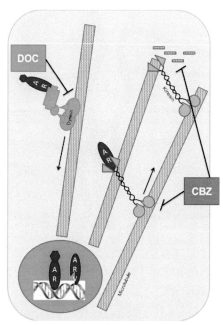

Figure 4, Sarah K. Martin and Natasha Kyprianou (See Page 138 of this volume.)

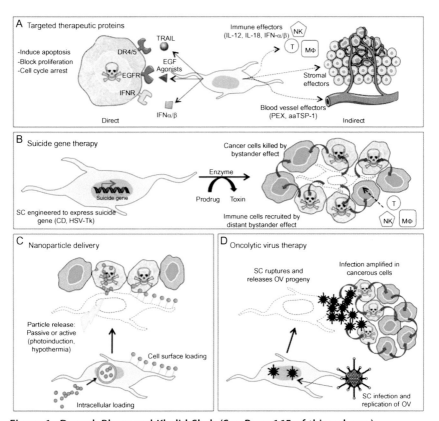

Figure 1, Deepak Bhere and Khalid Shah (See Page 165 of this volume.)

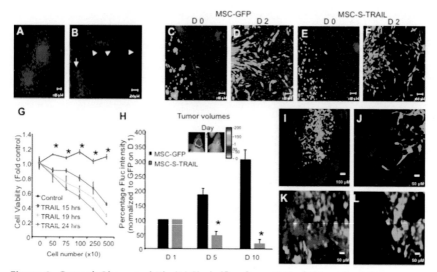

Figure 2, Deepak Bhere and Khalid Shah (See Page 178 of this volume.)

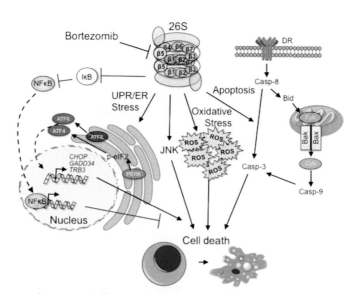

Figure 1, Nathan G. Dolloff (See Page 194 of this volume.)

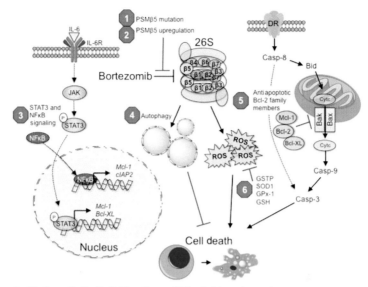

Figure 2, Nathan G. Dolloff (See Page 195 of this volume.)

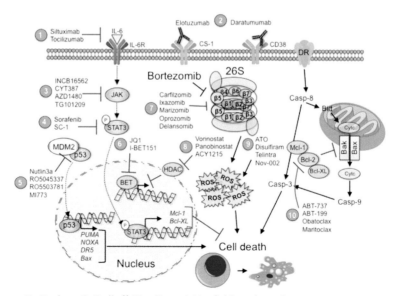

Figure 3, Nathan G. Dolloff (See Page 213 of this volume.)

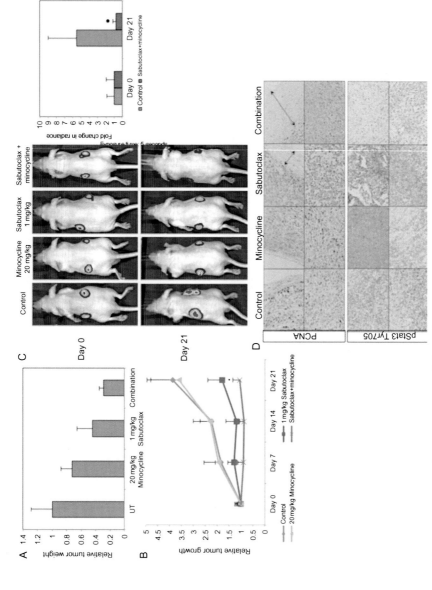

Figure 1, Bridget A. Quinn et al. (See Page 296 of this volume.)

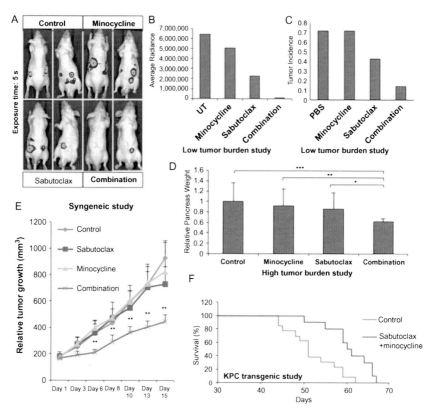

Figure 2, Bridget A. Quinn *et al.* (See Page 297 of this volume.)

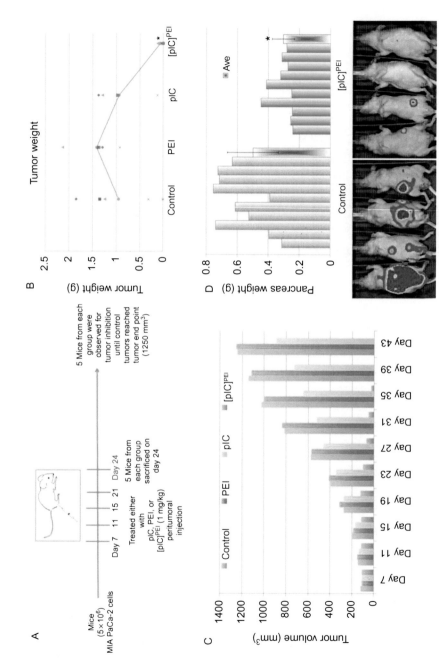

Figure 5, Bridget A. Quinn et al. (See Page 300 of this volume.)

CPI Antony Rowe
Eastbourne, UK
June 21, 2015